READINGS IN AMERICAN DEMOCRACY

Readings in American Democracy

EDITED BY
GERALD STOURZH
RALPH LERNER

AND

H. C. HARLAN

Second Edition

New York   OXFORD UNIVERSITY PRESS   1966

Copyright © 1959, 1966 by Oxford University Press, Inc.
This material has been previously published in a similar form by the American Foundation for Political Education: Copyright, 1957, 1958.
Library of Congress Catalogue Card Number: 66-11530
Printed in the United States of America

PREFACE TO THE SECOND EDITION

The present revision of *Readings in American Democracy* serves the same purposes as its previous forms. Originally designed to stimulate discussion and thinking by *adults* about the fundamental principles of American life, these readings have since been widely adopted for classroom use with undergraduates, and in a variety of institutions of higher learning.

Such widespread and continuing use can be looked upon as validating the educational approach used, and as a testimony to the timeless character of the issues raised. Changes in the current edition focus on problems of race relations and civil rights inherent in a federal form of democratic government, and seen in historical perspective. As in earlier forms, this revision aims to raise fundamental questions. It definitely does not attempt to suggest answers, or to reflect any particular viewpoint, other than a faith in the educational value of confrontation with basic issues.

Chicago, Illinois   H. C. Harlan
June 1965

PREFACE TO THE FIRST EDITION

The purpose of this volume of readings is threefold: to serve the college and university student as primary or supplementary course material, to offer the general reader an opportunity to acquaint himself systematically with the best thought on a wide range of the fundamental problems, and to function as the basic reading material for organized programs of adult education discussion groups.

Several unique features distinguish this book from other books of readings and affect the ways in which it can best be used. Its main characteristic is the plan of selection and editing. The articles are grouped in sections, each about 50 to 75 pages long, intended to be read as a unit. The selections have the continuity of a debate; opposing views are deliberately juxtaposed and the reader must judge the merits of each argument. The articles within each unit will be most instructive, therefore, if they are read in the order in which they appear. The sections are also best taken up in order, for the materials in later sections assume knowledge of the issues discussed in earlier sections.

This book has no index; although it contains a great deal of historical information, it is not a book for "looking up" isolated facts or events. It is a book of political argument. Although the readings present clear and forceful statements of a variety of positions—some quite partisan, others more detached and analytical—the collection as a whole is not meant to represent or support any particular viewpoint. Readers should not assume that any author in these pages is the spokesman of the editors. Nor is this volume intended to be comprehensive. The aim is rather to provoke thought on a limited number of highly significant problems and to provide a sound basis for further study.

These readings were originally prepared by members of the staff of the American Foundation for Political Education for use in discussion programs for the education of adults. During the years 1947–1958, the AFPE organized and conducted discussion groups in hundreds of communities, in cooperation with local educational institutions and civic agencies. Many of these programs continue, but in 1958 the AFPE ceased its practice of directly subsidizing the costs of local programs and has changed the nature of its activities. It now concentrates on the development of materials in different subject-matter fields and on the development of a variety of training programs. To conform to the changed focus of the work, the organization's name has been changed to the American Foundation for Continuing Education. The Foundation seeks to cooperate with persons and organizations interested in conducting discussion programs and welcomes requests for assistance or advice on leadership training or promotion. Inquiries may be addressed to the Executive Director, American Foundation for Continuing Education, 19 South La Salle Street, Chicago 3, Illinois.

THE EDITORS

Chicago, Illinois
April 1959

CONTENTS

I. THE GREAT TASK BEFORE US

1. Abraham Lincoln, *The Gettysburg Address* — 3

II. A SOCIETY OF FREE MEN

1. Michel-Guillaume de Crèvecoeur, *What Is an American?* — 7
2. Alexis de Tocqueville, *Democracy in America* — 11
3. William Graham Sumner, *What Social Classes Owe to Each Other* — 29
4. Granville Hicks, *How We Live Now in America* — 44
5. George F. Kennan, *Commencement, 1955* — 56

III. THE RIGHTS OF THE INDIVIDUAL

1. *The Declaration of Independence* — 67
2. *The Bill of Rights* — 71
3. George Bernard Shaw, *Natural Rights* — 73
4. Carl L. Becker, *Freedom of Speech and Press* — 75
5. *The Dennis Case* — 90

IV. THE WILL OF THE PEOPLE

1. Thomas Jefferson, *First Inaugural Address* — 121
2. Edmund Burke, *The Partnership of Generations* — 126
3. Thomas Jefferson, *The Rights of the Living Generation* — 127
4. T. B. Macaulay, *All Sail and No Anchor* — 131
5. Henry L. Mencken, *Rogues and the Mob* — 134
6. Robert M. MacIver, *The Genius of Democracy* — 140
7. Edward H. Carr, *From Individualism to Mass Democracy* — 152

V. THE DIFFUSION OF POWER

1. James Madison, *The Federalist, No. 51* — 167
2. Walter Bagehot, *Power and Responsibility* — 172
3. *A Government of Laws, Not of Men: The Supreme Court Fight of 1937* — 180
4. *Property Rights and National Security: Constitutional Conflict and the Seizure of the Steel Mills* — 197

VI. THE POLITICS OF DEMOCRACY

1. Edmund Burke, *Representing the People* — 223
2. James Madison, *The Federalist No. 10* — 227
3. James Fenimore Cooper, *On Party* — 235
4. Herbert Agar, *The Price of Union* — 238
5. Franklin D. Roosevelt, *Tweedledum and Tweedledee* — 247
6. James MacGregor Burns, *Republicans, Democrats: Who's Who?* — 249

VII. THE WELFARE STATE

1. C. A. R. Crosland, *The Case for Social Equality* — 259
2. Bertrand de Jouvenel, *The Ethics of Redistribution* — 266
3. Herbert H. Lehman, *Security and Freedom* — 280
4. Sir Alexander Gray, *Three Ways to Live Under the Welfare State* — 284
5. Franklin D. Roosevelt, *To Promote the General Welfare* — 293
6. Herbert Hoover, *The American System of Liberty* — 305

VIII. CIVIL RIGHTS AND FEDERALISM

1. *The Civil Rights Cases* — 313
2. *Plessy v. Ferguson* — 338
3. Burke Marshall, *Federalism and the Administration of Justice* — 345
4. *Brown v. Board of Education of Topeka* — 362
5. John F. Kennedy, *The Moral Crisis* — 367

IX. AMERICA IN A WORLD OF NATIONS

1. Thomas Paine, *An Asylum for Mankind* — 375
2. Carl Schurz, *Manifest Destiny* — 379
3. Albert J. Beveridge, *The Star of Empire* — 393
4. Woodrow Wilson, *Our Duty to the World* — 403
5. Carlos P. Romulo, *The American Dilemma: Democracy and Empire* — 409
6. George F. Kennan, *Our Duty to Ourselves* — 417

X. STATESMANSHIP IN A FREE DEMOCRACY

1. Stephen A. Douglas, *Slavery and the Right to Self-Government* — 427
2. Abraham Lincoln, *Slavery and the Constitution* — 433
3. Henry David Thoreau, *The Law of God and the Law of the Land* — 438
4. Lyman Bryson, *On Deceiving the Public for the Public Good* — 442
5. Alexis de Tocqueville, *The Predicament of Democracy* — 453
6. Dean Acheson, *Democratic Representation in an Age of Crisis* — 461

APPENDIX

The Constitution of the United States — 467

I

THE GREAT TASK BEFORE US

The Gettysburg Address

BY ABRAHAM LINCOLN

[November 19, 1863]

FOURSCORE and seven years ago our fathers brought forth upon this continent a new nation, conceived in liberty, and dedicated to the proposition that all men are created equal.

Now we are engaged in a great civil war, testing whether that nation, or any nation so conceived and so dedicated, can long endure. We are met on a great battlefield of that war. We have come to dedicate a portion of that field as a final resting-place for those who here gave their lives that that nation might live. It is altogether fitting and proper that we should do this.

But, in a larger sense, we cannot dedicate—we cannot consecrate —we cannot hallow—this ground. The brave men, living and dead, who struggled here, have consecrated it far above our poor power to add or detract. The world will little note nor long remember what we say here, but it can never forget what they did here. It is for us, the living, rather, to be dedicated here to the unfinished work which they who fought here have thus far so nobly advanced. It is rather for us to be here dedicated to the great task remaining before us—that from these honored dead we take increased devotion to that cause for which they gave the last full measure of devotion; that we here highly resolve that these dead shall not have died in vain; that this nation, under God, shall have a new birth of freedom; and that government of the people, by the people, for the people, shall not perish from the earth.

II

A SOCIETY OF FREE MEN

What Is an American?*

BY MICHEL-GUILLAUME JEAN DE CRÈVECOEUR

[1782]

I WISH I could be acquainted with the feelings and thoughts which must agitate the heart and present themselves to the mind of an enlightened Englishman, when he first lands on this continent. . . .

What a train of pleasing ideas this fair spectacle must suggest! it is a prospect which must inspire a good citizen with the most heartfelt pleasure. The difficulty consists in the manner of viewing so extensive a scene. He is arrived on a new continent; a modern society offers itself to his contemplation, different from what he had hitherto seen. It is not composed, as in Europe, of great lords who possess every thing, and of a herd of people who have nothing. Here are no aristocratical families, no courts, no kings, no bishops, no ecclesiastical dominion, no invisible power giving to a few a very visible one; no great manufacturers employing thousands, no great refinements of luxury. The rich and the poor are not so far removed from each other as they are in Europe.

Some few towns excepted, we are all tillers of the earth, from Nova Scotia to West Florida. We are a people of cultivators, scattered over an immense territory, communicating with each other by means of good roads and navigable rivers, united by the silken bands of mild government, all respecting the laws without dreading their power, because they are equitable. We are all animated with the spirit of industry, which is unfettered, and unrestrained, because each person works for himself. If he travels through our rural districts, he views not the hostile castle, and the haughty mansion, contrasted with the clay-built hut and miserable cabbin, where cattle and men help to keep each other warm, and swell in meanness, smoke, and indigence. A pleasing uniformity of decent competence appears throughout our habitations. The meanest of

---

* From Letter III of *Letters from an American Farmer*. Michel-Guillaume Jean de Crèvecoeur (1735–1813), a Frenchman who served under Montcalm, settled in the 1760's as a farmer in New York State. After the American Revolution he was French Consul in New York, returning to Europe in 1790.

our log-houses is a dry and comfortable habitation. Lawyer or merchant are the fairest titles our towns afford; that of a farmer is the only appellation of the rural inhabitants of our country. It must take some time ere he can reconcile himself to our dictionary, which is but short in words of dignity, and names of honour. There, on a Sunday, he sees a congregation of respectable farmers and their wives, all clad in neat homespun, well mounted, or riding in their own humble waggons. There is not among them an esquire, saving the unlettered magistrate. There he sees a parson as simple as his flock, a farmer who does not riot on the labour of others. We have no princes, for whom we toil, starve, and bleed: we are the most perfect society now existing in the world. Here man is free as he ought to be; nor is this pleasing equality so transitory as many others are. Many ages will not see the shores of our great lakes replenished with inland nations, nor the unknown bounds of North America entirely peopled. Who can tell how far it extends? Who can tell the millions of men who it will feed and contain? for no European foot has as yet travelled half the extent of this mighty continent!

The next wish of this traveller will be to know whence came all these people? they are a mixture of English, Scotch, Irish, French, Dutch, Germans, and Swedes. From this promiscuous breed, that race now called Americans have arisen. The eastern provinces must indeed be excepted, as being the unmixed descendants of Englishmen. I have heard many wish they had been more intermixed also: for my part, I am no wisher; and think it much better as it has happened. They exhibit a most conspicuous figure in this great and variegated picture; they too enter for a great share in the pleasing perspective displayed in these thirteen provinces. I know it is fashionable to reflect on them; but I respect them for what they have done; for the accuracy and wisdom with which they have settled their territory; for the decency of their manners; for their early love of letters; their ancient college, the first in this hemisphere; for their industry, which to me, who am but a farmer, is the criterion of every thing. There never was a people, situated as they are, who, with so ungrateful a soil, have done more in so short a time. Do you think that the monarchial ingredients which are more prevalent in other governments, have purged them from all foul stains? Their histories assert the contrary.

In this great American asylum, the poor of Europe have by some means met together, and in consequence of various causes; to what purpose should they ask one another, what countrymen they are? Alas, two thirds of them had no country. Can a wretch who wanders about, who works and starves, whose life is a continual scene of sore affliction or pinching penury; can that man call England or any other kingdom his country? A country that had no bread for him, whose fields procured him no harvest, who met with nothing but the frowns of the rich, the severity of the laws, with jails and punishments; who owned not a single foot of the extensive surface of this planet? No! urged by a variety of motives, here they came. Every thing has tended to regenerate them; new laws, a new mode of living, a new social system; here they are become men: in Europe they were as so many useless plants, wanting vegetative mould, and refreshing showers; they withered, and were mowed down by want, hunger, and war: but now, by the power of transplantation like all other plants, they have taken root and flourished! Formerly they were not numbered in any civil list of their country, except in those of the poor; here they rank as citizens. By what invisible power has this surprizing metamorphosis been performed? By that of the laws and that of their industry. The laws, the indulgent laws, protect them as they arrive, stamping on them the symbol of adoption; they receive ample rewards for their labours; these accumulated rewards procure them lands; those lands confer on them the title of freemen; and to that title every benefit is affixed which men can possibly require. This is the great operation daily performed by our laws. From whence proceed these laws? From our government. Whence that government? It is derived from the original genius and strong desire of the people, ratified and confirmed by government. This is the great chain which links us all. . . .

What attachment can a poor European emigrant have for a country where he had nothing? The knowledge of the language, the love of a few kindred as poor as himself, were the only cords that tied him: his country is now that which gives him land, bread, protection, and consequence: *Ubi panis ibi patria* is the motto of all emigrants. What then is the American, this new man? He is either an European, or the descendant of an European; hence that strange mixture of blood, which you will find in no other country.

I could point out to you a man, whose grandfather was an Englishman, whose wife was Dutch, whose son married a French woman, and whose present four sons have now four wives of different nations. *He* is an American, who, leaving behind him all his ancient prejudices and manners, receives new ones from the new mode of life he has embraced, the new government he obeys, and the new rank he holds. He becomes an American by being received in the broad lap of our great *Alma Mater*.

Here individuals of all nations are melted into a new race of men, whose labours and posterity will one day cause great change in the world. Americans are the western pilgrims, who are carrying along with them that great mass of arts, sciences, vigour, and industry, which began long since in the east; they will finish the great circle. The Americans were once scattered all over Europe; here they are incorporated into one of the finest systems of population which has ever appeared, and which will hereafter become distinct by the power of the different climates they inhabit. The American ought, therefore, to love this country much better than that wherein either he or his forefathers were born. Here the rewards of his industry follow with equal steps the progress of his labour; his labour is founded on the basis of nature, *self-interest;* can it want a stronger allurement? . . . The American is a new man, who acts upon new principles; he must therefore entertain new ideas and form new opinions. From involuntary idleness, servile dependance, penury, and useless labour, he has passed to toils of a very different nature, rewarded by ample subsistence.—This is an American. . . .

Democracy in America*
BY ALEXIS DE TOCQUEVILLE

[1840]

Second Book
Influence of Democracy on the Feelings of the Americans

Chapter I
Why Democratic Nations Show a More Ardent and Enduring Love of Equality Than of Liberty

THE FIRST and most intense passion that is produced by equality of condition is, I need hardly say, the love of that equality. My readers will therefore not be surprised that I speak of this feeling before all others.

Everybody has remarked that in our time, and especially in France, this passion for equality is every day gaining ground in the human heart. It has been said a hundred times that our contemporaries are far more ardently and tenaciously attached to equality than to freedom; but as I do not find that the causes of the fact have been sufficiently analyzed, I shall endeavor to point them out.

It is possible to imagine an extreme point at which freedom and equality would meet and blend. Let us suppose that all the people take a part in the government, and that each one of them has an

---

* Reprinted from Volume II, Book II, of *Democracy in America* by Alexis de Tocqueville, by permission of Alfred A. Knopf, Inc. Copyright 1945 by Alfred A. Knopf, Inc. Published by Vintage Books, Inc., 1956. Alexis de Tocqueville (1805–1859), a French nobleman and political thinker, and Foreign Minister of France in 1849, spent nine months in America in 1831–32. His *Democracy in America* was written as a result of this visit.

equal right to take part in it. As no one is different from his fellows, none can exercise a tyrannical power; men will be perfectly free because they are all entirely equal; and they will all be perfectly equal because they are entirely free. To this ideal state democratic nations tend. This is the only complete form that equality can assume upon earth; but there are a thousand others which, without being equally perfect, are not less cherished by those nations.

The principle of equality may be established in civil society without prevailing in the political world. There may be equal rights of indulging in the same pleasures, of entering the same professions, of frequenting the same places; in a word, of living in the same manner and seeking wealth by the same means, although all men do not take an equal share in the government. A kind of equality may even be established in the political world though there should be no political freedom there. A man may be the equal of all his countrymen save one, who is the master of all without distinction and who selects equally from among them all the agents of his power. Several other combinations might be easily imagined by which very great equality would be united to institutions more or less free or even to institutions wholly without freedom.

Although men cannot become absolutely equal unless they are entirely free, and consequently equality, pushed to its furthest extent, may be confounded with freedom, yet there is good reason for distinguishing the one from the other. The taste which men have for liberty and that which they feel for equality are, in fact, two different things; and I am not afraid to add that among democratic nations they are two unequal things.

Upon close inspection it will be seen that there is in every age some peculiar and preponderant fact with which all others are connected; this fact almost always gives birth to some pregnant idea or some ruling passion, which attracts to itself and bears away in its course all the feelings and opinions of the time; it is like a great stream towards which each of the neighboring rivulets seems to flow.

Freedom has appeared in the world at different times and under various forms; it has not been exclusively bound to any social condition, and it is not confined to democracies. Freedom cannot, therefore, form the distinguishing characteristic of democratic ages. The peculiar and preponderant fact that marks those ages as its own is the equality of condition; the ruling passion of men in those periods is the love of this equality. Do not ask what singular charm the men of democratic ages find in being equal, or what special reasons they may have for clinging so tenaciously to equality rather than to the other advantages that society holds out to them: equality is the distinguishing characteristic of the age they live in; that of itself is enough to explain that they prefer it to all the rest.

But independently of this reason there are several others which will at all times habitually lead men to prefer equality to freedom.

If a people could ever succeed in destroying, or even in diminishing the equality that prevails in its own body, they could do so only by long and laborious efforts. Their social condition must be modified, their laws abolished, their opinions superseded, their habits changed, their manners corrupted. But political liberty is more easily lost; to neglect to hold it fast is to allow it to escape. Therefore not only do men cling to equality because it is dear to them; they also adhere to it because they think it will last forever.

That political freedom in its excesses may compromise the tranquillity, the property, the lives of individuals is obvious even to narrow and unthinking minds. On the contrary, none but attentive and clear-sighted men perceive the perils with which equality threatens us, and they commonly avoid pointing them out. They know that the calamities they apprehend are remote and flatter themselves that they will only fall upon future generations, for which the present generation takes but little thought. The evils that freedom sometimes brings with it are immediate; they are apparent to all, and all are more or less affected by them. The evils that extreme equality may produce are slowly disclosed; they creep gradually into the social frame; they are seen only at intervals; and at the moment at which they become most violent, habit already causes them to be no longer felt.

The advantages that freedom brings are shown only by the lapse of time, and it is always easy to mistake the cause in which they originate. The advantages of equality are immediate, and they may always be traced from their source.

Political liberty bestows exalted pleasures from time to time upon a certain number of citizens. Equality every day confers a number of small enjoyments on every man. The charms of equality are every instant felt and are within the reach of all; the noblest hearts are not insensible to them, and the most vulgar souls exult in them. The passion that equality creates must therefore be at once strong and general. Men cannot enjoy political liberty unpurchased by some sacrifices, and they never obtain it without great exertions. But the pleasures of equality are self-proffered; each of the petty incidents of life seems to occasion them, and in order to taste them, nothing is required but to live.

Democratic nations are at all times fond of equality, but there are certain epochs at which the passion they entertain for it swells to the height of fury. This occurs at the moment when the old social system, long menaced, is overthrown after a severe internal struggle, and the barriers of rank are at length thrown down. At such times men pounce upon equality as their booty, and they cling to it as to some precious treasure which they fear to lose. The passion for equality penetrates on every side into men's hearts, expands there, and fills them entirely. Tell them not that by this blind surrender of themselves to an exclusive passion they risk their dearest interests; they are deaf. Show them not freedom escaping from their grasp while they are looking another way; they are blind, or rather they can discern but one object to be desired in the universe. . . .

I think that democratic communities have a natural taste for freedom; left to themselves, they will seek it, cherish it, and view any privation of it with regret. But for equality their passion is ardent, insatiable, incessant, invincible; they call for equality in freedom; and if they cannot obtain that, they still call for equality in slavery. They will endure poverty, servitude, barbarism, but they will not endure aristocracy.

This is true at all times, and especially in our own day. All men and all powers seeking to cope with this irresistible passion will be overthrown and destroyed by it. In our age freedom cannot be established without it, and despotism itself cannot reign without its support.

Chapter II

OF INDIVIDUALISM IN DEMOCRATIC COUNTRIES

I HAVE shown how it is that in ages of equality every man seeks for his opinions within himself; I am now to show how it is that in the same ages all his feelings are turned towards himself alone. *Individualism* is a novel expression, to which a novel idea has given birth. Our fathers were only acquainted with *égoïsme* (selfishness). Selfishness is a passionate and exaggerated love of self, which leads a man to connect everything with himself and to prefer himself to everything in the world. Individualism is a mature and calm feeling, which disposes each member of the community to sever himself from the mass of his fellows and to draw apart with his family and his friends, so that after he has thus formed a little circle of his own, he willingly leaves society at large to itself. Selfishness originates in blind instinct; individualism proceeds from erroneous judgment more than from depraved feelings; it originates as much in deficiencies of mind as in perversity of heart.

Selfishness blights the germ of all virtue; individualism, at first, only saps the virtues of public life; but in the long run it attacks and destroys all others and is at length absorbed in downright selfishness. Selfishness is a vice as old as the world, which does not belong to one form of society more than to another; individualism is of democratic origin, and it threatens to spread in the same ratio as the equality of condition.

Among aristocratic nations, as families remain for centuries in the same condition, often on the same spot, all generations become, as it were, contemporaneous. A man almost always knows his forefathers and respects them; he thinks he already sees his remote descendants and he loves them. He willingly imposes duties on himself towards the former and the latter, and he will frequently sacrifice his personal gratifications to those who went before and to those who will come after him. Aristocratic institutions, moreover, have the effect of closely binding every man to several of his fellow citizens. As the classes of an aristocratic people are strongly marked and permanent, each of them is regarded by its own members as a sort of lesser country, more tangible and more cherished than the country at large. As in aristocratic communities all the citizens occupy fixed positions, one above another, the result is that each of them always sees a man above himself whose patronage is necessary to him, and below himself another man whose co-operation he may claim. Men living in aristocratic ages are therefore almost always closely attached to something placed out of their own sphere, and they are often disposed to forget themselves. It is true that in these ages the notion of human fellowship is faint and that men seldom think of sacrificing themselves for mankind; but they often sacrifice themselves for other men. In democratic times, on the contrary, when the duties of each individual to the race are much more clear, devoted service to any one man becomes more rare; the bond of human affection is extended, but it is relaxed.

Among democratic nations new families are constantly springing up, others are constantly falling away, and all that remain change their condition; the woof of time is every instant broken and the track of generations effaced. Those who went before are soon forgotten; of those who will come after, no one has any idea: the interest of man is confined to those in close propinquity to himself. As each class gradually approaches others and mingles with them, its members become undifferentiated and lose their class identity for each other. Aristocracy had made a chain of all the members of the community, from the peasant to the king; democracy breaks that chain and severs every link of it.

As social conditions become more equal, the number of persons increases who, although they are neither rich nor powerful enough to exercise any great influence over their fellows, have nevertheless acquired or retained sufficient education and fortune to satisfy their own wants. They owe nothing to any man, they expect nothing from any man; they acquire the habit of always considering themselves as standing alone, and they are apt to imagine that their whole destiny is in their own hands.

Thus not only does democracy make every man forget his ancestors, but it hides his descendants and separates his contemporaries from him; it throws him back forever upon himself alone and threatens in the end to confine him entirely within the solitude of his own heart.

Chapter III

INDIVIDUALISM STRONGER AT THE CLOSE OF A DEMOCRATIC
REVOLUTION THAN AT OTHER PERIODS

THE PERIOD when the construction of democratic society upon the ruins of an aristocracy has just been completed is especially that at which this isolation of men from one another and the selfishness resulting from it most forcibly strike the observer. Democratic communities not only contain a large number of independent citizens, but are constantly filled with men who, having entered but yesterday upon their independent condition, are intoxicated with their new power. They entertain a presumptuous confidence in their own strength, and as they do not suppose that they can henceforward ever have occasion to claim the assistance of their fellow creatures, they do not scruple to show that they care for nobody but themselves.

An aristocracy seldom yields without a protracted struggle, in the course of which implacable animosities are kindled between the different classes of society. These passions survive the victory, and traces of them may be observed in the midst of the democratic confusion that ensues. Those members of the community who were at the top of the late gradations of rank cannot immediately forget their former greatness; they will long regard themselves as aliens in the midst of the newly composed society. They look upon all those whom this state of society has made their equals as oppressors, whose destiny can excite no sympathy; they have lost sight of their former equals and feel no longer bound to their fate by a common interest; each of them, standing aloof, thinks that he is reduced to care for himself alone. Those, on the contrary, who were formerly at the foot of the social scale and who have been brought up to the common level by a sudden revolution cannot enjoy their newly acquired independence without secret uneasiness; and if they meet with some of their former superiors on the same footing as themselves, they stand aloof from them with an expression of triumph and fear.

It is, then, commonly at the outset of democratic society that citizens are most disposed to live apart. Democracy leads men not to draw near to their fellow creatures; but democratic revolutions lead them to shun each other and perpetuate in a state of equality the animosities that the state of inequality created.

The great advantage of the Americans is that they have arrived at a state of democracy without having to endure a democratic revolution, and that they are born equal instead of becoming so.

Chapter IV

THAT THE AMERICANS COMBAT THE EFFECTS OF INDIVIDUALISM BY FREE INSTITUTIONS

DESPOTISM, which by its nature is suspicious, sees in the separation among men the surest guarantee of its continuance, and it usually makes every effort to keep them separate. No vice of the human heart is so acceptable to it as selfishness: a despot easily forgives his subjects for not loving him, provided they do not love one another. He does not ask them to assist him in governing the state; it is enough that they do not aspire to govern it themselves. He stigmatizes as turbulent and unruly spirits those who would combine their exertions to promote the prosperity of the community; and, perverting the natural meaning of words, he applauds as good citizens those who have no sympathy for any but themselves.

Thus the vices which depotism produces are precisely those which equality fosters. These two things perniciously complete and assist each other. Equality places men side by side, unconnected by any common tie; despotism raises barriers to keep them asunder; the former predisposes them not to consider their fellow creatures, the latter makes general indifference a sort of public virtue. . . .

The Americans have combated by free institutions the tendency of equality to keep men asunder, and they have subdued it. The legislators of America did not suppose that a general representation of the whole nation would suffice to ward off a disorder at once so natural to the frame of democratic society and so fatal; they also thought that it would be well to infuse political life into each portion of the territory in order to multiply to an infinite extent opportunities of acting in concert for all the members of the community and to make them constantly feel their mutual dependence. The plan was a wise one. The general affairs of a country engage the attention only of leading politicians, who assemble from time to time in the same places; and as they often lose sight of each other afterwards, no lasting ties are established between them. But if the object be to have the local affairs of a district conducted by the men who reside there, the same persons are always in contact, and they are, in a manner, forced to be acquainted and to adapt themselves to one another.

It is difficult to draw a man out of his own circle to interest him in the destiny of the state, because he does not clearly understand what influence the destiny of the state can have upon his own lot. But if it is proposed to make a road cross the end of his estate, he will see at a glance that there is a connection between this small public affair and his greatest private affairs; and he will discover, without its being shown to him, the close tie that unites private to general interest. Thus far more may be done by entrusting to the citizens the administration of minor affairs than by surrendering to them in the control of important ones, towards interesting them in the public welfare and convincing them that they constantly stand in need of one another in order to provide for it. A brilliant achievement may win for you the favor of a people at one stroke; but to earn the love and respect of the population that surrounds you, a long succession of little services rendered and of obscure good deeds, a constant habit of kindness, and an established reputation for disinterestedness will be required. Local freedom, then, which leads a great number of citizens to value the affection of their neighbors and of their kindred, perpetually brings men together and forces them to help one another in spite of the propensities that sever them.

In the United States the more opulent citizens take great care not to stand aloof from the people; on the contrary, they constantly keep on easy terms with the lower classes: they listen to them, they speak to them every day. They know that the rich in democracies always stand in need of the poor, and that in democratic times you attach a poor man to you more by your manner than by benefits conferred. The magnitude of such benefits, which sets off the difference of condition, causes a secret irritation to those who reap advantage from them, but the charm of simplicity of manners is almost irresistible; affability carries men away, and even want of polish is not always displeasing. This truth does not take root at once in the minds of the rich. They generally resist it as long as the democratic revolution lasts, and they do not acknowledge it immediately after that revolution is accomplished. They are very ready to do good to the people, but they still choose to keep them at arm's length; they think that is sufficient, but they are mistaken. They might spend fortunes thus without warming the hearts of the population around them; that population does not ask them for the sacrifice of their money, but of their pride.

It would seem as if every imagination in the United States were upon the stretch to invent means of increasing the wealth and satisfying the wants of the public. The best-informed inhabitants of each district constantly use their information to discover new truths that may augment the general prosperity; and if they have made any such discoveries, they eagerly surrender them to the mass of the people.

When the vices and weaknesses frequently exhibited by those who govern in America are closely examined, the prosperity of the people occasions, but improperly occasions, surprise. Elected magistrates do not make the American democracy flourish; it flourishes because the magistrates are elective.

It would be unjust to suppose that the patriotism and the zeal that every American displays for the welfare of his fellow citizens are wholly insincere. Although private interest directs the greater part of human actions in the United States as well as elsewhere, it does not regulate them all. I must say that I have often seen Americans make great and real sacrifices to the public welfare; and I have noticed a hundred instances in which they hardly ever failed to lend faithful support to one another. The free institutions which the inhabitants of the United States possess, and the political rights of which they make so much use, remind every citizen, and in a thousand ways, that he lives in society. They every instant impress upon his mind the notion that it is the duty as well as the interest of men to make themselves useful to their fellow creatures; and as he sees no particular ground of animosity to them, since he is never either their master or their slave, his heart readily leans to the side of kindness. Men attend to the interests of the public, first by necessity, afterwards by choice; what was intentional becomes an instinct, and by dint of working for the good of one's fellow citizens, the habit and the taste for serving them are at length acquired.

Many people in France consider equality of condition as one evil and political freedom as a second. When they are obliged to yield to the former, they strive at least to escape from the latter. But I contend that in order to combat the evils which equality may produce, there is only one effectual remedy: namely, political freedom.

Chapter XIX

What Causes Almost All Americans to Follow Industrial Callings

AGRICULTURE is perhaps, of all the useful arts, that which improves most slowly among democratic nations. Frequently, indeed, it would seem to be stationary, because other arts are making rapid strides towards perfection. On the other hand, almost all the tastes and habits that the equality of condition produces naturally lead men to commercial and industrial occupations.

Suppose an active, enlightened, and free man, enjoying a competency, but full of desires; he is too poor to live in idleness, he is rich enough to feel himself protected from the immediate fear of want, and he thinks how he can better his condition. This man has conceived a taste for physical gratifications, which thousands of his fellow men around him indulge in; he has himself begun to enjoy these pleasures, and he is eager to increase his means of satisfying these tastes more completely. But life is slipping away, time is urgent; to what is he to turn? The cultivation of the ground promises an almost certain result to his exertions, but a slow one; men are not enriched by it without patience and toil. Agriculture is therefore only suited to those who already have great superfluous wealth or to those whose penury bids them seek only a bare subsistence. The choice of such a man as we have supposed is soon made; he sells his plot of ground, leaves his dwelling, and embarks on some hazardous but lucrative calling.

Democratic communities abound in men of this kind; and in

proportion as the equality of conditions becomes greater, their multitude increases. Thus, democracy not only swells the number of working-men, but leads men to prefer one kind of labor to another; and while it diverts them from agriculture, it encourages their tastes for commerce and manufactures.[1]

This spirit may be observed even among the richest members of the community. In democratic countries, however opulent a man is supposed to be, he is almost always discontented with his fortune because he finds that he is less rich than his father was, and he fears that his sons will be less rich than himself. Most rich men in democracies are therefore constantly haunted by the desire of obtaining wealth, and they naturally turn their attention to trade and manufactures, which appear to offer the readiest and most efficient means of success. In this respect they share the instincts of the poor without feeling the same necessities; say, rather, they feel the most imperious of all necessities, that of not sinking in the world.

In aristocracies the rich are at the same time the governing power. The attention that they unceasingly devote to important public affairs diverts them from the lesser cares that trade and manufactures demand. But if an individual happens to turn his attention to business, the will of the body to which he belongs will immediately prevent him from pursuing it; for, however men may declaim against the rule of numbers, they cannot wholly escape it; and even among those aristocratic bodies that most obstinately refuse to acknowledge the rights of the national majority, a private majority is formed which governs the rest.

In democratic countries, where money does not lead those who

---

[1] It has often been remarked that manufacturers and merchants are inordinately addicted to physical gratifications, and this has been attributed to commerce and manufactures; but that, I apprehend, is to take the effect for the cause. The taste for physical gratifications is not imparted to men by commerce or manufactures, but it is rather this taste that leads men to engage in commerce and manufactures, as a means by which they hope to satisfy themselves more promptly and more completely. If commerce and manufactures increase the desire of well-being, it is because every passion gathers strength in proportion as it is cultivated, and is increased by all the efforts made to satiate it. All the causes that make the love of worldly welfare predominate in the heart of man are favorable to the growth of commerce and manufactures. Equality of conditions is one of those causes; it encourages trade, not directly, by giving men a taste for business, but indirectly, by strengthening and expanding in their minds a taste for well-being.

possess it to political power, but often removes them from it, the rich do not know how to spend their leisure. They are driven into active life by the disquietude and the greatness of their desires, by the extent of their resources, and by the taste for what is extraordinary, which is almost always felt by those who rise, by whatever means, above the crowd. Trade is the only road open to them. In democracies nothing is greater or more brilliant than commerce; it attracts the attention of the public and fills the imagination of the multitude; all energetic passions are directed towards it. Neither their own prejudices nor those of anybody else can prevent the rich from devoting themselves to it. The wealthy members of democracies never form a body which has manners and regulations of its own; the opinions peculiar to their class do not restrain them, and the common opinions of their country urge them on. Moreover, as all the large fortunes that are found in a democratic community are of commercial growth, many generations must succeed one another before their possessors can have entirely laid aside their habits of business.

Circumscribed within the narrow space that politics leaves them, rich men in democracies eagerly embark in commercial enterprise; there they can extend and employ their natural advantages, and, indeed, it is even by the boldness and the magnitude of their industrial speculations that we may measure the slight esteem in which productive industry would have been held by them if they had been born in an aristocracy.

A similar observation is likewise applicable to all men living in democracies, whether they are poor or rich. Those who live in the midst of democratic fluctuations have always before their eyes the image of chance; and they end by liking all undertakings in which chance plays a part. They are therefore all led to engage in commerce, not only for the sake of the profit it holds out to them, but for the love of the constant excitement occasioned by that pursuit.

The United States of America has only been emancipated for half a century from the state of colonial dependence in which it stood to Great Britain; the number of large fortunes there is small and capital is still scarce. Yet no people in the world have made such rapid progress in trade and manufactures as the Americans; they constitute at the present day the second maritime nation in the world, and although their manufactures have to struggle with almost insurmountable natural impediments, they are not prevented from making great and daily advances.

In the United States the greatest undertakings and speculations are executed without difficulty, because the whole population are engaged in productive industry, and because the poorest as well as the most opulent members of the commonwealth are ready to combine their efforts for these purposes. The consequence is that a stranger is constantly amazed by the immense public works executed by a nation which contains, so to speak, no rich men. The Americans arrived but as yesterday on the territory which they inhabit, and they have already changed the whole order of nature for their own advantage. They have joined the Hudson to the Mississippi and made the Atlantic Ocean communicate with the Gulf of Mexico, across a continent of more than five hundred leagues in extent which separates the two seas. The longest railroads that have been constructed up to the present time are in America.

But what most astonishes me in the United States is not so much the marvelous grandeur of some undertakings as the innumerable multitude of small ones. Almost all the farmers of the United States combine some trade with agriculture; most of them make agriculture itself a trade. It seldom happens that an American farmer settles for good upon the land which he occupies; especially in the districts of the Far West, he brings land into tillage in order to sell it again, and not to farm it; he builds a farmhouse on the speculation that, as the state of the country will soon be changed by the increase of population, a good price may be obtained for it.

Every year a swarm of people from the North arrive in the Southern states and settle in the parts where the cotton plant and the sugar-cane grow. These men cultivate the soil in order to make it produce in a few years enough to enrich them; and they already look forward to the time when they may return home to enjoy the competency thus acquired. Thus the Americans carry their businesslike qualities into agriculture, and their trading passions are displayed in that as in their other pursuits.

The Americans make immense progress in productive industry, because they all devote themselves to it at once; and for this same reason they are exposed to unexpected and formidable embarrassments. As they are all engaged in commerce, their commercial affairs are affected by such various and complex causes that it is impossible to foresee what difficulties may arise. As they are all more or less engaged in productive industry, at the least shock given to business all private fortunes are put in jeopardy at the same time, and the state is shaken. I believe that the return of these commercial panics is an endemic disease of the democratic nations of our age. It may be rendered less dangerous, but it cannot be cured, because it does not originate in accidental circumstances, but in the temperament of these nations.

Chapter XX

How an Aristocracy May be Created by Manufactures

I HAVE shown how democracy favors the growth of manufactures and increases without limit the numbers of the manufacturing classes; we shall now see by what side-road manufacturers may possibly, in their turn, bring men back to aristocracy.

It is acknowledged that when a workman is engaged every day upon the same details, the whole commodity is produced with greater ease, speed, and economy. It is likewise acknowledged that the cost of production of manufactured goods is diminished by the extent of the establishment in which they are made and by the amount of capital employed or of credit. These truths had long been imperfectly discerned, but in our time they have been demonstrated. They have been already applied to many very important kinds of manufactures, and the humblest will gradually be governed by them. I know of nothing in politics that deserves to fix the attention of the legislator more closely than these two new axioms of the science of manufactures.

When a workman is unceasingly and exclusively engaged in the fabrication of one thing, he ultimately does his work with singular dexterity; but at the same time he loses the general faculty of applying his mind to the direction of the work. He every day becomes more adroit and less industrious; so that it may be said of him that in proportion as the workman improves, the man is degraded. What can be expected of a man who has spent twenty years of his life in making heads for pins? And to what can that mighty human intelligence which has so often stirred the world be applied in him except it be to investigate the best method of making pins' heads? When a workman has spent a considerable portion of his existence in this manner, his thoughts are forever set upon the object of his daily toil; his body has contracted certain fixed habits, which it can never shake off; in a word, he no longer belongs to himself, but to the calling that he has chosen. It is in vain that laws and manners have been at pains to level all the barriers round such a man and to open to him on every side a thousand different paths to fortune; a theory of manufactures more powerful than customs and laws binds him to a craft, and frequently to a spot, which he cannot leave; it assigns to him a certain place in society, beyond which he cannot go; in the midst of universal movement it has rendered him stationary.

In proportion as the principle of the division of labor is more extensively applied, the workman becomes more weak, more narrow-minded, and more dependent. The art advances, the artisan recedes. On the other hand, in proportion as it becomes more manifest that the productions of manufactures are by so much the cheaper and better as the manufacture is larger and the amount of capital employed more considerable, wealthy and educated men come forward to embark in manufactures, which were heretofore abandoned to poor or ignorant handicraftsmen. The magnitude of the efforts required and the importance of the results to be obtained attract them. Thus at the very time at which the science of manufactures lowers the class of workmen, it raises the class of masters.

While the workman concentrates his faculties more and more upon the study of a single detail, the master surveys an extensive whole, and the mind of the latter is enlarged in proportion as that of the former is narrowed. In a short time the one will require nothing but physical strength without intelligence; the other stands in need of science, and almost of genius, to ensure success. This man resembles more and more the administrator of a vast empire; that man, a brute.

The master and the workman have then here no similarity, and their differences increase every day. They are connected only like the two rings at the extremities of a long chain. Each of them fills the station which is made for him, and which he does not leave; the one is continually, closely, and necessarily dependent upon the other and seems as much born to obey as that other is to command. What is this but aristocracy?

As the conditions of men constituting the nation become more and more equal, the demand for manufactured commodities becomes more general and extensive, and the cheapness that places these objects within the reach of slender fortunes becomes a great element of success. Hence there are every day more men of great opulence and education who devote their wealth and knowledge to manufactures and who seek, by opening large establishments and by a strict division of labor, to meet the fresh demands which are made on all sides. Thus, in proportion as the mass of the nation turns to democracy, that particular class which is engaged in manufactures becomes more aristocratic. Men grow more alike in the one, more different in the other; and inequality increases in the less numerous class in the same ratio in which it decreases in the community. Hence it would appear, on searching to the bottom, that aristocracy should naturally spring out of the bosom of democracy.

But this kind of aristocracy by no means resembles those kinds which preceded it. It will be observed at once that, as it applies exclusively to manufactures and to some manufacturing callings, it is a monstrous exception in the general aspect of society. The small aristocratic societies that are formed by some manufacturers in the midst of the immense democracy of our age contain, like the great aristocratic societies of former ages, some men who are very opulent and a multitude who are wretchedly poor. The poor have few means of escaping from their condition and becoming rich, but the rich are constantly becoming poor, or they give up business when they have realized a fortune. Thus the elements of which the class of the poor is composed are fixed, but the elements of which the class of the rich is composed are not so. To tell the truth, though there are rich men, the class of rich men does not exist; for these rich individuals have no feelings or purposes, no traditions or hopes, in common; there are individuals, therefore, but no definite class.

Not only are the rich not compactly united among themselves, but there is no real bond between them and the poor. Their relative position is not a permanent one; they are constantly drawn together or separated by their interests. The workman is generally dependent on the master, but not on any particular master; these two men meet in the factory, but do not know each other elsewhere; and while they come into contact on one point, they stand very far apart on all others. The manufacturer asks nothing of the workman but his labor; the workman expects nothing from him but his wages. The one contracts no obligation to protect nor the other to defend, and they are not permanently connected either by habit or by duty. The aristocracy created by business rarely settles in the midst of the manufacturing population which it directs; the object is not to govern that population, but to use it. An aristocracy thus constituted can have no great hold upon those whom it employs, and even if it succeeds in retaining them at one moment, they escape the next; it knows not how to will, and it cannot act.

The territorial aristocracy of former ages was either bound by law, or thought itself bound by usage, to come to the relief of its serving-men and to relieve their distresses. But the manufacturing aristocracy of our age first impoverishes and debases the men who serve it and then abandons them to be supported by the charity of the public. This is a natural consequence of what has been said before. Between the workman and the master there are frequent relations, but no real association.

I am of the opinion, on the whole, that the manufacturing aristocracy which is growing up under our eyes is one of the harshest that ever existed in the world; but at the same time it is one of the most confined and least dangerous. Nevertheless, the friends of democracy should keep their eyes anxiously fixed in this direction; for if ever a permanent inequality of conditions and aristocracy again penetrates into the world, it may be predicted that this is the gate by which they will enter.

What Social Classes Owe to Each Other*

BY WILLIAM GRAHAM SUMNER

[1883]

On a new philosophy: That poverty is the best policy.

IT IS COMMONLY ASSERTED that there are in the United States no classes, and any allusion to classes is resented. On the other hand, we constantly read and hear discussions of social topics in which the existence of social classes is assumed as a simple fact. "The poor," "the weak," "the laborers," are expressions which are used as if they had exact and well-understood definition. Discussions are made to bear upon the assumed rights, wrongs, and misfortunes of certain social classes; and all public speaking and writing consists, in a large measure, of the discussion of general plans for meeting the wishes of classes of people who have not been able to satisfy their own desires. These classes are sometimes discontented, and sometimes not. Sometimes they do not know that anything is amiss with them until the "friends of humanity" come to them with offers of aid. Sometimes they are discontented and envious. They do not take their achievements as a fair measure of their rights. They do not blame themselves or their parents for their lot, as compared with that of other people. Sometimes they claim that they have a right to everything of which they feel the need for their happiness on earth. To make such a claim against God or Nature would, of course, be only to say that we claim a right to live on earth if we can. But God and Nature have ordained the chances and conditions of life on earth once for all. The case cannot be re-opened. We cannot get a revision of the laws of hu-

---

* Selections from Chapters I-III and XI. William Graham Sumner (1840–1910) served for several years as an Episcopalian minister. From 1872 on he occupied the chair of political and social science at Yale University. He is the most significant American representative of a strict *laissez-faire* philosophy.

man life. We are absolutely shut up to the need and duty, if we would learn how to live happily, of investigating the laws of Nature, and deducing the rules of right living in the world as it is. These are very wearisome and commonplace tasks. They consist in labor and self-denial repeated over and over again in learning and doing. When the people whose claims we are considering are told to apply themselves to these tasks they become irritated and feel almost insulted. They formulate their claims as rights against society—that is, against some other men. In their view they have a right, not only to *pursue* happiness, but to *get* it; and if they fail to get it, they think they have a claim to the aid of other men—that is, to the labor and self-denial of other men—to get it for them. They find orators and poets who tell them that they have grievances, so long as they have unsatisfied desires.

Now, if there are groups of people who have a claim to other people's labor and self-denial, and if there are other people whose labor and self-denial are liable to be claimed by the first groups, then there certainly are "classes," and classes of the oldest and most vicious type. For a man who can command another man's labor and self-denial for the support of his own existence is a privileged person of the highest species conceivable on earth. Princes and paupers meet on this plane, and no other men are on it at all. On the other hand, a man whose labor and self-denial may be diverted from his maintenance to that of some other man is not a free man, and approaches more or less toward the position of a slave. Therefore we shall find that, in all the notions which we are to discuss, this elementary contradiction, that there are classes and that there are not classes, will produce repeated confusion and absurdity We shall find that, in our efforts to eliminate the old vices of class government, we are impeded and defeated by new products of the worst class theory. We shall find that all the schemes for producing equality and obliterating the organization of society produce a new differentiation based on the worst possible distinction—the right to claim and the duty to give one man's effort for another man's satisfaction. We shall find that every effort to realize equality necessitates a sacrifice of liberty.

It is very popular to pose as a "friend of humanity," or a "friend of the working classes." The character, however, is quite exotic in the United States. It is borrowed from England, where some men, otherwise of small account, have assumed it with great success and advantage. Anything which has a charitable sound and a kind-hearted tone generally passes without investigation, because it is disagreeable to assail it. Sermons, essays, and orations assume a conventional standpoint with regard to the poor, the weak, etc.; and it is allowed to pass as an unquestioned doctrine in regard to social classes that "the rich" ought to "care for the poor;" that Churches especially ought to collect capital from the rich and spend it for the poor; that parishes ought to be clusters of institutions by means of which one social class should perform its duties to another; and that clergymen, economists, and social philosophers have a technical and professional duty to devise schemes for "helping the poor." The preaching in England used all to be done to the poor—that they ought to be contented with their lot and respectful to their betters. Now, the greatest part of the preaching in America consists in injunctions to those who have taken care of themselves to perform their assumed duty to take care of others. Whatever may be one's private sentiments, the fear of appearing cold and hard-hearted causes these conventional theories of social duty and these assumptions of social fact to pass unchallenged.

Let us notice some distinctions which are of prime importance to a correct consideration of the subject which we intend to treat.

Certain ills belong to the hardships of human life. They are natural. They are part of the struggle with Nature for existence. We cannot blame our fellow-men for our share of these. My neighbor and I are both struggling to free ourselves from these ills. The fact that my neighbor has succeeded in this struggle better than I constitutes no grievance for me. Certain other ills are due to the malice of men, and to the imperfections or errors of civil institutions. These ills are an object of agitation, and a subject of discussion. The former class of ills is to be met only by manly effort and energy; the latter may be corrected by associated effort. The former class of ills is constantly grouped and generalized, and made the object of social schemes. We shall see, as we go on, what that means. The second class of ills may fall on certain social classes, and reform will take the form of interference by other classes in favor of that one. The last fact is, no doubt, the reason why people have been led, not noticing distinctions, to believe that the same method was applicable to the other class of ills. The distinction here made between the ills which belong to the struggle for existence and those which are due to the faults of human institutions is of prime importance.

It will also be important, in order to clear up our ideas about the notions which are in fashion, to note the relation of the economic to the political significance of assumed duties of one class to another. That is to say, we may discuss the question whether one class owes duties to another by reference to the economic effects which will be produced on the classes and society; or we may discuss the political expediency of formulating and enforcing rights and duties respectively between the parties. In the former case we might assume that the givers of aid were willing to give it, and we might discuss the benefit or mischief of their activity. In the other case we must assume that some at least of those who were forced to give aid did so unwillingly. Here, then, there would be a question of rights. The question whether voluntary charity is mischievous or not is one thing; the question whether legislation which forces one man to aid another is right and wise, as well as economically beneficial, is quite another question. Great confusion and consequent error is produced by allowing these two questions to become entangled in the discussion. Especially we shall need to notice the attempts to apply legislative methods of reform to the ills which belong to the order of Nature.

There is no possible definition of "a poor man." A pauper is a person who cannot earn his living; whose producing powers have fallen positively below his necessary consumption; who cannot, therefore, pay his way. A human society needs the active co-operation and productive energy of every person in it. A man who is present as a consumer, yet who does not contribute either by land, labor, or capital to the work of society, is a burden. On no sound political theory ought such a person to share in the political power of the State. He drops out of the ranks of workers and producers. Society must support him. It accepts the burden, but he must be cancelled from the ranks of the rulers likewise. So much for the pauper. About him no more need to be said. But he is not the "poor man." The "poor man" is an elastic term, under which any number of social fallacies may be hidden.

Neither is there any possible definition of "the weak." Some are weak in one way, and some in another; and those who are weak in one sense are strong in another. In general, however, it may be said that those whom humanitarians and philanthropists call the weak are the ones through whom the productive and conservative forces of society are wasted. They constantly neutralize and destroy the finest efforts of the wise and industrious, and are a deadweight on the society in all its struggles to realize any better things. Whether the people who mean no harm, but are weak in the essential powers necessary to the performance of one's duties in life, or those who are malicious and vicious, do the more mischief, is a question not easy to answer.

Under the names of the poor and the weak, the negligent, shiftless, inefficient, silly, and imprudent are fastened upon the industrious and prudent as a responsibility and a duty. On the one side, the terms are extended to cover the idle, intemperate, and vicious, who, by the combination, gain credit which they do not deserve, and which they could not get if they stood alone. On the other hand, the terms are extended to include wage-receivers of the humblest rank, who are degraded by the combination. The reader who desires to guard himself against fallacies should always scrutinize the terms "poor" and "weak" as used, so as to see which or how many of these classes they are made to cover.

The humanitarians, philanthropists, and reformers, looking at the facts of life as they present themselves, find enough which is sad and unpromising in the condition of many members of society. They see wealth and poverty side by side. They note great inequality of social position and social chances. They eagerly set about the attempt to account for what they see, and to devise schemes for remedying what they do not like. In their eagerness to recommend the less fortunate classes to pity and consideration they forget all about the rights of other classes; they gloss over all the faults of the classes in question, and they exaggerate their misfortunes and their virtues. They invent new theories of property, distorting rights and perpetrating injustice, as any one is sure to do who sets about the re-adjustment of social relations with the interests of one group distinctly before his mind, and the interests of all other groups thrown into the background. When I have read certain of these discussions I have thought that it must be quite disreputable to be respectable, quite dishonest to own property, quite unjust to go one's way and earn one's own living, and that the only really admirable person was the good-for-nothing. The man who by his own effort raises himself above poverty appears, in these discussions, to be of no account. The man who has done nothing to raise himself above poverty finds that the social doctors flock about him, bringing the capital which they have collected from the other class, and promising him the aid of the State to give him what the other had to work for. . . . On the theories of the social philosophers to whom I have referred, we should get a new maxim of judicious living: Poverty is the best policy. If you get wealth, you will have to support other people; if you do not get wealth, it will be the duty of other people to support you.

No doubt one chief reason for the unclear and contradictory theories of class relations lies in the fact that our society, largely controlled in all its organization by one set of doctrines, still contains survivals of old social theories which are totally inconsistent with the former. In the Middle Ages men were united by custom and prescription into associations, ranks, guilds, and communities of various kinds. These ties endured as long as life lasted. Consequently society was dependent, throughout all its details, on status, and the tie, or bond, was sentimental. In our modern state, and in the United States more than anywhere else, the social structure is based on contract, and status is of the least importance. Contract, however, is rational—even rationalistic. It is also realistic, cold, and matter-of-fact. A contract relation is based on a sufficient reason, not on custom or prescription. It is not permanent. It endures only so long as the reason for it endures. In a state based on contract sentiment is out of place in any public or common affairs. It is relegated to the sphere of private and personal relations, where it depends not at all on class types, but on personal acquaintance and personal estimates. The sentimentalists among us always seize upon the survivals of the old order. They want to save them and restore them. Much of the loose thinking also which troubles us in our social discussions arises from the fact that men do not distinguish the elements of status and of contract which may be found in our society.

Whether social philosophers think it desirable or not, it is out of the question to go back to status or to the sentimental relations which once united baron and retainer, master and servant, teacher and pupil, comrade and comrade. That we have lost some grace and elegance is undeniable. That life once held more poetry and romance is true enough. But it seems impossible that any one who has studied the matter should doubt that we have gained immeasurably, and that our farther gains lie in going forward, not in going backward. The feudal ties can never be restored. If they could be restored they would bring back personal caprice, favoritism, sycophancy, and intrigue. A society based on contract is a society of free and independent men, who form ties without favor or obligation, and co-operate without cringing or intrigue. A society based on contract, therefore, gives the utmost room and chance for individual development, and for all the self-reliance and dignity of a free man. That a society of free men, co-operating under contract, is by far the strongest society which has ever yet existed; that no such society has ever yet developed the full measure of strength of which it is capable; and that the only social improvements which are now conceivable lie in the direction of more complete realization of a society of free men united by contract, are points which cannot be controverted. It follows, however, that one man, in a free state, cannot claim help from, and cannot be charged to give help to, another. To understand the full meaning of this assertion it will be worth while to see what a free democracy is.

That a free man is a sovereign, but that a sovereign cannot take "tips."

A free man, a free country, liberty, and equality are terms of constant use among us. They are employed as watch-words as soon as any social questions come into discussion. It is right that they should be so used. They ought to contain the broadest convictions and most positive faiths of the nation, and so they ought to be available for the decision of questions of detail.

In order, however, that they may be so employed successfully and correctly it is essential that the terms should be correctly defined, and that their popular use should conform to correct definitions. No doubt it is generally believed that the terms are easily understood, and present no difficulty. Probably the popular notion is, that liberty means doing as one has a mind to, and that it is a metaphysical or sentimental good. A little observation shows that there is no such thing in this world as doing as one has a mind to. There is no man, from the tramp up to the President, the Pope, or the Czar, who can do as he has a mind to. There never has been any man, from the primitive barbarian up to a Humboldt or a Darwin, who could do as he had a mind to. The "Bohemian" who determines to realize some sort of liberty of this kind accomplishes his purpose only by sacrificing most of the rights and turning his back on most of the duties of a civilized man, while filching as much as he can of the advantages of living in a civilized state. Moreover, liberty is not a metaphysical or sentimental thing at all. It is positive, practical, and actual. It is produced and maintained by law and institutions, and is, therefore, concrete and historical. Sometimes we speak distinctively of civil liberty; but if there be any liberty other than civil liberty—that is, liberty under law—it is a mere fiction of the schoolmen, which they may be left to discuss.

Even as I write, however, I find in a leading review the following definition of liberty: Civil liberty is "the result of the restraint exercised by the sovereign people on the more powerful individuals and classes of the community, preventing them from availing themselves of the excess of their power to the detriment of the other classes." This definition lays the foundation for the result which it is apparently desired to reach, that "a government by the people can in no case become a paternal government, since its lawmakers are its mandatories and servants carrying out its will, and not its fathers or its masters." Here we have the most mischievous fallacy under the general topic which I am discussing distinctly formulated. In the definition of liberty it will be noticed that liberty is construed as the act of the sovereign people against somebody who must, of course, be differentiated from the sovereign people. Whenever "people" is used in this sense for anything less than the total population, man, woman, child, and baby, and whenever the great dogmas which contain the word "people" are construed under the limited definition of "people," there is always fallacy.

History is only a tiresome repetition of one story. Persons and classes have sought to win possession of the power of the State in order to live luxuriously out of the earnings of others. Autocracies, aristocracies, theocracies, and all other organizations for holding political power, have exhibited only the same line of action. It is the extreme of political error to say that if political power is only taken away from generals, nobles, priests, millionnaires, and scholars, and given to artisans and peasants, these latter may be trusted to do only right and justice, and never to abuse the power; that they will repress all excess in others, and commit none themselves. They will commit abuse, if they can and dare, just as others have done. The reason for the excesses of the old governing classes lies in the vices and passions of human nature—cupidity, lust, vindictiveness, ambition, and vanity. These vices are confined to no nation, class, or age. They appear in the church, the academy, the workshop, and the hovel, as well as in the army or the palace. . . . The only thing which has ever restrained these vices of human nature in those who had political power is law sustained by impersonal institutions. If political power be given to the masses who have not hitherto had it, nothing will stop them from abusing it but laws and institutions. To say that a popular government cannot be paternal is to give it a charter that it can do no wrong. The trouble is that a democratic government is in greater danger than any other of becoming paternal, for it is sure of itself, and ready to undertake anything, and its power is excessive and pitiless against dissentients.

What history shows is, that rights are safe only when guaranteed against all arbitrary power, and all class and personal interest. . . . There has been no liberty at all, save where a state has known how to break out, once for all, from this delusive round; to set barriers to selfishness, cupidity, envy, and lust, in *all* classes, from highest to lowest, by laws and institutions; and to create great organs of civil life which can eliminate, as far as possible, arbitrary and personal elements from the adjustment of interests and the definition of rights. Liberty is an affair of laws and institutions which bring rights and duties into equilibrium. It is not at all an affair of selecting the proper class to rule.

The notion of a free state is entirely modern. It has been developed with the development of the middle class, and with the growth of a commercial and industrial civilization. Horror at human slavery is not a century old as a common sentiment in a civilized state. The idea of the "free man," as we understand it, is the product of a revolt against mediaeval and feudal ideas; and our notion of equality, when it is true and practical, can be explained only by that revolt. It was in England that the modern idea found birth. It has been strengthened by the industrial and commercial development of that country. It has been inherited by all the English-speaking nations, who have made liberty real because they have inherited it, not as a notion, but as a body of institutions. . . .

The notion of civil liberty which we have inherited is that of *a status created for the individual by laws and institutions, the effect of which is that each man is guaranteed the use of all his own powers exclusively for his own welfare.* It is not at all a matter of elections, or universal suffrage, or democracy. All institutions are to be tested by the degree to which they guarantee liberty. It is not to be admitted for a moment that liberty is a means to social ends, and that it may be impaired for major considerations. Any one who so argues has lost the bearing and relation of all the facts and factors in a free state. A human being has a life to live, a career to run. He is a centre of powers to work, and of capacities to suffer. What his powers may be—whether they can carry him far or not; what his chances may be, whether wide or restricted; what his fortune may be, whether to suffer much or little—are questions of his personal destiny which he must work out and endure as he can; but for all that concerns the bearing of the society and its institutions upon that man, and upon the sum of happiness to which he can attain during his life on earth, the product of all history and all philosophy up to this time is summed up in the doctrine, that he should be left free to do the most for himself that he can, and should be guaranteed the exclusive enjoyment of all that he does. If the society—that is to say, in plain terms, if his fellow-men, either individually, by groups, or in a mass—impinge upon him otherwise than to surround him with neutral conditions of security, they must do so under the strictest responsibility to justify themselves. Jealousy and prejudice against all such interferences are high political virtues in a free man. It is not at all the function of the State to make men happy. They must make themselves happy in their own way, and at their own risk. The functions of the State lie entirely in the conditions or chances under which the pursuit of happiness is carried on, so far as those conditions or chances can be affected by civil organization. Hence, liberty for labor and security for earnings are the ends for which civil institutions exist, not means which may be employed for ulterior ends.

Now, the cardinal doctrine of any sound political system is, that rights and duties should be in equilibrium.... An immoral political system is created whenever there are privileged classes—that is, classes who have arrogated to themselves rights while throwing the duties upon others. In a democracy all have equal political rights. That is the fundamental political principle. A democracy, then, becomes immoral, if all have not equal political duties. This is unquestionably the doctrine which needs to be reiterated and inculcated beyond all others, if the democracy is to be made sound and permanent. Our orators and writers never speak of it, and do not seem often to know anything about it; but the real danger of democracy is, that the classes which have the power under it will assume all the rights and reject all the duties—that is, that they will use the political power to plunder those-who-have. Democracy, in order to be true to itself, and to develop into a sound working system, must oppose the same cold resistance to any claims for favor on the ground of poverty, as on the ground of birth and rank. It can no more admit to public discussion, as within the range of possible action, any schemes for coddling and helping wage-receivers than it could entertain schemes for restricting political power to wage-payers. It must put down schemes for making "the rich" pay for whatever "the poor" want, just as it tramples on the old theories that only the rich are fit to regulate society. One needs but to watch our periodical literature to see the danger that democracy will be construed as a system of favoring a new privileged class of the many and the poor.

Holding in mind, now, the notions of liberty and democracy as we have defined them, we see that it is not altogether a matter of fanfaronade when the American citizen calls himself a "sovereign." A member of a free democracy is, in a sense, a sovereign. He has no superior. He has reached his sovereignty, however, by a process of reduction and division of power which leaves him no inferior. It is very grand to call one's self a sovereign, but it is greatly to the purpose to notice that the political responsibilities of the free man have been intensified and aggregated just in proportion as political rights have been reduced and divided.... The free man who steps forward to claim his inheritance and endowment as a free and equal member of a great civil body must understand that his duties and responsibilities are measured to him by the same scale as his rights and his powers. He wants to be subject to no man. He wants to be equal to his fellows, as all sovereigns are equal. So be it; but he cannot escape the deduction that he can call no man to his aid. The other sovereigns will not respect his independence if he becomes dependent, and they cannot respect his equality if he sues for favors. The free man in a free democracy, when he cut off all the ties which might pull him down, severed also all the ties by which he might have made others pull him up. He must take all the consequences of his new status. He is, in a certain sense, an isolated man. The family tie does not bring to him disgrace for the misdeeds of his relatives, as it once would have done, but neither does it furnish him with the support which it once would have given. The relations of men are open and free, but they are also loose. A free man in a free democracy derogates from his rank if he takes a favor for which he does not render an equivalent.

A free man in a free democracy has no duty whatever toward other men of the same rank and standing, except respect, courtesy, and good-will. We cannot say that there are no classes, when we are speaking politically, and then say that there are classes, when we are telling A what it is his duty to do for B. In a free state every man is held and expected to take care of himself and his family, to make no trouble for his neighbor, and to contribute his full share to public interests and common necessities. If he fails in this he throws burdens on others. He does not thereby acquire rights against the others. On the contrary, he only accumulates obligations toward them; and if he is allowed to make his deficiencies a ground of new claims, he passes over into the position of a privileged or petted person—emancipated from duties, endowed with claims. This is the inevitable result of combining democratic political theories with humanitarian social theories. It would be aside from my present purpose to show, but it is worth noticing in passing, that one result of such inconsistency must surely be to undermine democracy, to increase the power of wealth in the democracy, and to hasten the subjection of democracy to plutocracy; for a man who accepts any share which he has not earned in another man's capital cannot be an independent citizen.

It is often affirmed that the educated and wealthy have an obligation to those who have less education and property, just because the latter have political equality with the former, and oracles and warnings are uttered about what will happen if the uneducated classes who have the suffrage are not instructed at the care and expense of the other classes. In this view of the matter universal suffrage is not a measure for *strengthening* the State by bringing to its support the aid and affection of all classes, but it is a new burden, and, in fact, a peril. Those who favor it represent it as a peril. This doctrine is politically immoral and vicious. When a community establishes universal suffrage, it is as if it said to each new-comer, or to each young man: "We give you every chance that any one else has. Now come along with us; take care of yourself, and contribute your share to the burdens which we all have to bear in order to support social institutions." Certainly, liberty, and universal suffrage, and democracy are not pledges of care and protection, but they carry with them the exaction of individual responsibility. The State gives equal rights and equal chances just because it does not mean to give anything else. It sets each man on his feet, and gives him leave to run, just because it does not mean to carry him. Having obtained his chances, he must take upon himself the responsibility for his own success or failure. It is a pure misfortune to the community, and one which will redound to its injury, if any man has been endowed with political power who is a heavier burden then than he was before; but it cannot be said that there is any new *duty* created for the good citizens toward the bad by the fact that the bad citizens are a harm to the State.

That it is not wicked to be rich; nay, even, that it is not wicked to be richer than one's neighbor.

... Undoubtedly there are cases of fraud, swindling, and other financial crimes; that is to say, the greed and selfishness of men are perpetual. They put on new phases, they adjust themselves to new forms of business, and constantly devise new methods of fraud and robbery, just as burglars devise new artifices to circumvent every new precaution of the lock-makers. The criminal law needs to be improved to meet new forms of crime, but to denounce financial devices which are useful and legitimate because use is made of them for fraud, is ridiculous and unworthy of the age in which we live.

... Any one who believes that any great enterprise of an industrial character can be started without labor must have little experience of life. Let any one try to get a railroad built, or to start a factory and win reputation for its products, or to start a school and win a reputation for it, or to found a newspaper and make it a success, or to start any other enterprise, and he will find what obstacles must be overcome, what risks must be taken, what perseverance and courage are required, what foresight and sagacity are necessary. Especially in a new country, where many tasks are waiting, where resources are strained to the utmost all the time, the judgment, courage, and perseverance required to organize new enterprises and carry them to success are sometimes heroic. Persons who possess the necessary qualifications obtain great rewards. They ought to do so. It is foolish to rail at them. Then, again, the ability to organize and conduct industrial, commercial, or financial enterprises is rare; the great captains of industry are as rare as great generals. The great weakness of all co-operative enterprises is in the matter of supervision. Men of routine or men who can do what they are told are not hard to find; but men who can think and plan and tell the routine men what to do are very rare. They are paid in proportion to the supply and demand of them. . . .

The aggregation of large fortunes is not at all a thing to be regretted. On the contrary, it is a necessary condition of many forms of social advance. If we should set a limit to the accumulation of wealth, we should say to our most valuable producers, "We do not want you to do us the services which you best understand how to perform, beyond a certain point." It would be like killing off our generals in war. A great deal is said, in the cant of a certain school, about "ethical views of wealth," and we are told that some day men will be found of such public spirit that, after they have accumulated a few millions, they will be willing to go on and labor simply for the pleasure of paying the taxes of their fellow-citizens. Possibly this is true. It is a prophecy. It is as impossible to deny it as it is silly to affirm it. For if a time ever comes when there are men of this kind, the men of that age will arrange their affairs accordingly. There are no such men now, and those of us who live now cannot arrange our affairs by what men will be a hundred generations hence. . . .

Wherefore we should love one another.

... Rights do not pertain to *results*, but only to *chances*. They pertain to the *conditions* of the struggle for existence, not to any of the results of it; to the *pursuit* of happiness, not to the possession of happiness. It cannot be said that each one has a right to have some property, because if one man had such a right some other man or men would be under a corresponding obligation to provide him with some property. Each has a right to acquire and possess property if he can. It is plain what fallacies are developed when we overlook this distinction. Those fallacies run through *all* socialistic schemes and theories. If we take rights to pertain to results, and then say that rights must be equal, we come to say that men have a right to be equally happy, and so on in all the details. Rights should be equal, because they pertain to chances, and all ought to have equal chances so far as chances are provided or limited by the action of society. This, however, will not produce equal results, but it is right just because it will produce unequal results—that is, results which shall be proportioned to the merits of individuals. We each owe it to the other to guarantee mutually the chance to earn, to possess, to learn, to marry, etc., etc., against any interference which would prevent the exercise of those rights by a person who wishes to prosecute and enjoy them in peace for the pursuit of happiness. If we generalize this, it means that all-of us ought to guarantee rights to each of us. ...

... The only help which is generally expedient, even within the limits of the private and personal relations of two persons to each other, is that which consists in helping a man to help himself. This always consists in opening the chances. A man of assured position can, by an effort which is of no appreciable importance to him, give aid which is of incalculable value to a man who is all ready to make his own career if he can only get a chance. The truest and deepest pathos in this world is not that of suffering but that of brave struggling. The truest sympathy is not compassion, but a fellow-feeling with courage and fortitude in the midst of noble effort.

How We Live Now in America*

BY GRANVILLE HICKS

[1953]

AROUND OUR HOUSE in upstate New York are lawn and flower gardens and a kitchen garden, but beyond this small area of cultivation the forest has taken over. Yet this was a farm for more than a hundred years, supporting four generations, and when Roxborough's population was at its peak, in the middle of the 19th century, it was one of hundreds of farms in the town. Today there aren't half a dozen men in the whole of Roxborough who get the major part of their livelihood from farming. We have our kitchen gardens, and some of us have hens, and here and there a family keeps a cow, but most of what we eat comes from the counters and shelves of supermarkets. Our clothing is selected from mail order catalogues or bought in city stores, and the spinning wheels and looms and hatchels of our ancestors are bait for antique dealers. The fuel we burn is delivered in trucks and wood for the fireplace is hard to come by and almost as dear as in the cities.

These are signs of a two-way revolution. The farming is now done elsewhere, and done with fewer and fewer men and more and more machines, under conditions that often approximate the conditions of the mass production industries. We live in the country, but we live on the produce of California and Texas and Florida and New Jersey, just as we would if we lived in Troy or Albany or New York City. And by and large we make our living out of industry, most of us directly, by working in the factories of the Capital District, most of the others indirectly, by catering to the summer people or serving the needs of the commuters. The small town has been absorbed by the great society.

---

* From *Commentary*, December, 1953. Reprinted by permission. Granville Hicks (b. 1901) is a writer and author of many books and articles on literary and social criticism.

The kind of life we lead is made possible by a series of machines, just as it would be if we lived in a city, the only difference being that most of the machines are owned by us individually instead of being parts of a gigantic impersonal apparatus. The apartment dweller takes for granted water and sanitation and light and heat and transportation, whereas these are matters that we have to take care of ourselves. Even at that, however, we are in much the same situation as millions of people who live in the housing developments that have multiplied on the fringes of the cities, and we are closer to the apartment dwellers than we are to the people who lived on these acres twenty-five years ago.

Counting cars, pumps, lawn mower, refrigerator, vacuum cleaner, and so forth, our family has at its disposal at least a dozen motors, and our neighbors have as many or more. For most people in the town the automobile is the indispensable machine, for it is what makes it possible for them not to have to choose between working in the city and living in the country. But a car is not merely a key to a job; for us in Roxborough, as for millions of Americans, it is the foundation of our social life, and we drive in it both for recreation and to get to the places where recreation, of the kind we happen to be seeking, can be found. For us, as for millions, it has altered the mores of courtship and sex, and it is with respect to the use of the car or the cars that a major part of family life has to be planned. Moreover, for those who know something about machinery, and they are numerous in Roxborough, it is a focus of intellectual interest and one of the livelier topics of conversation.

Next to the automobile, it is electricity that has revolutionized Roxborough. The first power lines were run into the town barely twenty-five years ago, against fanatical opposition. We ourselves, because we were a long way from the main line, spent several summers and two winters here without electricity. Kerosene lamps were no novelty to my wife and me, for we had grown up with them, and we remembered the excitement of getting electric lights. That was what electricity meant at the time of the First World War—safer, more adequate, more convenient lighting. But by 1937, when electricity was coaxed by subsidies up our road, it brought release from a dozen kinds of drudgery by way of the automatic water pump, the refrigerator, the washing machine, the electric flatiron, and, eventually, the electric range and the automatic oil burner. As the power lines have pushed farther and farther back into the hills since World War II, they have carried all these conveniences and new means of entertainment as well. When power fails, it is, many people say, the television that they miss most. If we are happy enough without TV, we do miss the record player and the FM radio.

As anybody can see, rural life more and more closely resembles urban life, and this is true not only for small towns like ours, which has become a semi-suburb, but also for the areas that raise the nation's food. In the dairy country of western New York, in the Corn Belt, in the Wheat Belt, in the fruit-raising sections of California and Florida, farmers not only have thousands of dollars worth of labor-saving machinery in their fields and barns; they have the conveniences and luxuries in their homes. Even in the poorer farm country the automobile and electricity are remaking the pattern of rural life.

Naturally the conveniences and luxuries have to be paid for, and they wouldn't be available if people couldn't buy them. The revolution, in other words, is economic as well as technological. During the worst years of the depression we were spending summers in Roxborough and winters in the city, and we could see that the victims of the depression fared better in Roxborough. The city jobless had nothing, but in Roxborough a man without a job could raise some food, cut some wood, do a little tinkering here and there. He was better off not only because he and his family ate more adequately but also because he had something to do. If a depression should come along now, we in Roxborough would still be better fixed than city people, but we should have more to lose than did the people who were living here in 1930. That is part of the price we pay for being absorbed by the great society.

What, on the other hand, we have gained will not be underestimated by someone who has been looking at the town for twenty years. The effect of prosperity is to be measured not merely in terms of motors and gadgets but also and especially in terms of human satisfactions. Men, well paid, able to live comfortably, have more self-respect. Women, released from part of their drudgery, with a little leisure and money enough to enjoy it, seem younger and prettier. Men, women, and children look better and obviously feel better, for they eat more nourishing food and can afford to go to doctors and dentists. If a TV set or a new car seems more important to a lot of people than dental repairs or a coat of paint on the house, most of them get around to the dentistry and the paint sooner or later.

As the town has become a semi-suburb, it has acquired a more varied population. Most of the newcomers are factory workers, just as most of the natives are, but we have acquired some white-collar workers and professionals. Yet class stratification hasn't taken place. Like any other town, Roxborough is full of cliques, but their composition is constantly shifting, and they seldom seem to be determined by the jobs or the incomes or the education of their members. And if you look at the cars parked outside the town hall when some community function is going on, all you can tell about the people inside is that most of them seem to be pretty prosperous.

The blurring of class lines is not peculiar to Roxborough; it is happening all over America. The breaking down of class distinctions, as has often been said, is most evident in the matter of clothes. . . . By and large it is considered improper in our society today to use clothes as a sign of social status or even, in the daily routine, of wealth. Clothes that served a symbolic purpose—the top hat, the frock coat, the starched shirt—have almost completely disappeared, and formal dress has become less common and less formal. Not only do Americans of all classes look much the same when they dress up; they are indistinguishable when they are not dressed up, and that is a larger and larger part of the time. If it takes a sharp eye to tell an original evening gown from an expensive imitation and the expensive imitation from a model that can be purchased by any shop girl, it is quite impossible to tell the rich girl from the poor one when both are wearing blue jeans. . . .

In the more stable societies of the past, when class differences were generally accepted, there were recognized ways in which they were symbolized. We Americans have never been wholly at ease with such symbols, and probably that is why at certain periods so much emphasis has been laid upon them. After the Civil War, when great fortunes were being made, the newly rich found it imperative to display their wealth, and we entered upon a period of what Thorstein Veblen called conspicuous consumption. When a man made a million, he built an ostentatious house that was too large for his family and staffed it with more servants than he needed. As his wealth grew, his carriages, his private car, his yacht, his clothes, his wife's clothes, his art collection, his philanthropies, everything about his life proclaimed the fact that he had money.

Today the conspicuous consumers are not the industrialists and financiers but the stars of movies and radio and the members of the expense-account aristocracy. And though many people envy these publicized figures their swimming pools and their night club glamor and extravagance, almost everyone is aware that their wealth doesn't mean much. . . .

Although there are fewer very poor and, relatively speaking, fewer very rich than there were twenty-five or fifty years ago, there are still grievous economic inequalities in this country, but they don't seem so important as they once did. This is partly because almost all of us are better off than we were, and therefore aren't so bitter against those who are still better off. It is partly because we know that the rich are heavily taxed and that their powers are curbed in many ways. And it is partly because the rich, instead of flaunting their economic superiority, seem a little embarrassed about it and try to assure us that they aren't any different from us. So we go about our various businesses with no feeling of inferiority and little resentment.

Most of the able-bodied men in Roxborough, as I have said, work in factories. They work forty or forty-four hours a week instead of the fifty-four or sixty or seventy-two hours that were common in the 19th century and persisted into the 20th. Many of them are engaged in work that taxes body and mind, but in this time of full employment a man's nose can't be kept to the grindstone all through the working day. Likely enough he gets a break for coffee, and if he goes to the toilet and sneaks a cigarette, he isn't going to be fired. When our Roxborough people come home they aren't exhausted, but are ready to work around the house—and practically every man in town has a dozen home repair jobs that his conscience and his wife are nagging him to finish—or go to a party or a square dance, or settle in for an evening with TV. Almost everybody is in debt, for no one makes enough to pay for all the things he and his wife and children want, but it is comparatively seldom that a car or a television set is repossessed. These people are getting a share of the good things of life, and they know it.

To the responsibilities and obligations of the great society most of them give little thought. They pay their union dues as automatically and, when they think about it, as resentfully as they pay their income taxes. For the most part they regard both government and unions as necessary evils, and they are more conscious of both than they are of the higher levels of management and ownership. They feel no animosity towards the remote big shots, for they are scarcely aware of their existence. Their gripes are over little things —the unreasonableness of a foreman, the stupidity of a fellow worker, the cost and quality of food in the cafeteria. Few of them, so far as I can tell, find acute satisfaction in their work, and yet no one feels that his job is meaningless drudgery. (When a man feels that way today, he goes and gets another job.) Any job that I hear talk about calls for the exercise of some skill that entitles a man to the respect of his companions and to self-respect. And even when the day's work is pretty tedious, a time comes when a man can go home to race motorcycles or paint the house or make love to his wife.

For the women of Roxborough the revolution has been even more of a success. To begin with, if they need or want to work, there are jobs for them and a great variety of jobs. . . .

What makes the big difference for most Roxborough women is the revolution that has taken place in their homes through the acquisition of labor-saving machinery. It has been a rapid revolution everywhere, and in Roxborough it has largely come about in ten or a dozen years. We have leapt almost in a single bound from an era in which a woman had to carry water, sweep and beat her rugs, scrub the family washing, and keep food in the spring house, into an era of multiple mechanical servants. Unless she has young children, a Roxborough woman enjoys hours of daily leisure, hours in which, of course, she may be very busy, but not under the old compulsions. . . .

Today, with so many better paid and otherwise more attractive jobs open to women, domestic labor has become a major luxury. Higher and higher in the social and financial scale we find that women are relying on machines, just as the Roxborough women do, to lighten their burdens and give them leisure.

If we try to add it all up, we can only conclude that this is a prosperous society, with a larger and larger proportion of the people sharing in the prosperity. Distinctions diminish between country and city, between rich and poor, between men and women, between old families and the children of immigrants, even—though so much more is to be desired—between black and white.

In *Democracy in America* Alexis de Tocqueville wrote:

The good things and the evils of life are more equally distributed in the world: great wealth tends to disappear, the number of small fortunes to increase; desires and gratifications are multiplied, but extraordinary prosperity and irremediable penury are alike unknown. The sentiment of ambition is universal, but the scope of ambition is seldom vast. . . . Human existence becomes longer, and property more secure; life is not adorned with brilliant trophies, but it is extremely easy and tranquil. Few pleasures are either very refined or very coarse; and highly polished manners are as uncommon as great brutality of taste. . . . Almost all extremes are softened or blunted: all that was most prominent is superseded by some mean term, at once less lofty and less low, less brilliant and less obscure, than what before existed in the world.

It was in 1831 that de Tocqueville visited the United States, but much of what he wrote fits our revolution. In the second half of the 19th century, with the expansion of industry, it appeared that he had been a poor prophet, for the differences between rich and poor were constantly becoming greater, and equality of opportunity seemed to be a myth. That was the theme of much of the agitation of the period: the promise of American life had been betrayed. But now, with the industrial revolution in a later phase, de Tocqueville's predictions seem to be coming true. If in our day we tend to overstate the degree of equality, as certainly he did in his, the contrast between what we have and what the world has generally known can scarcely be exaggerated.

Although he believed that the worldwide triumph of democracy was inevitable and was on the whole a good thing, de Tocqueville, it will be recalled, was not unqualifiedly enthusiastic about the new order. Immediately after the passage I have quoted he wrote, "When I survey this countless multitude of beings, shaped in each other's likeness, amid whom nothing rises and nothing falls, the sight of such universal uniformity saddens and chills me, and I am tempted to regret that state of society which has ceased to be."

The nightmare of "universal uniformity" has haunted critics of democracy ever since, and we still hear voices, some American, some European, complaining of the standardization of life in the United States.

American democracy is open to many criticisms, including many of those expressed by de Tocqueville more than a hundred years ago, but the charge of standardization will not hold water. As the experience of totalitarian countries ought to teach us, it takes a lot of doing to standardize human beings even in a slave state, and in a free state it is impossible. The individual may be standardized in a stable, stratified society, but in a social order that is constantly changing he cannot be. You cannot make people alike by giving them more and more opportunities to be different.

A quarter of a century ago, as mass production got under way, the assembly lines devised by efficiency engineers threatened what seemed to be an intolerable standardization of labor. Man was treated as a machine, not as a human being. But it soon became clear that it was more economical, as well as more humane, to treat machines as machines and men as men. Purely mechanical labor can always be done more cheaply by machines, freeing human beings for the more varied and more creative tasks that the machines cannot perform. Furthermore, through their unions and their individual acts of self-assertion, human beings find ways of resisting the tyranny and monotony of the assembly line. And so long as he has leisure, no man is likely to become a robot; even the man who performs one simple function all through the working day turns into a Jack-of-all-trades when he gets home.

The standardization of the product is the basis of modern industry, and sometimes the effects are good and sometimes they are bad, but even when they are bad, a cure for the evil is likely to appear. If the baking industry succeeds in foisting upon the nation a uniformly tasteless loaf of bread, smaller bakers emerge to furnish better bread for those who are willing to pay for it, and more and more housewives spend some of their leisure in doing their own baking. One supermarket is much like another, but any supermarket has a variety of goods on its shelves that would dumbfound the corner grocer of a generation ago. . . .

It is in mass entertainment that many critics have found the greatest threat of standardization—movies, radio, TV. Unquestionably these media have introduced an element of uniformity in American life: they give us our heroes, our catch phrases, our jokes; they influence our tastes and values. Possibly this degree of uniformity, holding together a widely dispersed people, is not a bad thing, even though it is accomplished on a low cultural level. But in any case what the critics forget is that these media increasingly cater to a diversity of tastes and that through such by-products as the long-playing record they have enriched the lives of those who happen not to want what the millions want.

The way we American spend our leisure time is the best proof that standardization is a myth. The critic will point out that hobbies are themselves constantly being standardized, and that is true.... But that does not do away with the fact of diversity. How do you suppose the people of Roxborough are entertaining themselves this weekend? They are not all watching TV, though probably more are doing that than any other one thing; they are not all out riding in their automobiles, though that, too, is one of the more popular forms of diversion; they are gardening, bird-watching, listening to symphonies, tinkering with machinery, horseback-riding, playing cards, dancing, even reading books. You cannot see them without knowing that these are people who have multitudinous choices and are happy in that fact.

Perhaps it is a law of industrial development that machines first enslave men and then liberate them. What we are seeing in the United States is an indication, if not proof, that at the higher technological levels industrialism can provide the abundance that once seemed a utopian pipe dream, and that on this abundance can be built a civilization in which opportunity is almost universal and special privilege at a minimum. Our revolution, like all revolutions, has entailed great suffering and brought many evils, but at least we see the possibility that the good can outweigh the bad.

Certainly few of my Roxborough neighbors seem dissatisfied with their lot. They know, to be sure, that they're not living in paradise: they and their families suffer from disease and accidents; husbands and wives sometimes get along badly, and children are often a problem; no day is without its worries and irritations. But this is the kind of society they want to live in—one in which jobs are plentiful and pay is good and there are lots of things you can buy with your money. Their outlook on life may be materialistic, as some Europeans charge, but they are easier-going about money than most Europeans of any nation or class. They enjoy being generous and unsuspicious in money matters, and there are few of them who would cheat you unless they thought you were getting ready to cheat them. It would be reckless to say that all spend their money wisely, but most of them get what is for them their money's worth.

Whatever their personal problems, there are only two social problems that concern them, but these concern them constantly and deeply: the possibility of depression and the possibility of war. Either, they fully realize, can destroy most of what they value in life. Depression is something most of those over twenty-five know something about at firsthand, for deprivation and fear in the early 30's left their mark even on the very young. As for war, it was only eight years ago that atom bombs fell on Hiroshima and Nagasaki, and I believe there are few of my neighbors who do not give daily thought to the chance of their falling on New York or Washington or Schenectady.

All America, of course, is in the same boat; either depression or war would be the end of the particular chapter we as a people are now writing. Since we have no assurance that we can avoid a depression, and are even more dubious about preventing war, we can only regard our experiment thus far as tentative. But it is not any the less significant because it may be interrupted—or, for that matter, finished for good and all.

Even if we leave depression and war out of account, we know that we pay a price for what we have, but most of my neighbors would say it is not too high a price. The price is to be defined chiefly in terms of the stresses and strains that are incident to the functioning of the great society. These stresses and strains are most palpable when you have to travel from Astor Place, say, to 125th Street on New York City's Lexington Avenue subway at five-thirty in an August heat wave, or when, in a November ice storm, you drive from a factory in the southeast section of Detroit to a housing project on the northwest boundary. Who can reckon the price in irritation and frustration, to say nothing of the threat to life and limb, that is paid by each one of millions of commuters in an age in which home is here and work is there? How many traumas, to say nothing of the accidents that get into the papers, are sustained in taking city and suburban families for a Sunday at the beach or in the country? It is no wonder that a fairly large number of people crack up. But fortunately the human species is tough, and, as most of my neighbors would say, we can take it.

The great society is necessarily a highly organized society, and organization is painful even when it is efficient and doubly so when it isn't. The "hurry up and wait" of the war is constantly reflected in civilian life. The instrument of organization is bureaucracy, and the *sine qua non* of bureaucracy is red tape. . . .

Our version of the great society operates partly through legal compulsions, partly through economic inducements, and partly through persuasion. Persuasion most commonly takes the form of advertising, and the advertising man is one of the distinctive products of our culture. Since mass production is possible only if a lot of people can be counted on to want the same thing at the same time, some form of advertising is indispensable, but the constant assault to which our sensibilities are subjected seems excessive. Fortunately most people develop an immunity to pressure and a skepticism regarding claims, but the atmosphere is polluted just the same.

The blatant voice of the advertising man reminds us that this is still, in some degree, a business civilization, and it is precisely the question of the degree that is important. Not only has the power of business been cut down since the 20's; the prestige of business has declined. For one thing, the expansion of government has created new careers outside of business, and many young people have found government service more attractive. As both government and organized labor have grown stronger, business has come to seem less important to the average citizen. The profit motive is still a vital force, but to many people "security" is a more potent word than "profit." Moreover, as prosperity becomes widespread, there seems to be less general acceptance of pecuniary standards. There are a few sharp dealers among my neighbors, and their capacity for making a fast buck is almost always spoken of with a kind of grudging admiration, but most people see no reason for trying to emulate them. They feel that if these local entrepreneurs want money as badly as all that, it's O.K., but they're not going to be bothered so long as they can get by. And they feel the same, as far as I can tell, about the big shots of business when they read about them in the papers.

No one is foolish enough to think that the state of affairs that exists in the United States is perfect. It is enough if we have here some indication that the industrial revolution is at long last being made to serve the interests of all or almost all of the people. We are approximating a way of life that has been the goal of human struggle for a long time. In this period of American hegemony in the West our friends may very well wonder whether we are wise enough to make good use of our power, but our follies ought not to blind them—or us—to the real significance of what has been happening in the United States. It is here, not in Russia, that the great revolution of our time is most advanced. Even if disaster overtakes our experiment tomorrow, the degree of success we have achieved ought to offer hope to mankind so long as mankind survives.

Commencement, 1955*

BY GEORGE F. KENNAN

[1955]

... THERE are certain traditional visions of self-fulfillment which, as I recall it, used to be commonly held out to young people, but which a person of my generation finds it difficult to accept in their entirety. I am referring here to the possibilities for happiness and self-realization in the purely personal and subjective field—in the development, that is, of the individual personality both within itself and in its relations of intimacy with other individuals. Former generations of Americans had a high degree of confidence in these possibilities. They had derived this, together with some of their political beliefs, from the romantic movement of Europe's nineteenth century, with its exalted idealization of the individual personality, and its corresponding portrayal of personal attachment between individuals as the greatest potential source of human happiness. This romanticism has survived to the present day, almost unmodified, in a portion of American folklore and particularly in our stupendous output of commercialized entertainment, except that here it always has an optimistic ring, whereas in Europe it was generally conceived as tragic.

But in this respect, too, time—I am afraid—has wrought its changes; I, for one, could not honestly encourage you to place on this personal world, important and promising as it may be, the same exalted hopes which our popular fiction and screen drama might suggest it to warrant. Others, I am confident, feel the same way.

Why is this? I am not sure any of us understand too well why this change has come about. Perhaps our modern sense of humor has something to do with it—for romanticism and humor are like

---

* From *Social Research* Vol. XXII, Summer, 1955. Reprinted by permission. George F. Kennan (b. 1904) is a member of the Institute of Advanced Studies at Princeton University. He served on active duty in the U. S. Foreign Service from 1927 to 1952, last as U. S. Ambassador to the Soviet Union.

oil and water. Then, too, the modern psychology has played its part. The psychologists have demonstrated certain things of which the ancients were excellently aware: namely, that greatness is never more than a partial attribute of the human personality (no man being, as they used to say, a hero to his valet); and that there is no such thing as a permanently uncomplicated relationship between any two human beings. Beyond this, the totalitarians, with their tortures and brainwashings and their shameful experimentation on the borderlands of human endurance, have also taught us some sad but unforgettable lessons about the fragility of the human personality—about the pathetic way in which its dignity and integrity are dependent on such elementary things as sleep and food and company and some minimal ability to gauge the future and plan for it.

These lessons need not weaken our concern for the elevation of the individual or our faith in the principle of love that stands at the center of so much of western religious thought. But they do make incumbent upon us, it seems to me, a greater modesty in the assessment of man's individual nature—a recognition that man is not super-man, never can be, never will be. It obliges us to recognize that a portion of our own nature must always stand in tragic conflict with the discipline of civilization in which we prefer to live which means that there will always be an element of tragedy in human affairs. It obliges us to recognize, finally, that the physical and spiritual sides of human affection, while they often go together for a time, have each its own law and its own necessity, and rarely combine to produce in permanence that state of transcendent happiness which our popular fiction and screen drama, following in the path of the romanticists, would have us believe is just around the corner.

The second of the reasons why I cannot hold out to you the same sort of assurance that used to be dispensed on these occasions is a very familiar one, about which people have thought and spoken too much, if anything, in these recent years. I am referring to the state of international insecurity in which we now live—in short, to the danger of war.

I am personally convinced that those whom we have recently come to regard as our adversaries have no more desire for such a war than you and I have. They want other things to which, thus far, we have found it impossible to agree; but they don't want another world war.

Now of course wars *can* come, and do, even when nobody wants them. But the fact that nobody does want one is a tremendously hopeful fact, to begin with; and it means that we have a better chance than many people suppose of avoiding war, if our policies are wise and moderate, coupled always with vigilance and with the maintenance of conciliatory, unprovocative strength.

Whether our policies *will* be that is a question of the degree to which a democracy will be able to develop maturity of statesmanship, which in turn means depth and subtlety of understanding, privacy of deliberation, and concentration on the long-term rather than the short-term effect.

These things come hard to a democracy. It is true that we have not done so badly, considering where we started, and when. But we are not yet out of the woods. The demagogues and philistines, though hushed and momentarily humbled, are still with us and their influence is still dominant in certain phases of our national behaviour. Some of our international opponents are arrogant, inexperienced, and irresponsible people. Thus the danger—partly external but partly from within ourselves—has not fully passed. And in the day of atomic weapons, this is a terrible danger, indeed.

We have no need to lose our sense of proportion or our love of life on this account. The enormous life expectancy that modern western man enjoys is, after all, something quite new and revolutionary. As little as two or three centuries ago, life in Europe was no doubt fully as hazardous, though for quite other reasons, as it is in our time; yet in those earlier centuries, people still contrived to lead full-blooded lives, rich in belief, in expression, in creativity.

Let us take a lesson from these forefathers, and learn to face life with the same unhesitating affirmation that they exhibited. We need not be forgetful of the danger; and we should certainly never cease to work patiently and wisely for the lowering of international tensions. But it need not be a source of morbid fatalism or pessimism to us if this danger or uncertainty cannot be circumvented all at once. It is important, of course, that our lives should not be tragically and brutally cut short by war before our work is done. But what is more important still is that in whatever span is given to us we should live richly and to the limit of our capacities.

For this, it is not essential that life be long; it is only necessary that it be real and have meaning.

This brings us to the last of the reservations I mentioned at the start of this discussion. It is the one which seems to me the most troublesome, the most difficult to describe, and the most serious—even more serious than the danger of war. I have in mind the question whether our American civilization of the year 1955 *is* well designed to produce the rich and meaningful life.

We have always gone on the theory that an increase in the amount of material goods and leisure time gave the individual a wider area of choice by which to make his own inner life rich; and it has been part of our traditional philosophy that the individual, confronted with this choice, would know how to use it wisely. These were the assumptions that lay, so to speak, at the end of that rainbow of material progress we followed through the first century and a half of our national history. But we are now approaching the end of this rainbow, in the sense that material plenty is really there for almost all of our citizens who are not too lazy or too improvident to seek it. We are getting enough evidence today to examine the validity of these underlying assumptions. And the results of this examination, so far as I am concerned, are not encouraging. I am *not* sure that the area of choice for the individual is really made wider by the conditions in which our material plenty is being achieved. I am *not* sure that the mass of our people know how to make good use of that choice, where it exists. And consequently, I question whether material abundance alone, as we have sought it heretofore and are rapidly coming to know it today, is really bringing us the results we hoped. I question whether it is making us a happier—and inwardly a richer—people.

I am at a loss to know how to argue that proposition in the few words to which I must limit myself. Nothing is harder than to discuss the inner world of a great people such as ours, with all its immense variety and complexity. No scientific proof is possible. All judgments are necessarily subjective, intuitive, and imperfect. Yet these questions must be asked. I can only say: look around you. Look at the state of our youth. Look at the faces you see behind the endless streams of windshields on our highways. Look at the state of our education, our recreational habits, and our cultural life. And then ask yourselves: are these people as happy as they ought to be in the face of their material abundance? Are they *that* much happier than people elsewhere who do not have this abundance? Are they *that* much happier than the Americans of earlier generations, for whom this abundance did not exist? Has their spiritual advance really kept pace with their material advance?

If you can answer these questions in the affirmative, then you are yourself a happy person; I give you my congratulations and absolve you of any moral obligation of attention to the remaining portion of this discussion, for we will be departing from different premises. But if, like myself, you find yourselves obliged to answer these questions with a decided negative, then you will have to bear with me in my concern and to join me in asking why all this is so, what it is that we have failed to take into account in our calculations of the past, and what this means to us for the future.

Two factors loom up in my mind as central ones in connection with this failure of material progress to bring us greater benefits. First, there is the disintegration of real community life almost everywhere, as a result of the revolutionary innovations in transportation and communication that the last half century have brought us. Secondly, there is the growing domination of cultural and recreational activity by commercial media, usually connected with the advertising profession, whose motivation has little, if anything, to do in the deeper sense with human welfare.

This last process, as you know, has a wide variety of distressing effects. By making advertising the main business of our newspapers, overhadowing physically and financially the other function, commercialism has affected in unfortunate ways the freshness, the independence, and the competitiveness of our press— once a mainstay of the vitality of our society. Periodical publication it already dominates. By its great appetite for such things as film and reprint rights and condensations, it is threatening to dominate the field of literary publishing, thus far a last refuge of genuine cultural values. Concerned to divert rather than to develop, the commercial cultural product has led to passivity of recreation—to spectatoritis—on an appalling scale, dulling the creative faculties of millions of people, undermining the very talent for active recreation. It has invaded the home. It has asserted its dominion over the minds of small children, crowding the school and appropriating to itself a major role in the actual educational process. With its characteristic staccato patterns, its lack of follow-through, and its endless abrupt transitions of theme, commercial entertainment has tended everywhere to weaken the faculty of concentration and to debauch the capacity for sustained and orderly thought. At the back of all this is usually, though not always, advertising. Thus in ever-increasing degree the right of monopolizing our attention, of absorbing our capacity for intellectual and emotional reaction, of shaping the habits and the imagery on which our thought depends, is being claimed by those whose primary interest in us is only the influencing of our conduct as purchasers at the shopping center.

I have no quarrel with the advertisers. I know that many of them are personally distressed about these very same things. I do not say that their business purposes are in any way evil or reprehensible. But I do say that these purposes are irrelevant to, and wholly out of accord with the importance of, those human reactions they are using as a means to their end. It seems to me preposterous that cultural and educational stimuli of such enormous importance, exercising so intimate and vital an influence on the inner world of our people—on their comprehension of life and their attitudes toward it—should be left in permanence to the conscience of an industry the nature of whose interest has so little relation to the things that are really at stake.

The effects of the other factor I mentioned—the disintegration of local community life—are scarcely less alarming. Here the automobile, ubiquitous and triumphant, has been the principal disintegrating agent. What we are faced with is, as you know, not only a disintegration of the local civic community as such but also a fragmentization of the family group, to a point where the home loses its integrity and becomes a fortuitous dormitory for strangers of different age groups. There could be no more bitter blow than this at the true sources of man's security—no surer guaranty of his bad behavior. And even if this were not at stake, the deterioration of the local civic community would be bad enough. The meaning of citizenship begins with the neighborhood; and when this type of community begins to lose its reality for people, then I for one have no great confidence that they will be good citizens in the wider frameworks of the state and the nation. Yet it is hard for the local community to retain its vitality when violent changes are constantly occurring in its composition and function; when the disorderly and uncontrolled development of our great urban areas is constantly rendering physical equipment obsolete and unsuited to new purposes to which commercial interests dictate that it should be put; when administrative boundaries no longer have even a remote relation to social realities; when residence is being separated from livelihood by ever-increasing distances and time intervals; when the reckless and growing dispersal of all facilities for living tends increasingly to pull people away from their homes, to fragmentize their interests, their allegiances, and their civic influence; and when, finally, it becomes increasingly difficult for the individual citizen to survey and apprehend the social context in which his life proceeds, and by which the real prospects for his happiness and that of his children are largely determined. These things are happening—and happening all over our country. We have been, as a nation, extraordinarily obtuse to the importance, from the standpoint of the deeper satisfactions of those who use them, of the sheer geographic arrangement of the facilities for life and work. As in the case of our educational and recreational facilities, we have lightheartedly resigned the control of these things into the hands of people who are perfectly worthy people and doing nothing reprehensible, but whose interests, being purely commercial, do not even take into account the need for the preservation of the health and integrity of the local community itself. Today, we see all around us the chaotic and depressing effects of this failure on our part to insist on public responsibility for the control of processes that are certainly matters of public concern in their effects.

These are all facts that limit in important ways that freedom of choice which, by tradition and by the theory of our society, the individual citizen was supposed to enjoy and by virtue of which he was supposed to have been the master of his own fate and his own happiness. And it is because of things like this that our American environment has become in certain ways a dangerous and unhealthy one—not conducive to the best development of the individual, either for his own sake or from the standpoint of his value as a citizen. This is why I think that some day we are going to have to come to a new social philosophy, which will go deeper into the true sources of man's prosperity than does our traditional attachment to free enterprise or does, for that matter, socialism, or communism, or the rationale of the European welfare state. This new philosophy will have to take account of the fact that the satisfying of man's material needs is only the beginning, and does not answer, but only opens up for the first time in all their real complexity and difficulty, the crucial questions as to what environmental conditions are most favorable to man's individual enjoyment of the experience of life and to the dignity of his relationship with other men.

These are the questions we have not yet learned to ask ourselves, as a political society. We are going to have to ask them, and to find answers to them, before we will dispose of such things as personal insecurity, urban blight, civic apathy, juvenile delinquency, and mass cultural vulgarity. Until we do this, we will continue to be, as we are now, not a bad people or a weak one or even a consciously unhappy one, but an endangered one—a people in danger, at least partially, of becoming sluggish intellectually, underdeveloped emotionally, creative only where commercial interest raises its capricious demands, filled with an inner restlessness and dissatisfaction, incapable of integrating our full strength and of bringing it to bear where it is most needed, dull and uninteresting to others and, what is worse, not terribly interesting to ourselves.

I exaggerate, of course. I know of no other way to make the point. I am aware of the immense resources of strength in our people. It is precisely because I am aware of them that I am so concerned that they should be released and not frustrated or stultified. I know that there are great areas of American life relatively unaffected by the conditions I have mentioned. I know that there are places where people are reacting with courage and imagination, in ways tremendously encouraging, to precisely these dangers. I can never move about in this great country, which I love as deeply as the next man, without being whip-sawed between discouragement and a hopeful excitement.

But I want to emphasize that where people *have* taken up the struggle against these powerful compulsions of our time, which our political system has failed to see or prevent and to which it has shown itself frighteningly indifferent, they have had to do so as individuals, acting on their own perceptions and their own initiative, swimming against the tide. And the burden of my message to you today is that you, too, will have to become conscious of the existence of these dangers and compulsions and to learn how to resist them, individually or in voluntary association with other people—you, too, will have to apply your will, your ingenuity, and your initiative—if you are to defend, against an unfavorable environment, the privacy and quietness of your inner world, if you are to give yourself a chance for creativity and self-fulfillment, if you are to retain around you those elements of true community, both personal and civic, without which even the strongest man finds it difficult to express himself, and without which the child—the adult of the future—can hardly develop at all.

As things are today, you will not get much help in meeting this problem. Government, by and large, will be able to help you very little. Press and radio will not help very much in it. Your greatest aid will be your education, which will give you the requisite understanding; but even education cannot give you the will or the courage. For all of that, you, like the rest of us, will be on your own. On your success in this obscure and perplexing battle, and on the success of millions like you, rests the future of the American soul. And there, more than in any statistics of economic expansion, lies the real future of America.

III

THE RIGHTS OF THE INDIVIDUAL

The Declaration of Independence

[1776]

WHEN, in the course of human events, it becomes necessary for one people to dissolve the political bands which have connected them with another, and to assume, among the powers of the earth, the separate and equal station to which the laws of nature and of nature's God entitle them, a decent respect to the opinions of mankind requires that they should declare the causes which impel them to the separation.

We hold these truths to be self-evident, that all men are created equal; that they are endowed by their Creator with certain unalienable rights; that among these, are life, liberty, and the pursuit of happiness. That, to secure these rights, governments are instituted among men, deriving their just powers from the consent of the governed; that, whenever any form of government becomes destructive of these ends, it is the right of the people to alter or to abolish it and to institute new government, laying its foundation on such principles and organizing its powers in such form, as to them shall seem most likely to effect their safety and happiness. Prudence, indeed, will dictate that governments long established should not be changed for light and transient causes; and accordingly all experience hath shown, that mankind are more disposed to suffer, while evils are sufferable, than to right themselves by abolishing the forms to which they are accustomed. But when a long train of abuses and usurpations, pursuing invariably the same object, evinces a design to reduce them under absolute despotism, it is their right, it is their duty, to throw off such government, and to provide new guards for their future security. Such has been the patient sufferance of these colonies; and such is now the necessity which constrains them to alter their former systems of government. The history of the present king of Great Britain is a history of repeated injuries and usurpations, all having in direct object the establishment of an absolute tyranny over these States. To prove this, let facts be submitted to a candid world.

He has refused his assent to laws the most wholesome and necessary for the public good.

He has forbidden his Governors to pass laws of immediate and pressing importance, unless suspended in their operation till his assent should be tained; and, when so suspended, he has utterly neglected to attend to them.

He has refused to pass other laws for the accommodation of large districts of people, unless those people would relinquish the right of representation in the legislature, a right inestimable to them and formidable to tyrants only.

He has called together legislative bodies at places unusual, uncomfortable, and distant from the depository of their public records, for the sole purpose of fatiguing them into compliance with his measures.

He has dissolved representative houses repeatedly, for opposing with manly firmness his invasions on the rights of the people.

He has refused for a long time, after such dissolutions, to cause others to be elected; whereby the legislative powers, incapable of annihilation, have returned to the people at large for their exercise; the State remaining in the meantime exposed to all the dangers of invasion from without, and convulsions within.

He has endeavored to prevent the population of these States; for that purpose obstructing the laws of naturalization of foreigners; refusing to pass others to encourage their migration hither, and raising the conditions of new appropriations of lands.

He has obstructed the administration of justice, by refusing his assent to laws for establishing judiciary powers.

He has made judges dependent on his will alone, for the tenure of their offices, and the amount and payment of their salaries.

He has erected a multitude of new offices, and sent hither swarms of officers to harass our people, and eat out their substance.

He has kept among us, in times of peace, standing armies without the consent of our legislatures.

He has affected to render the military independent of, and superior to, the civil power.

He has combined with others to subject us to a jurisdiction foreign to our constitution and unacknowledged by our laws; giving his assent to their acts of pretended legislation:

For quartering large bodies of armed troops among us;

For protecting them, by a mock trial, from punishment, for any murders which they should commit on the inhabitants of these States;

For cutting off our trade with all parts of the world;

For imposing taxes on us without our consent;

For depriving us, in many cases, of the benefits of trial by jury;

For transporting us beyond seas to be tried for pretended offences;

For abolishing the free system of English laws in a neighboring province, establishing therein an arbitrary government, and enlarging its boundaries so as to render it at once an example and fit instrument for introducing the same absolute rule into these colonies;

For taking away our charters, abolishing our most valuable laws, and altering fundamentally the forms of our governments;

For suspending our own legislatures, and declaring themselves invested with power to legislate for us in all cases whatsoever.

He has abdicated government here, by declaring us out of his protection and waging war against us.

He has plundered our seas, ravaged our coasts, burnt our towns, and destroyed the lives of our people.

He is at this time transporting large armies of foreign mercenaries to complete the works of death, desolation, and tyranny, already begun, with circumstances of cruelty and perfidy scarcely paralleled in the most barbarous ages and totally unworthy the head of a civilized nation.

He has constrained our fellow-citizens taken captive on the high seas, to bear arms against their country, to become the executioners of their friends and brethren, or to fall themselves by their hands.

He has excited domestic insurrections amongst us, and has endeavored to bring on the inhabitants of our frontiers, the merciless Indian savages, whose known rule of warfare is an undistinguished destruction of all ages, sexes, and conditions.

In every stage of these oppressions we have petitioned for redress, in the most humble terms: Our repeated petitions have been answered only by repeated injury. A prince, whose character is thus marked by every act which may define a tyrant is unfit to be the ruler of a free people.

Nor have we been wanting in attention to our British brethren. We have warned them from time to time of attempts by their legislature to extend an unwarrantable jurisdiction over us. We have reminded them of the circumstances of our emigration and settlement here. We have appealed to their native justice and magnanimity, and we have conjured them by the ties of our common kindred to disavow these usurpations which would inevitably interrupt our connections and correspondence. They too have been deaf to the voice of justice and consanguinity. We must, therefore, acquiesce in the necessity, which denounces our separation, and hold them, as we hold the rest of mankind, enemies in war, in peace friends.

We, therefore, the Representatives of the United States of America, in General Congress assembled, appealing to the Supreme Judge of the World for the rectitude of our intentions, do, in the name, and by authority of the good people of these colonies, solemnly publish and declare, That these United Colonies are, and of right ought to be, free and independent States; that they are absolved from all allegiance to the British crown, and that all political connection between them and the state of Great Britain, is, and ought to be, totally dissolved; and that, as free and independent states, they have full power to levy war, conclude peace, contract alliances, establish commerce, and to do all other acts and things which independent States may of right do. And for the support of this declaration, with a firm reliance on the protection of Divine Providence, we mutually pledge to each other our lives, our fortunes, and our sacred honor.

[Signatures omitted]

The Bill of Rights

[1791]

ARTICLE THE FIRST. Congress shall make no law respecting the establishment of religion, or prohibiting the free exercise thereof; or abridging the freedom of speech, or of the press; or the right of the people peaceably to assemble, and to petition the government for a redress of grievances.

ARTICLE THE SECOND. A well-regulated militia being necessary to the security of a free State, the right of the people to keep and bear arms shall not be infringed.

ARTICLE THE THIRD. No soldier shall, in time of peace, be quartered in any house without the consent of the owner, nor in time of war, but in a manner prescribed by law.

ARTICLE THE FOURTH. The right of the people to be secure in their persons, houses, papers, and effects, against unreasonable searches and seizures, shall not be violated, and no warrants shall issue, but upon probable cause, supported by oath or affirmation, and particularly describing the place to be searched, and the persons or things to be seized.

ARTICLE THE FIFTH. No person shall be held to answer for a capital, or otherwise infamous crime, unless on a presentment or indictment of a grand jury, except in cases arising in the land or naval forces, or in the militia, when in actual service in time of war or public danger; nor shall any person be subject for the same offence to be twice put in jeopardy of life or limb; nor shall be compelled in any criminal case to be a witness against himself, nor be deprived of life, liberty, or property, without due process of law; nor shall private property be taken for public use without just compensation.

ARTICLE THE SIXTH. In all criminal prosecutions, the accused shall enjoy the right to a speedy and public trial, by an impartial jury of the State and district wherein the crime shall have been committed, which district shall have been previously ascertained by law, and to be informed of the nature and cause of the accusation; to be confronted with the witnesses against him; to have compulsory process for obtaining witnesses in his favor, and to have the assistance of counsel for his defence.

Article the Seventh. In suits at common law, where the value in controversy shall exceed twenty dollars, the right of trial by jury shall be preserved, and no fact tried by a jury, shall be otherwise re-examined in any court of the United States than according to the rules of the common law.

Article the Eighth. Excessive bail shall not be required, nor excessive fines imposed, nor cruel and unusual punishments inflicted.

Article the Ninth. The enumeration in the Constitution of certain rights, shall not be construed to deny or disparage others retained by the people.

Article the Tenth. The powers not delegated to the United States by the Constitution, nor prohibited by it to the States, are reserved to the States respectively, or to the people.

Natural Rights*

BY G. BERNARD SHAW

[1913]

... ENGLISH UNIVERSITIES have for some time past encouraged an extremely foolish academic exercise which consists in disproving the existence of natural rights on the ground that they cannot be deduced from the principles of any known political system. If they could, they would not be natural rights but acquired ones. Acquired rights are deduced from political constitutions; but political constitutions are deduced from natural rights. When a man insists on certain liberties without the slightest regard to demonstrations that they are not for his own good, nor for the public good, nor moral, nor reasonable, nor decent, nor compatible with the existing constitution of society, then he is said to claim a natural right to that liberty. When, for instance, he insists, in spite of the irrefutable demonstrations of many able pessimists, from the author of the book of Ecclesiastes to Schopenhauer, that life is an evil, on living, he is asserting a natural right to live. When he insists on a vote in order that his country may be governed according to his ignorance instead of the wisdom of the Privy Council, he is asserting a natural right to self-government. When he insists on guiding himself at 21 by his own inexperience and folly and immaturity instead of by the experience and sagacity of his father, or the well stored mind of his grandmother, he is asserting a natural right to independence. Even if Home Rule were as unhealthy as an Englishman's eating, as intemperate as his drinking, as filthy as his smoking, as licentious as his domesticity, as corrupt as his elections, as murderously greedy as his commerce, as cruel as his prisons, and as merciless as his streets, Ireland's claim to self-government would still be as good as England's. King James the First proved so cleverly and conclusively that the satisfaction of natural rights was incompatible with good government that his courtiers

---

* From "Preface for Politicians" in *John Bull's Other Island*. Reprinted by permission of the Public Trustee and The Society of Authors, London. George Bernard Shaw (1856–1950) was an Irish playwright and critic.

called him Solomon. We, more enlightened, call him Fool, solely because we have learnt that nations insist on being governed by their own consent—or, as they put it, by themselves and for themselves—and that they will finally upset a good government which denies them this even if the alternative be a bad government which at least creates and maintains an illusion of democracy. America, as far as one can ascertain, is much worse governed, and has a much more disgraceful political history than England under Charles I; but the American Republic is the stabler government because it starts from a formal concession of natural rights, and keeps up an illusion of safeguarding them by an elaborate machinery of democratic election.

Freedom of Speech and Press*

BY CARL L. BECKER

[1944]

> Congress shall make no law . . . abridging the freedom of speech, or of the press; or the right of the people peaceably to assemble, and to petition the government for a redress of grievances.
>
> FIRST AMENDMENT TO THE CONSTITUTION

THE TENTH AMENDMENT to the Constitution states that "the powers not delegated by the Constitution to the United States, nor prohibited by it to the States, are reserved to the States respecively, *or to the people.*" For our purpose the qualifying phrase "to the people" is more important than the rest of the statement, because it discloses the fundamental principle on which all of our governments, state and Federal, are grounded—the principle that in the last analysis all powers are reserved to the people. We might then, without changing our system of government in any way, add another amendment to all of our constitutions: "The powers not delegated to the governments by the people, nor prohibited by them to the individual, are reserved to the individual." This is only another way of saying that our Government purports to be a government of free men, and that restraints on the individual's freedom to act and to speak for himself are self-imposed because defined in laws made by the will of the people.

But the will of the people is an intangible thing. "The people" comprises many people—many individuals with diverse and conflicting desires and wills. How can this conflict of wills be reconciled? The practical way in a republic is by majority vote. If the

---

* Reprinted from *Freedom and Responsibility in the American Way of Life* by Carl L. Becker, by permission of Alfred A. Knopf, Inc. Copyright 1945 by Alfred A. Knopf, Inc. Published by Vintage Books, Inc., 1955. Carl L. Becker (1873–1945), American historian, is the author of many studies of political history and the history of ideas. Among his works are *The Declaration of Independence* (1922) and *The Heavenly City of the Eighteenth Century Philosophers* (1932).

majority wills to prohibit, for example, the sale and manufacture of spirituous liquors, then the minority has to renounce that right; but how then can we say that the minority is not bound against its will or subject to restraints not self-imposed? Rousseau had a solution for this difficulty. He maintained that all members of society have agreed, by an original social compact, to submit their individual wills in particular measures to the general will. When, therefore, you and I cast our votes for or against a proposed measure, we are not really voting for or against the measure; we are merely voting to determine what the general will is in respect to that measure. If we vote against it, and it turns out that the majority voted for it, we are not defeated, but only enlightened; and since we now know that the majority is for the measure we are for it too, because we have, by the original social compact, already agreed to submit our individual wills to the general will as soon as we know what that will is.

Does all this sound like hocus-pocus—a dialectical run-around? It does, a little. But, after all, does it not describe well enough, although with a good deal of verbal gymnastics, what we actually do and think when we vote for or against this or that measure, or for or against this or that party? When we have an election, more especially a presidential election, the Democrats make a tremendous hulabaloo about the country's going to ruin if the Republicans win, and the Republicans make an equally alarming hullabaloo about the country's going to ruin "if that man" is elected again. But when the shouting is over and the votes are counted, what do we do then? All of us, defeated and despondent minority no less than triumphant and elated majority, take the count, accept the decision, and go about our business. And no one really thinks that the country is going to ruin, or that anyone is going to be deprived of his natural rights because the laws will be made for some time by a party he did not will to place in power. And why do we act and think in this way if it be not that all of us have agreed, not in the explicit terms of an original compact perhaps, but in terms of an implicit understanding consciously or unconsciously subscribed to, that the general will of the nation is to be determined by majority vote, and that all our rights and liberties can best be secured by submitting voluntarily to the will of the majority?

However that may be, and whatever fine-spun theories we may

devise to resolve or obscure the difficulty, there is no use blinking the fact that the will of the majority is not the same thing as the will of all. Majority rule works well only so long as the minority is willing to accept the will of the majority as the will of the nation and let it go at that. Generally speaking, the minority will be willing to let it go at that so long as it feels that its essential interests and rights are not fundamentally different from those of the current majority, and so long as it can, in any case, look forward with confidence to mustering enough votes within four or sixteen years to become itself the majority and so redress the balance. But if it comes to pass that a large minority feels that it has no such chance, that it is a fixed and permanent minority and that another group or class with rights and interests fundamentally hostile to its own is in permanent control, then government by majority vote ceases in any sense to be government by the will of the people for the good of all, and becomes government by the will of some of the people for their own interests at the expense of the others.

The founders of the republic were fully aware of this danger. Jefferson accepted the device of majority vote because it was the only practicable method of registering the will of the people; but he was not so blind as to think that the majority was always right, or that it would never abuse its power. "I believe," he said, "that there exists a right that is independent of force; . . . that justice is the fundamental law of society; that the majority, oppressing an individual, is guilty of a crime, abuses its strength, and by acting on the principle of the strongest breaks up the foundations of society." Jefferson and his contemporaries had faith in republican government because they believed that it provided the best security for the natural rights of men against the tyranny of kings and aristocrats. But they were fully aware that even in a republic the natural rights of men need to be safeguarded against another sort of tyranny—the tyranny of the majority. Against the tyranny of the majority, or at all events against hasty and ill-considered action by the majority, the founding fathers endeavored, therefore, to erect adequate safeguards.

One of these safeguards is to be found in the organization of government and the distribution of powers within it—an organization based on the grand negative principle of checks and balances. But there was certain rights of the individual that the founding fathers regarded as sacred and imprescriptible—rights that no government, even one founded on the will of the people, could ever justly deny. This idea, that natural right is superior to prescriptive law, is indeed fundamental in the political philosophy formulated in the eighteenth century to justify the liberal-democratic revolution of modern times. It was clearly set forth in the French Declaration of the Rights of Man and the Citizen. Jefferson expressed it in the Declaration of Independence by saying that governments are instituted among men to secure the natural and inalienable rights of man, and that when any government becomes destructive of these rights it is the right of the people to alter or to abolish it. This central idea, with much of the phraseology in which Jefferson expressed it, appears in all our constitutions. To this day, therefore, the natural rights philosophy is implicit in our system of government. It still stands, as one may say, in the entrance ways of all our constitutions, safeguarding the imprescriptible rights of man.

The most important of these rights, the one that is essential to all the rest and the foundation of democratic government as we understand it, is freedom of the mind—the right of every person, as the Connecticut Constitution puts it, to "speak, write, and publish his sentiments on all subjects, being responsible for the abuse of that liberty."

A good many years ago, when Mussolini was much admired for clearing the streets of beggars and making the trains run on time, a good lady said to me that she couldn't understand all this palaver about freedom of speech and of the press. Isn't everyone, she asked, always free to say what he thinks? Of course, she added, one must be prepared to take the consequences. I was unable on the spur of the moment to find an answer to that one. But it has since occurred to me that the good lady was more profoundly right than she realized. Democratic government rests on the assumption that the people are capable of governing themselves better than any one or any few can do it for them; but this in turn rests on the further assumption that if the people are free to think, speak, and publish their sentiments on any subject, the consequences will be good. Well, sometimes they are and sometimes not. If we have faith in democracy, we get over this inconvenient fact by saying that by and large and in the long run the consequences will be good. But the point is that if we accept democracy we must accord to everyone the right to think, speak, and publish his sentiments on any subject; and in that case we must indeed be prepared to take the consequences, whatever they may turn out to be.

It will be well, therefore, to examine this fundamental right with some care, in order to see, first, why Jefferson and his contemporaries were so profoundly convinced that the consequences would always and everywhere be good; and, secondly, whether we are still justified, in the light of a longer practical experience, in supposing that by and large and in the long run the consequences will at least be good enough to go on with.

2

... Eighteenth-century philosophers ... were profoundly convinced ... that if men were free to inquire about all things, to doubt and discuss all things, to form opinions on the basis of knowledge and evidence, and to utter their opinions freely, the competition of knowledge and opinion in the market of rational discourse would ultimately banish ignorance and superstition and enable men to shape their conduct and their institutions in conformity with the fundamental and invariable laws of nature and the will of God.

The freedom of the individual mind from the compulsion of church and state was thus a right that Voltaire and Jefferson and their contemporaries derived in the first instance from political experience and the teaching of history. But for them it had a higher validity than that. It was a fundamental article of faith in their religious or philosophical conception of God and nature and of the relation of man to both. For most eighteenth-century philosophers the laws of nature and the will of God were the same thing, or rather the laws of nature were an intended and adequate expression of the will of God. Natural law, said the French philosopher Volney, is "the constant and regular order of facts by which God rules the universe; the order which his wisdom presents to the sense and reason of men, to serve them as an equal and common rule of conduct, and to guide them, without distinction of race or sect, towards perfection and happiness." In the eighteenth century God the Father had become attenuated into God the First

Cause or Creator. Having at the beginning of things constructed the universe on a rational plan as a convenient habitation for mankind, the Creator had withdrawn from the immediate and arbitrary control of human affairs, leaving men to work out their own salvation as best they could. But this they could do very well, because the beneficent intentions of God were revealed, not in sacred scriptures, but in the great open book of nature, which all men endowed with the light of reason could read and interpret. The mysterious ways in which God moved to perform his wonders, so far from being known through official and dogmatic pronouncements of church and state, were to be progressively discovered by the free play of human reason upon accumulated and verifiable knowledge. The free play of human reason, given time enough, could therefore discover the invariable laws of nature and nature's God and, by bringing the ideas and the institutions of men into conformity with them, find the way, as Volney said, to perfection and happiness.

This conception of God and nature and of the relation of man to both provided the eighteenth-century philosophers with their faith in the worth and dignity of the individual man and the efficacy of human reason. The eighteenth century was the moment in history when men experienced the first flush and freshness of the idea that man is master of his own fate; the moment in history, also, when this emancipating idea, not yet brought to the harsh test of experience, could be accepted with unclouded optimism. Never had the universe seemed less mysterious, more simply constructed, more open and visible and eager to yield its secrets to common-sense questions. Never had the nature of man seemed less perverse, or the intelligence and will of men more pliable to rational persuasion. Never had social and political evils seemed so wholly the result of ignorance and superstition, or so easily corrected by the spread of knowledge and the construction of social institutions on a rational plan. The first task of political science was to discover the natural rights of man, the second to devise the form of government best suited to secure them. And for accomplishing this high task, for creating and maintaining a society founded on justice and equality, the essential freedom was freedom of the mind.

The extraordinary faith of the early prophets of democracy in the efficacy of the human reason and in the native disposition of men to be guided by it is well brought out by John Stuart Mill in reference to his father, a hard-headed man if there ever was one. "So complete," says Mill, "was my father's reliance on the influence of reason over the minds of mankind, whenever it was allowed to reach them, that he felt that all would be gained if the whole population were taught to read, if all sorts of opinions were allowed to be addressed to them by word and writing, and if by means of the suffrage they could nominate a legislature to give effect to the opinions they adopted." It was as simple as that.

In practice we find it somewhat less simple, no doubt; but to this day our faith in democracy, if we have any, has the same ideological basis as that of James Mill. Since primitive times virtually all religious or social systems have attempted to maintain themselves by forbidding free criticism and analysis either of existing institutions or of the doctrine that sustains them; of democracy alone is it the cardinal principle that free criticism and analysis by all and sundry is the highest virtue. In its inception modern democracy was, therefore, a stupendous gamble for the highest stakes. It offered long odds on the capacity and integrity of the human mind. It wagered all it had that the freest exercise of the human reason would never disprove the proposition that only by the freest exercise of the human reason can a tolerably just and rational society ever be created.

The play is still on, and we are still betting on freedom of the mind, but the outcome seems now somewhat more dubious than it did in Jefferson's time, because a century and a half of experience makes it clear that men do not in fact always use their freedom of speech and of the press in quite the rational and disinterested way they are supposed to. An examination of freedom of the mind in practice should enable us, therefore, to estimate the odds for and against the theory somewhat more accurately than Jefferson and his contemporaries were able to do.

3

The democratic doctrine of freedom of speech and of the press, whether we regard it as a natural and inalienable right or not, rests upon certain assumptions. One of these is that men desire to know the truth and will be disposed to be guided by it. Another is that the sole method of arriving at the truth in the long run is by the free competition of opinion in the open market. Another is that, since men will inevitably differ in their opinions, each man must be permitted to urge, freely and even strenuously, his own opinion, provided he accords to others the same right. And the final assumption is that from this mutual toleration and comparison of diverse opinions the one that seems the most rational will emerge and be generally accepted.

The classic expression of this procedure and this attitude of mind is the saying attributed, incorrectly it may be, to Voltaire: "I disagree absolutely with what you say, but I will defend to the death your right to say it." For me these famous words always call up, at first, an agreeable picture—the picture of two elderly gentlemen, in powdered wigs and buckled shoes, engaged over their toddy in an amiable if perhaps somewhat heated discussion about the existence of the deity. But when I try to fit the phrase into the free competition of opinion as it actually works out in our present democratic society, the picture fades out into certain other pictures, some even more agreeable, others much less so. The more agreeable picture might be that of Pierre and Marie Curie working day and night for four years in their leaky laboratory for the sole purpose of discovering the truth about pitchblende. The less agreeable picture might be that of some business tycoon placing self above service by purveying misinformation about the value of certain stocks which he wishes to palm off on the public. Or it might be the picture of a newspaper editor blue-penciling a story altogether true and needing to be known, because it does not have a sufficiently sensational "news value." Or it might be the picture of some fruity-throated radio announcer availing himself every day of his inalienable right of misrepresenting the merits of a certain toothpaste. Or else the picture of a Congressional committee exercising its freedom of the press to denounce as Communists certain worthy men and women who are not Communists by any reasonable definition of Communism, because they have exercised their freedom of speech to say a good word for the labor unions or the Spanish Loyalists, or because they have subscribed for and read the *Nation,* the *New Republic,* or the *Daily Worker.*

These instances may serve to make vivid the fact that freedom of speech does not travel exclusively on a one-way street marked "Search for Truth." It often enough travels on a one-way street marked "Private Profit," or on another marked "Anything to Win the Election." Most often, no doubt, it travels every which way on the broad unmarked highway of diverse human activities. This is only to say that the right of free speech cannot be considered to any good purpose apart from the concrete situations in which it is exercised; and in all such situations the relevant questions are, who is exercising the right, by what means, and for what purposes? . . . All of our constitutions recognize that there may be abuses of the liberty that need to be defined and prohibited by law. The relevant question always is, what abuses are sufficiently grave to be prohibited by law? And the most relevant and difficult question of all is, in limiting the right because it is abused, at what point precisely does the limitation of the right become a greater evil than the abuse because it threatens to destroy the right altogether?

The classic instance of this dilemma arises in connection with the Communists and the Fascists—a dilemma that may be stated in the following way:

Democratic government rests on the right of the people to govern themselves, and therefore on the right of all citizens to advocate freely in speech and writing a modification of the existing form of government—the right of advocating, let us say, the abolition of the House of Representatives, or the election of a president for life. Well, the Communists and the Fascists avail themselves of the right of free speech to advocate the abolition of the democratic form of government altogether, and they maintain that since it cannot be done by persuasion and voting it should be done by force. Is this an abuse of the right of free speech? Is the democratic right of free speech to be accorded to those whose avowed aim is to destroy democratic government and free speech as a part of it? Are we expected to be loyal to the principle of free speech to the point where, writhing in pain among its worshipers, it commits suicide? That is certainly asking a lot.

It is asking too much so long as we remain in the realm of logical discourse. The program of the Fascists, and of the Communists in so far at least as the preliminaries of political reform are concerned, is based on an appeal to force rather than to persuasion. Their own principles teach us that it is logical for them to resist suppression, but merely impudent for them to resent it. Very well, then, since that is their program, let us stop talking, appeal to force, and see which is the stronger. Freedom of speech is for those who are *for* it—for those who are willing to accept it and abide by it as a political method; and I can see no reason why a democratic government should not defend its existence by force against internal as well as against external enemies whose avowed aim is to destroy it.

But strict logic is a poor counselor of political policy, and I can see no practical virtue in a syllogistic solution of the problem of Communist or Fascist propaganda. Freedom of speech can neither be suppressed by argument nor supported by suppressing argument. The real danger is not that Communists and Fascists will destroy our democratic government by free speaking, but that our democratic government, through its own failure to cure social evils, will destroy itself by breeding Communists and Fascists. If we can by the democratic method sufficiently alleviate social injustice, freedom of speech will sufficiently justify itself; if not, freedom of speech will in any case be lost in the shuffle.

It is in connection with social injustices that the question of free speech raises practical rather than theoretical problems. These problems concern the abuses of free speech committed by those who employ freedom of speech, not to destroy democratic government, but to serve their own interests by distorting the truth or betraying the public interest. That such abuses exist no one denies, and that some of them should be prohibited by law no one has ever denied. No one believes in the freedom of speech that issues in libel and slander. The practical question is, are our laws against slander and libel well adapted to meet the modern ingenious methods of insidious within-the-law vilification?

Or, to take a different case, no one can deny that much of our modern advertising is essentially dishonest; and it can hardly be maintained that to lie freely and all the time for private profit is not to abuse the right of free speech, whether it be a violation of the law or not. But again the practical question is, how much lying for private profit is to be permitted by law? Vendors of toothpaste say every day that their particular brands will cure bad breath, restore to the teeth the original brilliance of the enamel, and thereby enable anyone to recover the lost affection of husband or wife or boy or girl friend. What to do about it? Well, better let it go. The law cannot do everything; it must assume the people have some intelligence; and if the toothpaste is harmless, and the people are so ignorant as not to know the obvious facts of life (one of which is that high-powered advertising is a kind of mental test of the gullibility factor of the people), well, it is too bad maybe, but it is not a responsibility that the law can wisely assume. Liberty is the right of anyone to do whatever does not injure others; and it does not injure anyone to use the permitted brands of toothpaste, even if salt water or powdered chalk would do him just as much good besides being much less expensive.

It is quite another matter, however, if the systematic lying for private profit or personal advantage does injure others. If the toothpaste destroys the enamel instead of leaving it as it was; if the cosmetic, instead of being harmless, poisons the skin; if published misinformation induces ignorant or gullible people to invest their money in worthless stocks—in such and many similar cases the freedom of the liar to "speak, write, and publish his sentiments" needs to be restrained by law. The number of such laws has increased, is increasing, and will undoubtedly continue to increase. There can be no natural and inalienable right to lie systematically for private profit or personal advantage; and if the individual will not assume the responsibility for being reasonably honest, the government must assume the responsibility of restraining his freedom of speaking and acting for dishonest purposes and to the injury of others. But this raises a question of fundamental importance for the maintenance of democracy and of the freedom of speech and action that are inseparable from it. The question is, how far will it be necessary to go in making laws for curbing the dishonest and protecting the ignorant and gullible? How much ignorance, gullibility, and dishonesty can there be without making it impossible, by any number of laws designed to protect the ignorant and curb the dishonest, to preserve anything that can rightly be called democracy?

Democratic government is self-government, and self-government, if it be more than an empty form, is something far more than the popular election of representatives to make laws regulating everything and thereby to relieve the people of the responsibility for what they do. Whatever the form of government may be, it is not self-government unless the people are mostly intelligent enough and honest enough to do of their own accord what is right and necessary with a minimum of legal compulsion and restraint. Self-government works best, of course, in a small community in which everyone knows everyone else and in which the relations of men are therefore mostly direct and personal. It works well enough in a new and sparsely settled community in which there is room enough for everyone, in which the people do not get in each other's way too much whatever they do, and in which there is a fair chance for every man to get on in life as well as his ability and industry permit.

But we no longer live in such a community. We live in a highly complex and economically integrated community in which the relations of men are largely indirect and impersonal, and the life and fortunes of every citizen are profoundly affected, in ways that are not apparent and cannot be foreseen or avoided, by what others unknown to him are doing or planning to do. In such a community it is increasingly difficult for the honest to be intelligent and informed about what is going on and who is getting away with it, and therefore increasingly easy for the intelligent and informed to push their interests dishonestly under cover of the general ignorance. In such a community, it is obvious, there must be more laws for curbing the dishonest and for protecting the ignorant and the gullible. But it is equally obvious that self-government cannot be maintained by laws alone, and that if there are too many ignorant and gullible and dishonest people in the community the process of curbing the dishonest and protecting the ignorant may easily reduce the sphere of individual responsibility, and therefore of individual freedom, to the point where self-government in any real sense of the word ceases to exist.

The dishonest, undercover promotion of selfish interests on the part of the intelligent and the informed few thrives on the ignorance and gullibility of the many, and both are intimately related to the means by which information and misinformation can be communicated. In the eighteenth century it was taken for granted by the prophets of liberal democracy that if all men were free to "speak, write, and publish" their sentiments, the means of doing so would be freely available. Any group of citizens could meet in public assembly and argue to their hearts' content. Any man could establish a newspaper and every week air his opinions on all questions. Any man could, in the spirit of Brutus or Publicola, write a piece and get it published in the newspaper, or at slight cost in a penny pamphlet. Any man, if sufficiently high-brow, could write and publish a book. And all intelligent citizens could without too much effort attend the public forums, read the newspapers and many of the most talked-of books and pamphlets, and thereby, such was the theory, keep well abreast of what was being thought and said about public affairs in his community, and so play his proper part in molding public opinion and legislation.

It was not, of course, even in the eighteenth century, quite so simple as that; but today it is far less simple than it was in the eighteenth century. The average citizen still enjoys the inestimable right of freedom of speech and of the press. Any man can express his sentiments without first looking furtively over his shoulder to see if a government spy is in the offing; any man can, so far as the law is concerned, print a newspaper or a book without first submitting it to an official censor. This is the fundamentally important privilege; and no cataloguing of incidental violations of the right can obscure the fact that through the press and the radio detailed information about events, and the most diverse opinions, are with little let or hindrance daily and hourly presented to the people.

Daily and hourly presented to the people—it is this submerging flood of information and misinformation that makes the situation today so much less simple than it was in the eighteenth century. The means of gathering and communicating information about all that is being said and done and thought all over the world have become so perfected that no man can possibly take in, much less assimilate, more than a very small part of it. No man, unless he makes a full-time job of it, can hope to keep abreast of what is being said and done in the world at large, or even in his own country. The average citizen, although free to form and express his opinion, therefore plays a minor role in molding public opinion. His role is not to initiate, but passively to receive information and misinformation and diverse opinions presented to him by those who have access to the means of communication.

The propagation of information and opinion, to be effective under modern conditions, must be organized; and its promoters will have an indifferent success unless they resort to mass production and mass distribution of their wares. The chief instruments of propaganda are the press and the broadcasting stations. No one who does not command a great deal of capital can establish a broadcasting station. Much less, but still a good deal, of capital is required to establish a publishing company or a newspaper. . . . Any man can of course write a letter and get it published in a newspaper; but the chief instruments of propaganda are not readily available to the average citizen. They can be effectively used only by the Government, political parties, and party leaders, wealthy men and business corporations, associations organized for the promotion of specific causes, and the writers of books that publishers find it worth while to publish.

In our society free and impartial discussion, from which the truth is supposed to emerge, is permitted and does exist. But the thinking of the average citizen and his opinion about public affairs is in very great measure shaped by a wealth of unrelated information and by the most diverse ideas that the selective process of private economic enterprise presents to him for consideration—information the truth of which he cannot verify; ideas formulated by persons unknown to him, and too often inspired by economic, political, religious, or other interests that are never avowed.

4

As Jefferson and his contemporaries did, we still believe that self-government is the best form of government and that freedom of the mind is the most important of the rights that sustain it. We are less sure than they were that a beneficent intelligence designed the world on a rational plan for man's special convenience. We are aware that the laws of nature, and especially the laws of human nature, are less easily discovered and applied than they supposed. We have found it more difficult to define the essential rights of man and to secure them by simple institutional forms than they anticipated. We have learned that human reason is not the infallible instrument for recording the truth that they supposed it to be, and that men themselves are less amenable to rational persuasion. Above all we have learned that freedom of speech and of the press may be used to convey misinformation and distort the truth for personal advantage as well as to express and communicate it for the public good. But although we no longer have the unlimited and solvent backing of God or nature, we are still betting that freedom of the mind will never disprove the proposition that only through freedom of the mind can a reasonably just society ever be created.

We may win our bet, but we shall win it only on certain hard conditions. The conditions are that the people by and large be sufficiently informed to hold and express intelligent opinions on public affairs, and sufficiently honest and public-spirited to subordinate purely selfish interests to the general welfare. In so far as the intelligent and informed systematically employ freedom of speech and of the press for personal and antisocial ends, in so far as the mass of the people are so ignorant and ill-informed as to be capable of being fooled all of the time, freedom of speech and of the press loses its chief virtue and self-government is undermined. Self-government, and the freedom of speech and of the press that sustains it, can be maintained by law only in a formal sense; if they are to be maintained in fact the people must have sufficient intelligence and honesty to maintain them with a minimum of legal compulsion.

This heavy responsibility is the price of freedom, and it can be paid only by a people that has a high degree of integrity and intelligence. Neither intelligence nor integrity can be imposed by law. But the native intelligence, and perhaps the integrity of the people, can be reinforced by law, more especially by laws providing for schools and universities. . . .

The Dennis Case

Eugene Dennis *et al.* v. United States

(341 U.S. 494)

[1951]

Mr. Chief Justice Vinson announced the judgment of the Court and an opinion in which Mr. Justice Reed, Mr. Justice Burton and Mr. Justice Minton join.

Petitioners were indicted in July, 1948, for violation of the conspiracy provisions of the Smith Act, 54 Stat. 671, 18 U.S.C. (1946 ed.) sec. 11, during the period of April, 1945, to July, 1948. . . . A verdict of guilty as to all the petitioners was returned by the jury on October 14, 1949. The Court of Appeals affirmed the convictions, 183 F. 2d 201. We granted certiorari, 340 U.S. 863, limited to the following two questions: (1) Whether either Sec. 2 or Sec. 3 of the Smith Act, inherently or as construed and applied in the instant case, violates the First Amendment and other provisions of the Bill of Rights; (2) whether either Sec. 2 or Sec. 3 of the Act, inherently or as construed and applied in the instant case, violates the First and Fifth Amendments because of indefiniteness.

Sections 2 and 3 of the Smith Act . . . provide as follows:

"Sec. 2.

"(a) It shall be unlawful for any person—

"(1) to knowingly or willfully advocate, abet, advise, or teach the duty, necessity, desirability, or propriety of overthrowing or destroying any government in the United States by force or violence, or by the assassination of any officer of such government;

"(2) with intent to cause the overthrow or destruction of any government in the United States, to print, publish, edit, issue, circulate, sell, distribute, or publicly display any written or printed matter advocating, advising, or teaching the duty, necessity, desirability, or propriety of overthrowing or destroying any government in the United States by force or violence;

(3) to organize or help to organize any society, group, or assembly of persons who teach, advocate, or encourage the overthrow or destruction of any government in the United States by force or violence; or to be or become a member of, or affiliate with, any such society, group, or assembly of persons, knowing the purpose thereof.

"(b) For the purpose of this section, the term 'government in the United States' means the Government of the United States, the government of any State, Territory, or possession of the United States, the government of the District of Columbia, or the government of any political subdivision of any of them.

"SEC. 3. It shall be unlawful for any person to attempt to commit, or to conspire to commit, any of the acts prohibited by the provisions of . . . this title."

The indictment charged the petitioners with wilfully and knowingly conspiring (1) to organize as the Communist Party of the United States of America a society, group and assembly of persons who teach and advocate the overthrow and destruction of the Government of the United States by force and violence, and (2) knowingly and wilfully to advocate and teach the duty and necessity of overthrowing and destroying the Government of the United States by force and violence. The indictment further alleged that Sec. 2 of the Smith Act proscribes these acts and that any conspiracy to take such action is a violation of Sec. 3 of the Act. . . .

The obvious purpose of the statute is to protect existing Government, not from change by peaceable, lawful and constitutional means, but from change by violence, revolution and terrorism. That it is within the *power* of the Congress to protect the Government of the United States from armed rebellion is a proposition which requires little discussion. Whatever theoretical merit there may be to the argument that there is a "right" to rebellion against dictatorial governments is without force where the existing structure of the government provides for peaceful and orderly change. We reject any principle of governmental helplessness in the face of preparation for revolution, which principle, carried to its logical conclusion, must lead to anarchy. No one could conceive that it is not within the power of Congress to prohibit acts intended to overthrow the Government by force and violence. The question with which we are concerned here is not whether Congress has such *power*, but whether the *means* which it has employed conflict with the First and Fifth Amendments to the Constitution.

One of the bases for the contention that the means which Congress has employed are invalid takes the form of an attack on the face of the statute on the grounds that by its terms it prohibits academic discussion of the merits of Marxism-Leninism, that it stifles ideas and is contrary to all concepts of a free speech and a free press. Although we do not agree that the language itself has that significance, we must bear in mind that it is the duty of the federal courts to interpret federal legislation in a manner not inconsistent with the demands of the Constitution. *American Communications Assn.* v. *Douds,* 339 U.S. 382, 407 (1950)....

The very language of the Smith Act negates the interpretation which petitioners would have us impose on that Act. It is directed at advocacy, not discussion. Thus, the trial judge properly charged the jury that they could not convict if they found that petitioners did "no more than pursue peaceful studies and discussions or teaching and advocacy in the realm of ideas." He further charged that it was not unlawful "to conduct in an American college and university a course explaining the philosophical theories set forth in the books which have been placed in evidence." Such a charge is in strict accord with the statutory language, and illustrates the meaning to be placed on those words. Congress did not intend to eradicate the free discussion of political theories, to destroy the traditional rights of Americans to discuss and evaluate ideas without fear of governmental sanction. Rather Congress was concerned with the very kind of activity in which the evidence showed these petitioners engaged.

But although the statute is not directed at the hypothetical cases which petitioners have conjured, its application in this case has resulted in convictions for the teaching and advocacy of the overthrow of the Government by force and violence, which, even though coupled with the intent to accomplish that overthrow, contains an element of speech. For this reason, we must pay special heed to the demands of the First Amendment marking out the boundaries of speech.

We pointed out in *Douds, supra,* that the basis of the First Amendment is the hypothesis that speech can rebut speech, propaganda will answer propaganda, free debate of ideas will result in the wisest governmental policies. It is for this reason that this

Court has recognized the inherent value of free discourse. An analysis of the leading cases in this Court which have involved direct limitations on speech, however, will demonstrate that both the majority of the Court and the dissenters in particular cases have recognized that this is not an unlimited, unqualified right, but that the societal value of speech must, on occasion, be subordinated to other values and considerations.

No important case involving free speech was decided by this Court prior to *Schenck* v. *United States,* 249 U.S. 47 (1919). Indeed, the summary treatment accorded an argument based upon an individual's claim that the First Amendment protected certain utterances indicates that the Court at earlier dates placed no unique emphasis upon that right. It was not until the classic dictum of Justice Holmes in the *Schenck* case that speech *per se* received that emphasis in a majority opinion. That case involved a conviction under the Criminal Espionage Act, 40 Stat. 217. The question the Court faced was whether the evidence was sufficient to sustain the conviction. Writing for a unanimous Court, Justice Holmes stated that the "question in every case is whether the words used are used in such circumstances and are of such a nature as to create a clear and present danger that they will bring about the substantive evils that Congress has a right to prevent." 249 U.S. at 52. . . .

The rule we deduce . . . is that where an offense is specified by a statute in nonspeech or nonpress terms, a conviction relying upon speech or press as evidence of violation may be sustained only when the speech or publication created a "clear and present danger" of attempting or accomplishing the prohibited crime, *e.g.,* interference with enlistment. . . .

Speech is not an absolute, above and beyond control by the legislature when its judgment, subject to review here, is that certain kinds of speech are so undesirable as to warrant criminal sanction. Nothing is more certain in modern society than the principle that there are no absolutes, that a name, a phrase, a standard has meaning only when associated with the considerations which gave birth to the nomenclature. See *Douds,* 339 U.S. at 397. To those who would paralyze our Government in the face of impending threat by encasing it in a semantic straitjacket we must reply that all concepts are relative.

In this case we are squarely presented with the application of the "clear and present danger" test, and must decide what that phrase imports. We first note that many of the cases in which this Court has reversed convictions by use of this or similar tests have been based on the fact that the interest which the State was attempting to protect was itself too insubstantial to warrant restriction of speech. . . . Overthrow of the Government by force and violence is certainly a substantial enough interest for the Government to limit speech. Indeed, this is the ultimate value of any society, for if a society cannot protect its very structure from armed internal attack, it must follow that no subordinate value can be protected. If, then, this interest may be protected, the literal problem which is presented is what has been meant by the use of the phrase "clear and present danger" of the utterances bringing about the evil within the power of Congress to punish.

Obviously, the words cannot mean that before the Government may act, it must wait until the *putsch* is about to be executed, the plans have been laid and the signal is awaited. If Government is aware that a group aiming at its overthrow is attempting to indoctrinate its members and to commit them to a course whereby they will strike when the leaders feel the circumstances permit, action by the Government is required. The argument that there is no need for Government to concern itself, for Government is strong, it possesses ample powers to put down a rebellion, it may defeat the revolution with ease needs no answer. For that is not the question. Certainly an attempt to overthrow the Government by force, even though doomed from the outset because of inadequate numbers or power of the revolutionists, is a sufficient evil for Congress to prevent. The damage which such attempts create both physically and politically to a nation makes it impossible to measure the validity in terms of the probability of success, or the immediacy of a successful attempt. In the instant case the trial judge charged the jury that they could not convict unless they found that petitioners intended to overthrow the Government "as speedily as circumstances would permit." This does not mean, and could not properly mean, that they would not strike until there was certainty of success. What was meant was that the revolutionists would strike when they thought the time was ripe. We must therefore reject the contention that success or probability of success is the criterion. . . .

Chief Judge Learned Hand, writing for the majority below, interpreted the phrase as follows: "In each case [courts] must ask whether the gravity of the 'evil,' discounted by its improbability, justifies such invasion of free speech as is necessary to avoid the danger." 183 F.2d at 212. We adopt this statement of the rule. As articulated by Chief Judge Hand, it is as succinct and inclusive as any other we might devise at this time. It takes into consideration those factors which we deem relevant, and relates their significances. More we cannot expect from words.

Likewise, we are in accord with the court below, which affirmed the trial court's finding that the requisite danger existed. The mere fact that from the period 1945 to 1948 petitioners' activities did not result in an attempt to overthrow the Government by force and violence is of course no answer to the fact that there was a group that was ready to make the attempt. The formation by petitioners of such a highly organized conspiracy, with rigidly disciplined members subject to call when the leaders, these petitioners, felt that the time had come for action, coupled with the inflammable nature of world conditions, similar uprisings in other countries, and the touch-and-go nature of our relations with countries with whom petitioners were in the very least ideologically attuned, convince us that their convictions were justified on this score. And this analysis disposes of the contention that a conspiracy to advocate, as distinguished from the advocacy itself, cannot be constitutionally restrained, because it comprises only the preparation. It is the existence of the conspiracy which creates the danger. . . . If the ingredients of the reaction are present, we cannot bind the Government to wait until the catalyst is added. . . .

We hold that sections 2 (a) (1), 2 (a) (3) and 3 of the Smith Act, do not inherently, or as construed or applied in the instant case, violate the First Amendment and other provisions of the Bill of Rights, or the First and Fifth Amendments because of indefiniteness. Petitioners intended to overthrow the Government of the United States as speedily as the circumstances would permit. Their conspiracy to organize the Communist Party and to teach and advocate the overthrow of the Government of the United States by force and violence created a "clear and present danger" of an attempt to overthrow the Government by force and violence. They were properly and constitutionally convicted for violation of the Smith Act. The judgments of conviction are *Affirmed.* . . .

MR. JUSTICE FRANKFURTER, concurring in affirmance of the judgment. . . .

In enacting a statute which makes it a crime for the defendants to conspire to do what they have been found to have conspired to do, did Congress exceed its constitutional power?

Few questions of comparable import have come before this Court in recent years. The appellants maintain that they have a right to advocate a political theory, so long, at least, as their advocacy does not create an immediate danger of obvious magnitude to the very existence of our present scheme of society. On the other hand, the Government asserts the right to safeguard the security of the Nation by such a measure as the Smith Act. Our judgment is thus solicited on a conflict of interests of the utmost concern to the well-being of the country. . . . If adjudication is to be a rational process we cannot escape a candid examination of the conflicting claims with full recognition that both are supported by weighty title-deeds.

There come occasions in law, as elsewhere, when the familiar needs to be recalled. Our whole history proves even more decisively than the course of decisions in this Court that the United States has the powers inseparable from a sovereign nation. "America has chosen to be, in many respects, and to many purposes, a nation; and for all these purposes, her government is complete; to all these objects, it is competent." Chief Justice Marshall in *Cohens v. Virginia,* 6 Wheat. 264, 414. The right of a government to maintain its existence—self-preservation—is the most pervasive aspect of sovereignty. "Security against foreign danger," wrote Madison, "is one of the primitive objects of civil society." The Federalist, No. 41. The constitutional power to act upon this basic principle has been recognized by this Court at different periods and under diverse circumstances. "To preserve its independence, and give security against foreign aggression and encroachment, is the highest duty of every nation, and to attain these ends nearly all other considerations are to be subordinated. It matters not in what form such aggression and encroachment come. . . . The government, possessing the powers which are to be exercised for protection and security, is clothed with authority to determine the occasion on which the powers shall be called forth. . . ." *The Chinese Exclusion Case*, 130 U.S. 581, 606 . . .

But even the all-embracing power and duty of self-preservation is not absolute. Like the war power, which is indeed an aspect of the power of self-preservation, it is subject to applicable constitutional limitations. . . . Our Constitution has no provision lifting restrictions upon governmental authority during periods of emergency, although the scope of a restriction may depend on the circumstances in which it is invoked.

The First Amendment is such a restriction. It exacts obedience even during periods of war; it is applicable when war clouds are not figments of the imagination no less than when they are. The First Amendment categorically demands that "Congress shall make no law respecting an establishment of religion, or prohibiting the free exercise thereof; or abridging the freedom of speech, or of the press; or the right of the people peaceably to assemble, and to petition the Government for a redress of grievances." The right of a man to think what he pleases, to write what he thinks, and to have his thoughts made available for others to hear or read has an engaging ring of universality. The Smith Act and this conviction under it no doubt restrict the exercise of free speech and assembly. Does that, without more, dispose of the matter? . . .

"The law is perfectly well settled," this Court said over fifty years ago, "that the first ten amendments to the Constitution, commonly known as the Bill of Rights, were not intended to lay down any novel principles of government, but simply to embody certain guaranties and immunities which we had inherited from our English ancestors, and which had from time immemorial been subject to certain well-recognized exceptions arising from the necessities of the case. In incorporating these principles into the fundamental law there was no intention of disregarding the exceptions, which continued to be recognized as if they had been formally expressed." *Robertson* v. *Baldwin*, 165 U.S. 275, 281. . . . Absolute rules would inevitably lead to absolute exceptions, and such exceptions would eventually corrode the rules. The demands of free speech in a democratic society as well as the interest in national security are better served by candid and informed weighing of the competing interests, within the confines of the judicial process, than by announcing dogmas too inflexible for the non-Euclidian problems to be solved.

But how are competing interests to be assessed? Since they are not subject to quantitative ascertainment, the issue necessarily resolves itself into asking, who is to make the adjustment?—who is to balance the relevant factors and ascertain which interest is in the circumstances to prevail? Full responsibility for the choice cannot be given to the courts. Courts are not representative bodies. They are not designed to be a good reflex of a democratic society. Their judgment is best informed, and therefore most dependable, within narrow limits. Their essential quality is detachment, founded on independence. History teaches that the independence of the judiciary is jeopardized when courts become embroiled in the passions of the day and assume primary responsibility in choosing between competing political, economic and social pressures.

Primary responsibility for adjusting the interests which compete in the situation before us of necessity belongs to the Congress.... We must scrupulously observe the narrow limits of judicial authority even though self-restraint is alone set over us. Above all we must remember that this Court's power of judicial review is not "an exercise of the powers of a super-legislature." Mr. Justice Brandeis and Mr. Justice Holmes, dissenting in *Burns Baking Co.* v. *Bryan,* 264 U.S. 504, 534....

... There is ample justification for a legislative judgment that the conspiracy now before us is a substantial threat to national order and security. If the Smith Act is justified at all, it is justified precisely because it may serve to prohibit the type of conspiracy for which these defendants were convicted....

... The phrase "clear and present danger," in its origin, "served to indicate the importance of freedom of speech to a free society but also to emphasize that its exercise must be compatible with the preservation of other freedoms essential to a democracy and guaranteed by our Constitution." *Pennekamp* v. *Florida,* 328 U.S. 331, 350, 352-353 (concurring). It were far better that the phrase be abandoned than that it be sounded once more to hide from the believers in an absolute right of free speech the plain fact that the interest in speech, profoundly important as it is, is no more conclusive in judicial review than other attributes of democracy or than a determination of the people's representatives that a measure is necessary to assure the safety of government itself.

Not every type of speech occupies the same position on the scale of values. There is no substantial public interest in permitting certain kinds of utterances: "the lewd and obscene, the profane, the libelous, and the insulting or 'fighting' words—those which by their very utterance inflict injury or tend to incite an immediate breach of the peace." *Chaplinsky* v. *New Hampshire*, 315 U.S. 568, 572. We have frequently indicated that the interest in protecting speech depends on the circumstances of the occasion. See *Niemotko* v. *Maryland*, 340 U.S. at 275-283. It is pertinent to the decision before us to consider where on the scale of values we have in the past placed the type of speech now claiming constitutional immunity.

The defendants have been convicted of conspiring to organize a party of persons who advocate the overthrow of the Government by force and violence. The jury has found that the object of the conspiracy is advocacy as "a rule or principle of action," "by language reasonably and ordinarily calculated to incite persons to such action," and with the intent to cause the overthrow "as speedily as circumstances would permit."

On any scale of values which we have hitherto recognized, speech of this sort ranks low.

Throughout our decisions there has recurred a distinction between the statement of an idea which may prompt its hearers to take unlawful action, and advocacy that such action be taken. The distinction has its root in the conception of the common law that a person who procures another to do an act is responsible for that act as though he had done it himself. . . . We frequently have distinguished protected forms of expression from statements which "incite to violence and crime and threaten the overthrow of organized government by unlawful means." *Stromberg* v. *California*, 283 U.S. at 369. . . .

It is true that there is no divining rod by which we may locate "advocacy." Exposition of ideas readily merges into advocacy. The same Justice who gave currency to application of the incitement doctrine in this field dissented four times from what he thought was its misapplication. As he said in the *Gitlow* dissent, "Every idea is an incitement." 268 U.S. at 673. Even though advocacy of overthrow deserves little protection, we should hesitate to prohibit it if we thereby inhibit the interchange of rational ideas so essential to representative government and free society.

But there is underlying validity in the distinction between advocacy and the interchange of ideas, and we do not discard a useful tool because it may be misused. That such a distinction could be used unreasonably by those in power against hostile or unorthodox views does not negate the fact that it may be used reasonably against an organization wielding the power of the centrally controlled international Communist movement. The object of the conspiracy before us is clear enough that the chance of error in saying that the defendants conspired to advocate rather than to express ideas is slight. Mr. Justice Douglas quite properly points out that the conspiracy before us is not a conspiracy to overthrow the Government. But it would be equally wrong to treat it as a seminar in political theory.

These general considerations underlie decision of the case before us.

On the one hand is the interest in security. The Communist Party was not designed by these defendants as an ordinary political party. For the circumstances of its organization, its aims and methods, and the relation of the defendants to its organization and aims we are concluded by the jury's verdict. The jury found that the Party rejects the basic premise of our political system—that change is to be brought about by nonviolent constitutional process. The jury found that the Party advocates the theory that there is a duty and necessity to overthrow the Government by force and violence. It found that the Party entertains and promotes this view, not as a prophetic insight or as a bit of unworldly speculation, but as a program for winning adherents and as a policy to be translated into action. . . .

. . . In determining whether application of the statute to the defendants is within the constitutional powers of Congress, we are not limited to the facts found by the jury. We must view such a question in the light of whatever is relevant to a legislative judgment. We may take judicial notice that the Communist doctrines which these defendants have conspired to advocate are in the ascendency in powerful nations who cannot be acquitted of unfriendliness to the institutions of this country. We may take account of evidence brought forward at this trial and elsewhere, much of which has long been common knowledge. In sum, it would amply justify a legislature in concluding that recruitment of additional members for the Party would create a substantial danger to national security.

In 1947, it has been reliably reported, at least 60,000 members were enrolled in the Party. Evidence was introduced in this case that the membership was organized in small units, linked by an intricate chain of command, and protected by elaborate precautions designed to prevent disclosure of individual identity. There are no reliable data tracing acts of sabotage or espionage directly to these defendants. But a Canadian Royal Commission appointed in 1946 to investigate espionage reported that it was "overwhelmingly established" that "the Communist movement was the principal base within which the espionage network was recruited." The most notorious spy in recent history was led into the service of the Soviet Union through Communist indoctrination. Evidence supports the conclusion that members of the Party seek and occupy positions of importance in political and labor organizations. Congress was not barred by the Constitution from believing that indifference to such experience would be an exercise not of freedom but of irresponsibility.

On the other hand is the interest in free speech. The right to exert all governmental powers in aid of maintaining our institutions and resisting their physical overthrow does not include intolerance of opinions and speech that cannot do harm although opposed and perhaps alien to dominant, traditional opinion. The treatment of its minorities, especially their legal position, is among the most searching tests of the level of civilization attained by a society. It is better for those who have almost unlimited power of government in their hands to err on the side of freedom. We have enjoyed so much freedom for so long that we are perhaps in danger of forgetting how much blood it cost to establish the Bill of Rights.

Of course no government can recognize a "right" of revolution, or a "right" to incite revolution if the incitement has no other purpose or effect. But speech is seldom restricted to a single purpose, and its effects may be manifold. A public interest is not wanting in granting freedom to speak their minds even to those who advocate the overthrow of the Government by force. For, as the evidence in this case abundantly illustrates, coupled with such advocacy is criticism of defects in our society. Criticism is the spur to reform; and Burke's admonition that a healthy society must reform in order to conserve has not lost its force. Astute observers have remarked that one of the characteristics of the American Republic is indifference to fundamental criticism. Bryce, The American Commonwealth, c. 84. It is a commonplace that there may be a grain of truth in the most uncouth doctrine, however false and repellent the balance may be. Suppressing advocates of overthrow inevitably will also silence critics who do not advocate overthrow but fear that their criticism may be so construed. No matter how clear we may be that the defendants now before us are preparing to overthrow our Government at the propitious moment, it is self-delusion to think that we can punish them for their advocacy without adding to the risks run by loyal citizens who honestly believe in some of the reforms these defendants advance. It is a sobering fact that in sustaining the conviction before us we can hardly escape restriction on the interchange of ideas.

We must not overlook the value of that interchange. Freedom of expression is the well-spring of our civilization—the civilization we seek to maintain and further by recognizing the right of Congress to put some limitation upon expression. Such are the paradoxes of life. For social development of trial and error, the fullest possible opportunity for the free play of the human mind is an indispensable prerequisite. The history of civilization is in considerable measure the displacement of error which once held sway as official truth by beliefs which in turn have yielded to other truths. Therefore the liberty of man to search for truth ought not to be fettered, no matter what orthodoxies he may challenge. Liberty of thought soon shrivels without freedom of expression. Nor can truth be pursued in an atmosphere hostile to the endeavor or under dangers which are hazarded only by heroes.

"The interest, which [the First Amendment] guards, and which gives it its importance, presupposes that there are no orthodoxies —religious, political, economic, or scientific—which are immune from debate and dispute. Back of that is the assumption—itself an orthodoxy, and the one permissible exception—that truth will be most likely to emerge, if no limitations are imposed upon utterances that can with any plausibility be regarded as efforts to present grounds for accepting or rejecting propositions whose truth the utterer asserts, or denies." *International Brotherhood of Electrical Workers* v. *Labor Board,* 181 F. 2d 34, 40. In the last analysis it is on the validity of this faith that our national security is staked.

It is not for us to decide how we would adjust the clash of interests which this case presents were the primary responsibility for reconciling it ours. Congress has determined that the danger created by advocacy of overthrow justifies the ensuing restriction on freedom of speech. The determination was made after due deliberation, and the seriousness of the congressional purpose is attested by the volume of legislation passed to effectuate the same ends.

Can we then say that the judgment Congress exercised was denied it by the Constitution? Can we establish a constitutional doctrine which forbids the elected representatives of the people to make this choice? Can we hold that the First Amendment deprives Congress of what it deemed necessary for the Government's protection?

To make validity of legislation depend on judicial reading of events still in the womb of time—a forecast, that is, of the outcome of forces at best appreciated only with knowledge of the topmost secrets of nations—is to charge the judiciary with duties beyond its equipment. . . .

. . . The distinction which the Founders drew between the Court's duty to pass on the power of Congress and its complementary duty not to enter directly the domain of policy is fundamental. But in its actual operation it is rather subtle, certainly to the common understanding. Our duty to abstain from confounding policy with constitutionality demands perceptive humility as well as self-restraint in not declaring unconstitutional what in a judge's private judgment is unwise and even dangerous.

Even when moving strictly within the limits of constitutional adjudication, judges are concerned with issues that may be said to involve vital finalities. The too easy transition from disapproval of what is undesirable to condemnation as unconstitutional, has led some of the wisest judges to question the wisdom of our scheme in lodging such authority in courts. But it is relevant to remind that in sustaining the power of Congress in a case like this nothing irrevocable is done. The democratic process at all events is not impaired or restricted. Power and responsibility remain with the people and immediately with their representation. All the Court says is that Congress was not forbidden by the Constitution to pass this enactment and a prosecution under it may be brought against a conspiracy such as the one before us.

The wisdom of the assumptions underlying the legislation and prosecution is another matter. In finding that Congress has acted within its power, a judge does not remotely imply that he favors the implications that lie beneath the legal issues. . . .

No one is better equipped than George F. Kennan to speak on the meaning of the menace of Communism and the spirit in which we should meet it.

"If our handling of the problem of Communist influence in our midst is not carefully moderated—if we permit it, that is, to become an emotional preoccupation and to blind us to the more important positive tasks before us—we can do a damage to our national purpose beyond comparison greater than anything that threatens us today from the Communist side. The American Communist party is today, by and large, an external danger. It represents a tiny minority in our country; it has no real contact with the feelings of the mass of our people; and its position as the agency of a hostile foreign power is clearly recognized by the overwhelming mass of our citizens.

"But the subjective emotional stresses and temptations to which we are exposed in our attempt to deal with this domestic problem are not an external danger: they represent a danger within ourselves—a danger that something may occur in our own minds and souls which will make us no longer like the persons by whose efforts this republic was founded and held together, but rather like the representatives of that very power we are trying to combat: intolerant, secretive, suspicious, cruel, and terrified of internal dissension because we have lost our own belief in ourselves and in the power of our ideals. The worst thing that our Communists could do to us, and the thing we have most to fear from their activities, is that we should become like them.

"That our country is beset with external dangers I readily concede. But these dangers, at their worst, are ones of physical destruction, of the disruption of our world security, of expense and inconvenience and sacrifice. These are serious, and sometimes terrible things, but they are all things that we can take and still remain Americans.

"The internal danger is of a different order. America is not just territory and people. There is lots of territory elsewhere, and there are lots of people; but it does not add up to America. America is something in our minds and our habits of outlook which causes us to believe in certain things and to behave in certain ways, and by which, in its totality, we hold ourselves distinguished from others. If that once goes there will be no America to defend. And that can go too easily if we yield to the primitive human instinct to escape from our frustrations into the realms of mass emotion and hatred and to find scapegoats for our difficulties in individual fellow-citizens who are, or have at one time been, disoriented or confused." George F. Kennan, Where do You Stand on Communism? New York Times Magazine, May 27, 1951.

Civil liberties draw at best only limited strength from legal guaranties. Preoccupation by our people with the constitutionality, instead of with the wisdom of legislation or of executive action, is preoccupation with a false value. Even those who would most freely use the judicial brake on the democratic process by invalidating legislation that goes deeply against their grain, acknowledge, at least by paying lip service, that constitutionality does not exact a sense of proportion or the sanity of humor or an absence of fear. Focusing attention on constitutionality tends to make constitutionality synonymous with wisdom. When legislation touches freedom of thought and freedom of speech, such a tendency is a formidable enemy of the free spirit. Much that should be rejected as illiberal, because repressive and envenoming, may well be not unconstitutional. The ultimate reliance for the deepest needs of civilization must be found outside their vindication in courts of law; apart from all else, judges, howsoever they may conscientiously seek to discipline themselves against it, unconsciously are too apt to be moved by the deep undercurrents of public feeling. A persistent, positive translation of the liberating faith into the feelings and thoughts and actions of men and women is the real protection against attempts to strait-jacket the human mind. Such temptations will have their way, if fear and hatred are not exorcized. The mark of a truly civilized man is confidence in the strength and security derived from the inquiring mind. We may be grateful for such honest comforts as it supports, but we must be unafraid of its uncertainties. Without open minds there can be no open society. And if society be not open the spirit of man is mutilated and becomes enslaved. . . .

MR. JUSTICE JACKSON, concurring.

This prosecution is the latest of never-ending, because never successful, quests for some legal formula that will secure an existing order against revolutionary radicalism. It requires us to reappraise, in the light of our own times and conditions, constitutional doctrines devised under other circumstances to strike a balance between authority and liberty.

Activity here charged to be criminal is conspiracy—that defendants conspired to teach and advocate, and to organize the Communist Party to teach and advocate, overthrow and destruction of the Government by force and violence. There is no charge of actual violence or attempt at overthrow.

The principal reliance of the defense in this Court is that the conviction cannot stand under the Constitution because the conspiracy of these defendants presents no "clear and present danger" of imminent or foreseeable overthrow. . . .

Communism . . . appears today as a closed system of thought representing Stalin's version of Lenin's version of Marxism. As an ideology, it is not one of spontaneous protest arising from American working-class experience. It is a complicated system of assumptions, based on European history and conditions, shrouded in an obscure and ambiguous vocabulary, which allures our ultra-sophisticated intelligentsia more than our hard-headed working people. From time to time it champions all manner of causes and grievances and makes alliances that may add to its foothold in government or embarrass the authorities.

The Communist Party, nevertheless, does not seek its strength primarily in numbers. Its aim is a relatively small party whose strength is in selected, dedicated, indoctrinated, and rigidly disciplined members. From established policy it tolerates no deviation and no debate. It seeks members that are, or may be, secreted in strategic posts in transportation, communications, industry, government, and especially in labor unions where it can compel employers to accept and retain its members. It also seeks to infiltrate and control organizations of professional and other groups. Through these placements in positions of power it seeks a leverage over society that will make up in power of coercion what it lacks in power of persuasion. The Communists have no scruples against sabotage, terrorism, assassination, or mob disorder; but violence is not with them, as with the anarchists, an end in itself. The Communist Party advocates force only when prudent and profitable. Their strategy of stealth precludes premature or uncoordinated outbursts of violence, except, of course, when the blame will be placed on shoulders other than their own. They resort to violence as to truth, not as a principle but as an expedient. Force or violence, as they would resort to it, may never be necessary, because infiltration and deception may be enough.

Force would be utilized by the Communist Party not to destroy government but for its capture. The Communist recognizes that an established government in control of modern technology cannot be overthrown by force until it is about ready to fall of its own weight. Concerted uprising, therefore, is to await that contingency and revolution is seen, not as a sudden episode, but as the consummation of a long process.

The United States, fortunately, has experienced Communism only in its preparatory stages and for its pattern of final action must look abroad. Russia, of course, was the pilot Communist revolution, which to the Marxist confirms the Party's assumptions and points its destiny. But Communist technique in the overturn of a free government was disclosed by the *coup d'etat* in which they seized power in Czechoslovakia. There the Communist Party during its preparatory stage claimed and received protection for its freedoms of speech, press, and assembly. Pretending to be but another political party, it eventually was conceded participation in government, where it entrenched reliable members chiefly in control of police and information services. When the government faced a foreign and domestic crisis, the Communist Party had established a leverage strong enough to threaten civil war. In a period of confusion the Communist plan unfolded and the underground organization came to the surface throughout the country in the form chiefly of labor "action committees." Communist officers of the unions took over transportation and allowed only persons with party permits to travel. Communist printers took over the newspapers and radio and put out only party-approved versions of events. Possession was taken of telegraph and telephone systems and communications were cut off wherever directed by party heads. Communist unions took over the factories, and in the cities a partisan distribution of food was managed by the Communist organization. A virtually bloodless abdication by the elected government admitted the Communists to power, whereupon they instituted a reign of oppression and terror, and ruthlessly denied to all others the freedoms which had sheltered their conspiracy.

The foregoing is enough to indicate that, either by accident or design, the Communist stratagem outwits the anti-anarchist pattern of statute aimed against "overthrow by force and violence" if qualified by the doctrine that only "clear and present danger" of accomplishing that result will sustain the prosecution.

The "clear and present danger" test was an innovation by Mr. Justice Holmes in the *Schenck* case, reiterated and refined by him and Mr. Justice Brandeis in later cases, all arising before the era of World War II revealed the subtlety and efficacy of modernized revolutionary techniques used by totalitarian parties. In those cases, they were faced with convictions under so-called criminal syndicalism statutes aimed at anarchists but which, loosely construed, had been applied to punish socialism, pacifism, and left-wing ideologies, the charges often resting on far-fetched inferences which, if true, would establish only technical or trivial violations. They proposed "clear and present danger" as a test for the sufficiency of evidence in particular cases.

I would save it, unmodified, for application as a "rule of reason" in the kind of case for which it was devised. When the issue is criminality of a hot-headed speech on a street corner, or circulation of a few incendiary pamphlets, or parading by some zealots behind a red flag, or refusal of a handful of school children to salute our flag, it is not beyond the capacity of the judicial process to gather, comprehend, and weigh the necessary materials for decision whether it is a clear and present danger of substantive evil or a harmless letting off of steam. It is not a prophecy, for the danger in such cases has matured by the time of trial or it was never present. The test applies and has meaning where a conviction is sought to be based on a speech or writing which does not directly or explicitly advocate a crime but to which such tendency is sought to be attributed by construction or by implication from external circumstances. The formula in such cases favors freedoms that are vital to our society, and, even if sometimes applied too generously, the consequences cannot be grave. But its recent expansion has extended, in particular to Communists, unprecedented immunities. Unless we are to hold our Government captive in a judge-made verbal trap, we must approach the problem of a well-organized, nation-wide conspiracy, such as I have described realistically as our predecessors faced the trivialities that were being prosecuted until they were checked with a rule of reason.

I think reason is lacking for applying that test to this case.

If we must decide that this Act and its application are constitutional only if we are convinced that petitioner's conduct creates a "clear and present danger" of violent overthrow, we must appraise imponderables, including international and national phenomena which baffle the best informed foreign offices and our most experienced politicians. We would have to foresee and predict the effectiveness of Communist propaganda, opportunities for infiltration, whether, and when, a time will come that they consider propitious for action, and whether and how fast our existing government will deteriorate. And we would have to speculate as to whether an approaching Communist *coup* would not be anticipated by a nationalistic fascist movement. No doctrine can be sound whose application requires us to make a prophecy of that sort in the guise of a legal decision. The judicial process simply is not adequate to a trial of such far-flung issues. The answers given would reflect our own political predilections and nothing more.

The authors of the clear and present danger test never applied it to a case like this, nor would I. If applied as it is proposed here, it means that the Communist plotting is protected during its period of incubation; its preliminary stages of organization and preparation are immune from the law; the Government can move only after imminent action is manifest, when it would, of course, be too late.

The highest degree of constitutional protection is due to the individual acting without conspiracy. But even an individual cannot claim that the Constitution protects him in advocating or teaching overthrow of government by force or violence. I should suppose no one would doubt that Congress has power to make such attempted overthrow a crime. But the contention is that one has the constitutional right to work up a public desire and will to do what it is a crime to attempt. I think direct incitement by speech or writing can be made a crime, and I think there can be a conviction without also proving that the odds favored its success by 99 to 1, or some other extremely high ratio. . . .

What really is under review here is a conviction of **conspiracy**, after a trial for conspiracy, on an indictment charging conspiracy, brought under a statute outlawing conspiracy. With due respect to my colleagues, they seem to me to discuss anything under the sun except the law of conspiracy. One of the dissenting opinions even appears to chide me for "invoking the law of conspiracy." As that is the case before us, it may be more amazing that its reversal can be proposed without even considering the law of conspiracy.

The Constitution does not make conspiracy a civil right. The Court has never before done so and I think it should not do so now. Conspiracies of labor unions, trade associations, and news agencies have been condemned, although accomplished, evidenced and carried out, like the conspiracy here, chiefly by letter-writing, meetings, speeches and organizations. Indeed, this Court seems, particularly in cases where the conspiracy has economic ends, to be applying its doctrines with increasing severity. While I consider criminal conspiracy a dragnet device capable of perversion into an instrument of injustice in the hands of a partisan or complacent judiciary, it has an established place in our system of law, and no reason appears for applying it only to concerted action claimed to disturb interstate commerce and withholding it from those claimed to undermine our whole Government.

The basic rationale of the law of conspiracy is that a conspiracy may be an evil in itself, independently of any other evil it seeks to accomplish. . . .

So far does this doctrine reach that it is well settled that Congress may make it a crime to conspire with others to do what an individual may lawfully do on his own. This principle is illustrated in conspiracies that violate the antitrust laws as sustained and applied by this Court. Although one may raise the prices of his own products, and many, acting without concert, may do so, the moment they conspire to that end they are punishable. The same principle is applied to organized labor. Any workman may quit his work for any reason, but concerted actions to the same end are in some circumstances forbidden. . . .

There is lamentation in the dissents about the injustice of conviction in the absence of some overt act. Of course, there has been no general uprising against the Government, but the record is replete with acts to carry out the conspiracy alleged, acts such as always are held sufficient to consummate the crime where the statute requires an overt act.

But the shorter answer is that no overt act is or need be required. The Court, in antitrust cases, early upheld the power of Congress to adopt the ancient common law that makes conspiracy itself a crime. . . . It is not to be supposed that the power of Congress to protect the Nation's existence is more limited than its power to protect interstate commerce.

Also, it is urged that since the conviction is for conspiracy to teach and advocate, and to organize the Communist Party to teach and advocate, the First Amendment is violated, because freedoms of speech and press protect teaching and advocacy regardless of what is taught or advocated. I have never thought that to be the law.

I do not suggest that Congress could punish conspiracy to advocate something, the doing of which it may not punish. Advocacy or exposition of the doctrine of communal property ownership, or any political philosophy unassociated with advocacy of its imposition by force or seizure of government by unlawful means could not be reached through conspiracy prosecution. But it is not forbidden to put down force or violence, it is not forbidden to punish its teaching or advocacy, and the end being punishable, there is no doubt of the power to punish conspiracy for the purpose.

The defense of freedom of speech or press has often been raised in conspiracy cases, because, whether committed by Communists, by businessmen, or by common criminals, it usually consists of words written or spoken, evidenced by letters, conversations, speeches or documents. Communication is the essence of every conspiracy, for only by it can common purpose and concert of action be brought about or be proved. However, when labor unions raised the defense of free speech against a conspiracy charge, we unanimously said:

"It rarely has been suggested that the constitutional freedom for speech and press extends its immunity to speech or writing used as an integral part of conduct in violation of a valid criminal statute. We reject the contention now. . . ." *Giboney* v. *Empire Storage & Ice Co.*, 336 U.S. 490, 498.

When our constitutional provisions were written, the chief forces recognized as antagonists in the struggle between authority and liberty were the Government on the one hand and the individual citizen on the other. It was thought that if the state could be kept in its place the individual could take care of himself.

In more recent times these problems have been complicated by the intervention between the state and the citizen of permanently organized, well-financed, semisecret and highly disciplined political organizations. Totalitarian groups here and abroad perfected the technique of creating private paramilitary organizations to coerce both the public government and its citizens. These organizations assert as against our Government all of the constitutional rights and immunities of individuals and at the same time exercise over their followers much of the authority which they deny to the Government. The Communist Party realistically is a state within a state, an authoritarian dictatorship within a republic. It demands these freedoms, not for its members, but for the organized party. It denies to its own members at the same time the freedom to dissent, to debate, to deviate from the party line, and enforces its authoritarian rule by crude purges, if nothing more violent.

The law of conspiracy has been the chief means at the Government's disposal to deal with the growing problems created by such organizations. I happen to think it is an awkward and inept remedy, but I find no constitutional authority for taking this weapon from the Government. There is no constitutional right to "gang up" on the Government.

While I think there was power in Congress to enact this statute and that, as applied in this case, it cannot be held unconstitutional, I add that I have little faith in the long-range effectiveness of this conviction to stop the rise of the Communist movement. Communism will not go to jail with these Communists. No decision by this Court can forestall revolution whenever the existing government fails to command the respect and loyalty of the people and sufficient distress and discontent is allowed to grow up among the masses. Many failures by fallen governments attest that no government can long prevent revolution by outlawry. Corruption, ineptitude, inflation, oppressive taxation, militarization, injustice, and loss of leadership capable of intellectual initiative in domestic or foreign affairs are allies on which the Communists count to bring opportunity knocking to their door. Sometimes I think they may be mistaken. But the Communists are not building just for today—the rest of us might profit by their example.

MR. JUSTICE DOUGLAS, dissenting.

If this were a case where those who claimed protection under the First Amendment were teaching the techniques of sabotage, the assassination of the President, the filching of documents from public files, the planting of bombs, the art of street warfare, and the like, I would have no doubts. The freedom to speak is not absolute; the teaching of methods of terror and other seditious conduct should be beyond the pale along with obscenity and immorality. This case was argued as if those were the facts. The argument imported much seditious conduct into the record. That is easy and it has popular appeal, for the activities of Communists in plotting and scheming against the free world are common knowledge. But the fact is that no such evidence was introduced at the trial. There is a statute which makes a seditious conspiracy unlawful. Petitioners, however, were not charged with a "conspiracy to overthrow" the Government. They were charged with a conspiracy to form a party and groups and assemblies of people who teach and advocate the overthrow of our Government by force or violence and with a conspiracy to advocate and teach its overthrow by force and violence. It may well be that indoctrination in the techniques of terror to destroy the Government would be indictable under either statute. But the teaching which is condemned here is of a different character.

So far as the present record is concerned, what petitioners did was to organize people to teach and themselves teach the Marxist-Leninist doctrine contained chiefly in four books: Foundations of Leninism by Stalin (1924), The Communist Manifesto by Marx and Engels (1848), State and Revolution by Lenin (1917), History of the Communist Party of the Soviet Union (B) (1939).

Those books are to Soviet Communism what Mein Kampf was to Nazism. If they are understood, the ugliness of Communism is revealed, its deceit and cunning are exposed, the nature of its activities becomes apparent, and the chances of its success less likely. That is not, of course, the reason why petitioners chose these books for their classrooms. They are fervent Communists to whom these volumes are gospel. They preached the creed with the hope that some day it would be acted upon.

The opinion of the Court does not outlaw these texts nor condemn them to the fire, as the Communists do literature offensive to their creed. But if the books themselves are not outlawed, if they can lawfully remain on library shelves, by what reasoning does their use in a classroom become a crime? It would not be a crime under the Act to introduce these books to a class, though that would be teaching what the creed of violent overthrow of the government is. The Act, as construed, requires the element of intent—that those who teach the creed believe in it. The crime then depends not on what is taught but on who the teacher is. That is to make freedom of speech turn not on *what is said,* but on the *intent* with which it is said. Once we start down that road we enter territory dangerous to the liberties of every citizen.

There was a time in England when the concept of constructive treason flourished. Men were punished not for raising a hand against the king but for thinking murderous thoughts about him. The Framers of the Constitution were alive to that abuse and took steps to see that the practice would not flourish here. Treason was defined to require overt acts—the evolution of a plot against the country into an actual project. The present case is not one of treason. But the analogy is close when the illegality is made to turn on intent, not on the nature of the act. We then start probing men's minds for motive and purpose; they become entangled in the law not for what they did but for *what they thought;* they get convicted not for what they said but for the purpose with which they said it.

Intent, of course, often makes the difference in the law. An act otherwise excusable or carrying minor penalties may grow to an abhorrent thing if the evil intent is present. We deal here, however, not with ordinary acts but with speech, to which the Constitution has given a special sanction.

The vice of treating speech as the equivalent of overt acts of a treasonable or seditious character is emphasized by a concurring opinion, which by invoking the law of conspiracy makes speech do service for deeds which are dangerous to society. The doctrine of conspiracy has served diverse and oppressive purposes and in its broad reach can be made to do great evil. But never until today has anyone seriously thought that the ancient law of conspiracy could constitutionally be used to turn speech into seditious conduct. Yet that is precisely what is suggested. I repeat that we deal here with speech alone, not with speech *plus* acts of sabotage or unlawful conduct. Not a single seditious act is charged in the indictment. To make a lawful speech unlawful because two men conceive it is to raise the law of conspiracy to appalling proportions. That course is to make a radical break with the past and to violate one of the cardinal principles of our constitutional scheme.

Free speech has occupied an exalted position because of the high service it has given our society. Its protection is essential to the very existence of a democracy. The airing of ideas releases pressures which otherwise might become destructive. When ideas compete in the market for acceptance, full and free discussion exposes the false and they gain few adherents. Full and free discussion even of ideas we hate encourages the testing of our own prejudices and preconceptions. Full and free discussion keeps a society from becoming stagnant and unprepared for the stresses and strains that work to tear all civilizations apart.

Full and free discussion has indeed been the first article of our faith. We have founded our political system on it. It has been the safeguard of every religious, political, philosophical, economic, and racial group amongst us. We have counted on it to keep us from embracing what is cheap and false; we have trusted the common sense of our people to choose the doctrine true to our genius and to reject the rest. This has been the one single outstanding tenet that has made our institutions the symbol of freedom and equality. We have deemed it more costly to liberty to suppress a despised minority than to let them vent their spleen. We have above all else feared the political censor. We have wanted a land where our people can be exposed to all the diverse creeds and cultures of the world.

There comes a time when even speech loses its constitutional immunity. Speech innocuous one year may at another time fan such destructive flames that it must be halted in the interests of the safety of the Republic. That is the meaning of the clear and present danger test. When conditions are so critical that there will be no time to avoid the evil that the speech threatens, it **is time** to call a halt. Otherwise, free speech which is the strength of the Nation will be the cause of its destruction.

Yet free speech is the rule, not the exception. The restraint to be constitutional must be based on more than fear, on more than passionate opposition against the speech, on more than a revolted dislike for its contents. There must be some immediate injury to society that is likely if speech is allowed. The classic statement of these conditions was made by Mr. Justice Brandeis in his concurring opinion in *Whitney v. California*, 274 U.S. 357, 376-377,

"Fear of serious injury cannot alone justify suppression of free speech and assembly. Men feared witches and burnt women. It is the function of speech to free men from the bondage of irrational fears. To justify suppression of free speech there must be reasonable ground to fear that serious evil will result if free speech is practiced. There must be reasonable ground to believe that the danger apprehended is imminent. There must be reasonable ground to believe that the evil to be prevented is a serious one. Every denunciation of existing law tends in some measure to increase the probability that there will be violation of it. Condonation of a breach enhances the probability. Expressions of approval add to the probability. Propagation of the criminal state of mind by teaching syndicalism increases it. Advocacy of law-breaking heightens it still further. But even advocacy of violation, however reprehensible morally, is not a justification for denying free speech where the advocacy falls short of incitement and there is nothing to indicate that the advocacy would be immediately acted on. The wide difference between advocacy and incitement, between preparation and attempt, between assembling and conspiracy, must be borne in mind. In order to support a finding of a clear and present danger it must be shown either that immediate serious violence was to be expected or was advocated, or that the past conduct furnished reason to believe that such advocacy was then contemplated.

"Those who won our independence by revolution were not cowards. They did not fear political change. They did not exalt order at the cost of liberty. To courageous, self-reliant men, with confidence in the power of free and fearless reasoning applied through the processes of popular government, no danger flowing from speech can be deemed clear and present, unless the incidence of the evil apprehended is so imminent that it may befall before there is opportunity for full discussion. *If there be time to expose through discussion the falsehood and fallacies to avert the evil by the processes of education, the remedy to be applied is more speech, not enforced silence.*" (Italics added)....

The nature of Communism as a force on the world scene would, of course, be relevant to the issue of clear and present danger of petitioners' advocacy within the United States. But the primary consideration is the strength and tactical position of petitioners and their converts in this country. On that there is no evidence in the record. If we are to take judicial notice of the threat of Communists within the nation, it should not be difficult to conclude that *as a political party* they are of little consequence. Communists in this country have never made a respectable or serious showing in any election. I would doubt that there is a village, let alone a city or county or state which the Communists could carry. Communism in the world scene is no bogey-man; but Communists as a political faction or party in this country plainly is. Communism has been so thoroughly exposed in this country that it has been crippled as a political force. Free speech has destroyed it as an effective political party. It is inconceivable that those who went up and down this country preaching the doctrine of revolution which petitioners espouse would have any success. In days of trouble and confusion when bread lines were long, when the unemployed walked the streets, when people were starving, the advocates of a short-cut by revolution might have a chance to gain adherents. But today there are no such conditions. The country is not in despair; the people know Soviet Communism; the doctrine of Soviet revolution is exposed in all of its ugliness and the American people want none of it.

How it can be said that there is a clear and present danger that this advocacy will succeed is, therefore, a mystery. Some nations less resilient than the United States, where illiteracy is high and where democratic traditions are only budding, might have to take drastic steps and jail these men for merely speaking their creed. But in America they are miserable merchants of unwanted ideas; their wares remain unsold. The fact that their ideas are abhorrent does not make them powerful.

The political impotence of the Communists in this country does not, of course, dispose of the problem. Their numbers; their positions in industry and government; the extent to which they have in fact infiltrated the police, the armed services, transportation, stevedoring, power plants, munitions works, and other critical places—these facts all bear on the likelihood that their advocacy of the Soviet theory of revolution will endanger the Republic. But the record is silent on these facts. . . .

... Free speech—the glory of our system of government—should not be sacrificed on anything less than plain and objective proof of danger that the evil advocated is imminent. On this record no one can say that petitioners and their converts are in such a strategic position as to have even the slightest chance of achieving their aims.

The First Amendment provides that "Congress shall make no law ... abridging the freedom of speech." The Constitution provides no exception. This does not mean, however, that the Nation need hold its hand until it is in such weakened condition that there is no time to protect itself from incitement to revolution. Seditious conduct can always be punished. But the command of the First Amendment is so clear that we should not allow Congress to call a halt to free speech except in the extreme case of peril from the speech itself. The First Amendment makes confidence in the common sense of our people and in their maturity of judgment the great postulate of our democracy. Its philosophy is that violence is rarely, if ever, stopped by denying civil liberties to those advocating resort to force. The First Amendment reflects the philosophy of Jefferson "that it is time enough for the rightful purposes of civil government for its officers to interfere when principles break out into overt acts against peace and good order." The political censor has no place in our public debates. Unless and until extreme and necessitous circumstances are shown our aim should be to keep speech unfettered and to allow the processes of law to be invoked only when the provocateurs among us move from speech to action.

Vishinsky wrote in 1948 in The Law of the Soviet State, "In our state, naturally there can be no place for freedom of speech, press, and so on for the foes of socialism."

Our concern should be that we accept no such standard for the United States. Our faith should be that our people will never give support to these advocates of revolution, so long as we remain loyal to the purposes for which our Nation was founded.

IV

THE WILL OF THE PEOPLE

First Inaugural Address*

BY THOMAS JEFFERSON

[1801]

Friends and Fellow Citizens:–

Called upon to undertake the duties of the first executive office of our country, I avail myself of the presence of that portion of my fellow citizens which is here assembled, to express my grateful thanks for the favor with which they have been pleased to look toward me, to declare a sincere consciousness that the task is above my talents, and that I approach it with those anxious and awful presentiments which the greatness of the charge and the weakness of my powers so justly inspire. A rising nation, spread over a wide and fruitful land, traversing all the seas with the rich productions of their industry, engaged in commerce with nations who feel power and forget right, advancing rapidly to destinies beyond the reach of mortal eye—when I contemplate these transcendent objects, and see the honor, the happiness, and the hopes of this beloved country committed to the issue and the auspices of this day, I shrink from the contemplation, and humble myself before the magnitude of the undertaking. Utterly indeed, should I despair, did not the presence of many whom I here see remind me, that in the other high authorities provided by our constitution, I shall find resources of wisdom, of virtue, and of zeal, on which to rely under all difficulties. To you, then, gentlemen, who are charged with the sovereign functions of legislation, and to those associated with you, I look with encouragement for that guidance and support which may enable us to steer with safety the vessel in which we are all embarked amid the conflicting elements of a troubled world.

During the contest of opinion through which we have passed, the animation of discussion and of exertions has sometimes worn an aspect which might impose on strangers unused to think freely and to speak and to write what they think; but this being now

---

* Thomas Jefferson (1743–1826) was the third President of the United States. He served from 1801 to 1809.

decided by the voice of the nation, announced according to the rules of the constitution, all will, of course, arrange themselves under the will of the law, and unite in common efforts for the common good. All, too, will bear in mind this sacred principle, that though the will of the majority is in all cases to prevail, that will, to be rightful, must be reasonable; that the minority possess their equal rights, which equal laws must protect, and to violate which would be oppression. Let us, then, fellow citizens, unite with one heart and one mind. Let us restore to social intercourse that harmony and affection without which liberty and even life itself are but dreary things. And let us reflect that having banished from our land that religious intolerance under which mankind so long bled and suffered, we have yet gained little if we countenance a political intolerance as despotic, as wicked, and capable of as bitter and bloody persecutions. During the throes and convulsions of the ancient world, during the agonizing spasms of infuriated man, seeking through blood and slaughter his long-lost liberty, it was not wonderful that the agitations of the billows should reach even this distant and peaceful shore; that this should be more felt and feared by some and less by others; that this should divide opinions as to measures of safety. But every difference of opinion is not a difference of principle. We have called by different names brethren of the same principle. We are all republicans—we are all federalists. If there be any among us who would wish to dissolve this Union or to change its republican form, let them stand undisturbed as monuments of the safety with which error of opinion may be tolerated where reason is left free to combat it. I know, indeed, that some honest men fear that a republican government cannot be strong; that this government is not strong enough. But would the honest patriot, in the full tide of successful experiment, abandon a government which has so far kept us free and firm, on the theoretic and visionary fear that this government, the world's best hope, may by possibility want energy to preserve itself? I trust not. I believe this, on the contrary, the strongest government on earth. I believe it is the only one where every man, at the call of the laws, would fly to the standard of the law, and would meet invasions of the public order as his own personal concern. Sometimes it is said that man cannot be trusted with the government of himself. Can he, then, be trusted with the government of others?

Or have we found angels in the forms of kings to govern him? Let history answer this question.

Let us, then, with courage and confidence pursue our own federal and republican principles, our attachment to our union and representative government. Kindly separated by nature and a wide ocean from the exterminating havoc of one quarter of the globe; too high-minded to endure the degradations of the others; possessing a chosen country, with room enough for our descendants to the hundredth and thousandth generation; entertaining a due sense of our equal right to the use of our own faculties, to the acquisitions of our industry, to honor and confidence from our fellow citizens, resulting not from birth but from our actions and their sense of them; enlightened by a benign religion, professed, indeed, and practiced in various forms, yet all of them including honesty, truth, temperance, gratitude, and the love of man; acknowledging and adoring an overruling Providence, which by all its dispensations proves that it delights in the happiness of man here and his greater happiness hereafter; with all these blessings, what more is necessary to make us a happy and prosperous people? Still one thing more, fellow citizens—a wise and frugal government, which shall restrain men from injuring one another, which shall leave them otherwise free to regulate their own pursuits of industry and improvement, and shall not take from the mouth of labor the bread it has earned. This is the sum of good government, and this is necessary to close the circle of our felicities.

About to enter, fellow citizens, on the exercise of duties which comprehend everything dear and valuable to you, it is proper that you should understand what I deem the essential principles of our government, and consequently those which ought to shape its administration. I will compress them within the narrowest compass they will bear, stating the general principle, but not all its limitations. Equal and exact justice to all men, of whatever state or persuasion, religious or political; peace, commerce, and honest friendship, with all nations—entangling alliances with none; the support of the state governments in all their rights, as the most competent administrations for our domestic concerns and the surest bulwarks against anti-republican tendencies; the preservation of the general government in its whole constitutional vigor, as the sheet anchor of our peace at home and safety abroad; a jealous care of the right of election by the people—a mild and safe corrective of abuses which are lopped by the sword of the revolution where peaceable remedies are unprovided; absolute acquiescence in the decisions of the majority—the vital principle of republics, from which there is no appeal but to force, the vital principle and immediate parent of despotism; a well-disciplined militia—our best reliance in peace and for the first moments of war, till regulars may relieve them; the supremacy of the civil over the military authority; economy in the public expense, that labor may be lightly burdened; the honest payment of our debts and sacred preservation of the public faith; encouragement of agriculture, and of commerce as its handmaid; the diffusion of information and the arraignment of all abuses at the bar of public reason; freedom of religion; freedom of the press; freedom of person under the protection of the habeas corpus; and trial by juries impartially selected—these principles form the bright constellation which has gone before us, and guided our steps through an age of revolution and reformation. The wisdom of our sages and the blood of our heroes have been devoted to their attainment. They should be the creed of our political faith—the text of civil instruction—the touchstone by which to try the services of those we trust; and should we wander from them in moments of error or alarm, let us hasten to retrace our steps and to regain the road which alone leads to peace, liberty, and safety.

I repair, then, fellow citizens, to the post you have assigned me. With experience enough in subordinate offices to have seen the difficulties of this, the greatest of all, I have learned to expect that it will rarely fall to the lot of imperfect man to retire from this station with the reputation and the favor which bring him into it. Without pretensions to that high confidence reposed in our first and great revolutionary character, whose preeminent services had entitled him to the first place in his country's love, and destined for him the fairest page in the volume of faithful history, I ask so much confidence only as may give firmness and effect to the legal administration of your affairs. I shall often go wrong through defect of judgment. When right, I shall often be thought wrong by those whose positions will not command a view of the whole ground. I ask your indulgence for my own errors, which will never be intentional; and your support against the errors of others, who may condemn what they would not if seen in all its parts. The approbation implied by your suffrage is a consolation to me for the past; and my future solicitude will be to retain the good opinion of those who have bestowed it in advance, to conciliate that of others by doing them all the good in my power, and to be instrumental to the happiness and freedom of all.

Relying, then, on the patronage of your good will, I advance with obedience to the work, ready to retire from it whenever you become sensible how much better choice it is in your power to make. And may that Infinite Power which rules the destinies of the universe, lead our councils to what is best, and give them a favorable issue for your peace and prosperity.

The Partnership of Generations*

BY EDMUND BURKE

[1790]

... SOCIETY is indeed a contract. Subordinate contracts for objects of mere occasional interest may be dissolved at pleasure—but the state ought not to be considered as nothing better than a partnership agreement in a trade of pepper and coffee, calico or tobacco, or some other such low concern, to be taken up for a little temporary interest, and to be dissolved by the fancy of the parties. It is to be looked on with other reverence; because it is not a partnership in things subservient only to the gross animal existence of a temporary and perishable nature. It is a partnership in all science; a partnership in all art; a partnership in every virtue, and in all perfection. As the ends of such a partnership cannot be obtained in many generations, it becomes a partnership not only between those who are living, but between those who are living, those who are dead, and those who are to be born. Each contract of each particular state is but a clause in the great primaeval contract of eternal society, linking the lower with the higher natures, connecting the visible and invisible world, according to a fixed compact sanctioned by the inviolable oath which holds all physical and moral natures, each in their appointed place. This law is not subject to the will of those, who by an obligation above them, and infinitely superior, are bound to submit their will to that law. The municipal corporations of that universal kingdom are not morally at liberty at their pleasure, and on their speculations of a contingent improvement, wholly to separate and tear asunder the bands of their subordinate community, and to dissolve it into an unsocial, uncivil, unconnected chaos of elementary principles.

* From *Reflections on the Revolution in France*. Edmund Burke (1729?–1797), one of the greatest English political thinkers, served for many years as Member of the British House of Commons. His writings directed against the French Revolution have made him the foremost philosopher of conservatism.

The Rights of the Living Generation

BY THOMAS JEFFERSON

1. To James Madison, Paris, September 6, 1789:

Dear Sir,—I sit down to write to you without knowing by what occasion I shall send my letter. I do it, because a subject comes into my head, which I would wish to develop a little more than is practicable in the hurry of the moment of making up general despatches.

The question, whether one generation of men has a right to bind another, seems never to have been started either on this or our side of the water. Yet it is a question of such consequences as not only to merit decision, but place also among the fundamental principles of every government. The course of reflection in which we are immersed here, on the elementary principles of society, has presented this question to my mind; and that no such obligation can be transmitted, I think very capable of proof. I set out on this ground, which I suppose to be self-evident, that the *earth belongs in usufruct to the living;* that the dead have neither powers nor rights over it. The portion occupied by any individual ceases to be his when himself ceases to be, and reverts to the society. If the society has formed no rules for the appropriation of its lands in severality, it will be taken by the first occupants, and these will generally be the wife and children of the decedent. If they have formed rules of appropriation, those rules may give it to the wife and children, or to some one of them, or to the legatee of the deceased. So they may give it to its creditor. But the child, the legatee or creditor, takes it, not by natural right, but by a law of the society of which he is a member, and to which he is subject. . . .

What is true of every member of the society, individually, is true of them all collectively; since the rights of the whole can be no more than the sum of the rights of the individuals. To keep our ideas clear when applying them to a multitude, let us suppose a whole generation of men to be born on the same day, to attain mature age on the same day, and to die on the same day, leaving a succeeding generation in the moment of attaining their mature age, all together. Let the ripe age be supposed of twenty-one years, and their period of life thirty-four years more, that being the average term given by the bills of mortality to persons of twenty-one years of age. Each successive generation would, in this way, come and go off the stage at a fixed moment, as individuals do now. Then I say, the earth belongs to each of these generations during its course, fully and in its own right. The second generation receives it clear of the debts and incumbrances of the first, the third of the second, and so on. For if the first could charge it with a debt, then the earth would belong to the dead and not to the living generation. Then, no generation can contract debts greater than may be paid during the course of its own existence. At twenty-one years of age, they may bind themselves and their lands for thirty-four years to come; at twenty-two, for thirty-three; at twenty-three, for thirty-two; and at fifty-four for one year only; because these are the terms of life which remain to them at the respective epochs. . . .

What is true of generations succeeding one another at fixed epochs, as has been supposed for clearer conception, is true for those renewed daily, as in the actual course of nature. As a majority of the contracting generation will continue in being thirty-four years, and a new majority will then come into possession, the former may extend their engagement to that term, and no longer. The conclusion then, is, that neither the representatives of a nation, nor the whole nation itself assembled, can validly engage debts beyond what they may pay in their own time, that is to say, within thirty-four years of the date of the engagement. . . .

On similar ground it may be proved, that no society can make a perpetual constitution, or even a perpetual law. The earth belongs always to the living generation: they may manage it, then, and what proceeds from it, as they please, during their usufruct. They are masters, too, of their own persons, and consequently may govern them as they please. But persons and property make the sum of the objects of government. The constitution and the laws of their predecessors are extinguished then, in their natural course, with those whose will gave them being. This could preserve that being, till it ceased to be itself, and no longer. Every constitution, then, and every law, naturally expires at the end of thirty-four years. If it be enforced longer, it is an act of force, and not of right. It may be said, that the succeeding generation exercising, in fact, the power of repeal, this leaves them as free as if the constitution or law had been expressly limited to thirty-four years only. In the first place, this objection admits the right, in proposing an equivalent. But the power of repeal is not an equivalent. It might be, indeed, if every form of government were so perfectly contrived, that the will of the majority could always be obtained, fairly and without impediment. But this is true of no form. The people cannot assemble themselves; their representation is unequal and vicious. Various checks are opposed to every legislative proposition. Factions get possession of the public councils, bribery corrupts them, personal interests lead them astray from the general interests of their constituents; and other impediments arise, so as to prove to every practical man, that a law of limited duration is much more manageable than one which needs a repeal.

2. To Samuel Kercheval, Monticello, July 12, 1816.

. . . Some men look at constitutions with sanctimonious reverence, and deem them like the ark of the covenant, too sacred to be touched. They ascribe to the men of the preceding age a wisdom more than human, and suppose what they did to be beyond amendment. I knew that age well; I belonged to it, and labored with it. It deserved well of its country. It was very like the present, but without the experience of the present; and forty years of experience in government is worth a century of book-reading; and this they would say themselves, were they to rise from the dead. I am certainly not an advocate for frequent and untried changes in laws and constitutions. I think moderate imperfections had better be borne with; because, when once known, we accommodate ourselves to them, and find practical means of correcting their ill effects. But I know also, that laws and institutions must go hand in hand with the progress of the human mind. As that becomes more developed, more enlightened, as new discoveries are made, new truths disclosed, and manners and opinions change with the change of circumstances, institutions must advance also, and keep pace with the times. We might as well require a man to wear still the coat which fitted him when a boy, as civilized society to remain ever under the regimen of their barbarous ancestors. It is this preposterous idea which has lately deluged Europe in blood. Their monarchs, instead of wisely yielding to the gradual change of circumstances, of favoring progressive accommodation to progressive improvement, have clung to old abuses, entrenched themselves behind steady habits, and obliged their subjects to seek through blood and violence rash and ruinous innovations, which, had they been referred to the peaceful deliberations and collected wisdom of the nation, would have been put into acceptable and salutary forms. Let us follow no such examples, nor weakly believe that one generation is not as capable as another of taking care of itself, and of ordering its own affairs. Let us, as our sister States have done, avail ourselves of our reason and experience, to correct the crude essays of our first and unexperienced, although wise, virtuous, and well-meaning councils. And lastly, let us provide in our constitution for its revision at stated periods. What these periods should be, nature herself indicates. By the European tables of mortality, of the adults living at any one moment of time, a majority will be dead in about nineteen years. At the end of that period then, a new majority is come into place; or, in other words, a new generation. Each generation is as independent of the one preceding, as that was of all which had gone before. It has then, like them, a right to choose for itself the form of government it believes most promotive of its own happiness; consequently, to accommodate to the circumstances in which it finds itself, that received from its predecessors; and it is for the peace and good of mankind, that a solemn opportunity of doing this every nineteen or twenty years, should be provided by the Constitution; so that it may be handed on, with periodical repairs, from generation to generation, to the end of time, if anything human can so long endure. . . .

All Sail and No Anchor*

BY THOMAS B. MACAULAY

[1857]

... I NEVER, in Parliament, in conversation, or even on the hustings,—a place where it is the fashion to court the populace,—uttered a word indicating an opinion that the supreme authority in a state ought to be intrusted to the majority of citizens told by the head, in other words, to the poorest and most ignorant part of society. I have long been convinced that institutions purely democratic must, sooner or later, destroy liberty, or civilisation, or both. In Europe, where the population is dense, the effect of such institutions would be almost instantaneous. What happened lately in France is an example. In 1848 a pure democracy was established there. During a short time there was reason to expect a general spoliation, a national bankruptcy, a new partition of the soil, a maximum of prices, a ruinous load of taxation laid on the rich for the purpose of supporting the poor in idleness. Such a system would, in twenty years, have made France as poor and barbarous as the France of the Carlovingians. Happily the danger was averted; and now there is a despotism, a silent tribune, an enslaved press. Liberty is gone, but civilisation has been saved. I have not the smallest doubt that, if we had a purely democratic government here, the effect would be the same. Either the poor would plunder the rich, and civilisation would perish, or order and property would be saved by a strong military government, and liberty would perish. You may think that your country enjoys an exemption from these evils. I will frankly own to you that I am of a very different opinion. Your fate I believe to be certain, though it is deferred by a physical cause. As long as you have a boundless extent of fertile and unoccupied land, your labouring population will be far more at ease than the labouring population

---

* From a letter to Hanry S. Randall, Jefferson's biographer, in H. M. Lydenberg, *What Did Macaulay Say About America?*, New York, 1925. Thomas Babington (Lord) Macaulay (1800–1859) was one of England's foremost historians and exponents of 19th century liberalism.

of the old world; and, while that is the case, the Jeffersonian polity may continue to exist without causing any fatal calamity. But the time will come when New England will be as thickly peopled as old England. Wages will be as low, and will fluctuate as much with you as with us. You will have your Manchesters and Birminghams, and in those Manchesters and Birminghams, hundreds of thousands of artisans will assuredly be sometimes out of work. Then your institutions will be fairly brought to the test. Distress every where makes the labourer mutinous and discontented, and inclines him to listen with eagerness to agitators who tell him that it is a monstrous iniquity that one man should have a million while another cannot get a full meal. In bad years there is plenty of grumbling here, and sometimes a little rioting. But it matters little. For here the sufferers are not the rulers. The supreme power is in the hands of a class, numerous indeed, but select; of an educated class, of a class which is, and knows itself to be, deeply interested in the security of property and maintenance of order. Accordingly, the malcontents are firmly, yet gently, restrained. The bad time is got over without robbing the wealthy to relieve the indigent. The springs of national prosperity soon begin to flow again: work is plentiful: wages rise; and all is tranquillity and cheerfulness. I have seen England pass three or four times through such critical seasons as I have described. Through such seasons the United States will have to pass, in the course of the next century, if not of this. How will you pass through them. I heartily wish you a good deliverance. But my reason and my wishes are at war; and I cannot help foreboding the worst. It is quite plain that your government will never be able to restrain a distressed and discontented majority. For with you the majority is the government, and has the rich, who are always a minority, absolutely at its mercy. The day will come when, in the State of New York, a multitude of people, none of whom has had more than half a breakfast, or expects to have more than half a dinner, will choose a Legislature. Is it possible to doubt what sort of a Legislature will be chosen? On one side is a statesman preaching patience, respect for vested rights, strict observance of public faith. On the other is a demagogue ranting about the tyranny of capitalists and usurers, and asking why anybody should be permitted to drink Champagne and to ride in a carriage, while thousands of honest folks are in want of necessaries. Which of the two candidates is likely to be preferred by a working man who hears his children cry for more bread? I seriously apprehend that you will, in some such season of adversity as I have described, do things which will prevent prosperity from returning; that you will act like people who should in a year of scarcity, devour all the seed corn, and thus make the next year a year, not of scarcity, but of absolute famine. There will be, I fear, spoliation. The spoliation will increase the distress. The distress will produce fresh spoliation. There is nothing to stop you. Your Constitution is all sail and no anchor. As I said before, when a society has entered on this downward progress, either civilisation or liberty must perish. Either some Caesar or Napoleon will seize the reins of government with a strong hand; or your republic will be as fearfully plundered and laid waste by barbarians in the twentieth Century as the Roman Empire was in the fifth;— with this difference, that the Huns and Vandals who ravaged the Roman Empire came from without, and that your Huns and Vandals will have been engendered within your own country by your own institutions.

Rogues and the Mob*

BY HENRY L. MENCKEN

[1926]

THE LOWLY Christian I have limned is not only the glory of democratic states, but also their boss. Sovereignty is in him, sometimes both actually and legally, but always actually. Whatever he wants badly enough, he can get. If he is misled by mountebanks and swindled by scoundrels it is only because his credulity and imbecility cover a wider area than his simple desires. The precise form of the government he suffers under is of small importance. Whether it be called a constitutional monarchy, as in England, or a representative republic, as in France, or a pure democracy, as in some of the cantons of Switzerland, it is always essentially the same. There is, first, the mob, theoretically and in fact the ultimate judge of all ideas and the source of all power. There is, second, the camorra of self-seeking minorities, each seeking to inflame, delude and victimize it. The political process thus becomes a mere battle of rival rogues. But the mob remains quite free to decide between them. It may even, under the hand of God, decide for a minority that happens, by some miracle, to be relatively honest and enlightened. If, in common practice, it sticks to the thieves, it is only because their words are words it understands and their ideas are ideas it cherishes. It has the power to throw them off at will, and even at whim, and it also has the means.

A great deal of paper and ink has been wasted discussing the difference between representative government and direct democracy. . . . It is generally held that representative government, as practically encountered in the world, is full of defects, some of them amounting to organic disease. Not only does it take the initiative in lawmaking out of the hands of the plain people, and leave them only the function of referees; it also raises certain obvious

---

* Reprinted from *Notes on Democracy* by Henry L. Mencken, by permission of Alfred A. Knopf, Inc. Copyright, 1926 by Alfred A. Knopf, Inc. Henry L. Mencken (1880–1955) was a literary, social, and political critic. Among his many books is *The American Language* (1918).

obstacles to their free exercise of that function. Scattered as they are, and unorganized save in huge, unworkable groups, they are unable, it is argued, to formulate their virtuous desires quickly and clearly, or to bring to the resolution of vexed questions the full potency of their native sagacity. Worse, they find it difficult to enforce their decisions, even when they have decided. Every Liberal knows this sad story, and has shed tears telling it. The remedy he offers almost always consists of a resort to what he calls a purer democracy. That is to say, he proposes to set up the recall, the initiative and referendum, or something else of the sort, and so convert the representative into a mere clerk or messenger. The final determination of all important public questions, he argues, ought to be in the hands of the voters themselves. They alone can muster enough wisdom for the business, and they alone are without guile. The cure for the evils of democracy is more democracy.

All this, of course, is simply rhetoric. Every time anything of the kind is tried it fails ingloriously. Nor is there any evidence that it has ever succeeded elsewhere, today or in the past. Certainly no competent historian believes that the citizens assembled in a New England town-meeting actually formulated *en masse* the transcendental and immortal measures that they adopted, nor even that they contributed anything of value to the discussion thereof. The notion is as absurd as the parallel notion, long held by philologues of defective powers of observation, that the popular ballads surviving from earlier ages were actually composed by the folk. The ballads, in point of fact, were all written by concrete poets, most of them not of the folk; the folk, when they had any hand in the business at all, simply acted as referees, choosing which should survive. In exactly the same way the New England town-meeting was led and dominated by a few men of unusual initiative and determination, some of them genuinely superior, but most of them simply demagogues and fanatics. The citizens in general heard the discussion of rival ideas, and went through the motions of deciding between them, but there is no evidence that they ever had all the relevant facts before them or made any effort to unearth them, or that appeals to their reason always, or even usually, prevailed over appeals to their mere prejudice and superstition....

The truth is that the difference between representative democracy and direct democracy is a great deal less marked than political sentimentalists assume. Under both forms the sovereign mob must employ agents to execute its will, and in either case the agents may have ideas of their own, based upon interests of their own, and the means at hand to do and get what they will. Moreover, their very position gives them a power of influencing the electors that is far above that of any ordinary citizen: they become politicians *ex officio,* and usually end by selling such influence as remains after they have used all they need for their own ends. Worse, both forms of democracy encounter the difficulty that the generality of citizens, no matter how assiduously they may be instructed, remain congenitally unable to comprehend many of the problems before them, or to consider all of those they do comprehend in an unbiased and intelligent manner. Thus it is often impossible to ascertain their views in advance of action, or even, in many cases, to determine their conclusions *post hoc.* . . . The great masses of Americans of today, though they are theoretically competent to decide all the larger matters of national policy, and have certain immutable principles, of almost religious authority, to guide them, actually look for leading to professional politicians, who are influenced in turn by small but competent and determined minorities, with special knowledge and special interests. . . .

The American people, true enough, are sheep. Worse, they are donkeys. Yet worse, to borrow from their own dialect, they are goats. They are thus constantly bamboozled and exploited by small minorities of their own number, by determined and ambitious individuals, and even by exterior groups. . . . But all the while they have the means in their hands to halt the obscenity whenever it becomes intolerable, and now and then, raised transiently to a sort of intelligence, they do put a stop to it. There are no legal or other bars to the free functioning of their will, once it emerges into consciousness, save only such bars as they themselves have erected, and these they may remove whenever they so desire. No external or super-legal power stands beyond their reach, exercising pressure upon them; they recognize no personal sovereign with inalienable rights and no class with privileges above the common law; they are even kept free, by a tradition as old as the Republic itself, of foreign alliances which would condition their autonomy. Thus their sovereignty, though it is limited in its everyday exercise by self-imposed constitutional checks and still more by restraints which lie in the very nature of government, whatever its form, is probably just as complete in essence as that of the most absolute monarch who ever hanged a peasant or defied the Pope.

What is too often forgotten, in discussing the matter, is the fact that no such monarch was ever actually free, at all times and under all conditions. In the midst of his most charming tyrannies he had still to bear it in mind that his people, oppressed too much, could always rise against him, and that he himself, though a king *von Gottes Gnaden*,* was yet biologically only a man, with but one gullet to slit; and if the people were feeble or too craven to be dangerous, then there was always His Holiness of Rome to fear or other agents of the King of Kings; and if these ghostly mentors, too, were silent, then he had to reckon with his ministers, his courtiers, his soldiers, his doctors, and his women. . . . It seems to me that the common people, under such a democracy as that which now prevails in the United States, are more completely sovereign, in fact as well as in law, than any of these ancient despots. They may be seduced and enchained by a great variety of prehensile soothsayers, just as Henry VIII was seduced and enchained by his wives, but, like Henry again, they are quite free to throw off their chains whenever they please, and to chop off the heads of their seducers. . . .

Nor is there much force or relevancy in the contention that democracy is incomplete in the United States (as in England, France, Germany and all other democratic countries) because certain classes of persons are barred from full citizenship, sometimes for reasons that appear to be unsound. To argue thus is to argue against democracy itself, for if the majority has not the right to decide what qualifications shall be necessary to participate in its sovereignty, then it has no sovereignty at all. What one usually finds, on examining any given case of class disfranchisement, is that the class disfranchised is not actively eager, as a whole, for the ballot, and that its lack of interest in the matter is at least presumptive evidence of its general political incompetence. . . .

There is little reason for believing that the extension of the franchise to the classes that still remain in the dark would make

---

* Editor's Note: German for "by the Grace of God."

government more delicately responsive to the general will. Such classes, as a matter of fact, are now so few and so small in numbers in all of the Western nations that they may be very conveniently disregarded. It is as if doctors of philosophy, members of the Society of the Cincinnati or men who could move their ears were disfranchised. In the United States, true enough, there is one disfranchised group that is much larger, to wit, that group of Americans whose African descent is visible to the naked eye and at a glance. But even in this case, the reality falls much below the appearance. The more intelligent American Negroes vote in spite of the opposition of the poor whites, their theological brothers and economic rivals, and not a few of them actually make their livings as professional politicians, even in the South. . . .

Moreover, even those who are actually disfranchised, say in the rural wastes of the South, may remove their disability by the simple device of moving away, as, in fact, hundreds of thousands have done. Their disfranchisement is thus not intrinsic and complete, but merely a function of their residence, like that of all persons, white or black, who live in the District of Columbia, and so it takes on a secondary and trivial character, as hay-fever, in the pathological categories, takes on a secondary and trivial character by yielding to a change of climate. Moreover, it is always extra-legal, and thus remains dubious: the theory of the fundamental law is that the coloured folk may and do vote. This theory they could convert into a fact at any time by determined mass action. The Nordics might resist that action, but they could not halt it: there would be another Civil War if they tried to do so, and they would be beaten a second time. If the blacks in the backwaters of the South keep away from the polls today it is only because they do not esteem the ballot highly enough to risk the dangers that go with trying to use it. That fact, it seems to me, convicts them of unfitness for citizenship in a democratic state, for the loftiest of all the rights of the citizen, by the democratic dogma, is that of the franchise, and whoever is not willing to fight for it, even at the cost of his last drop of gore, is surely not likely to exercise it with a proper sense of consecration after getting it. No one argues that democracy is destroyed in the United States by the fact that millions of white citizens, perfectly free under the law and the local *mores* of their communities to vote, nevertheless fail to do so. The difference between these negligent whites and the disfranchised Negroes is only superficial. Both have a clear legal right to the ballot; if they neglect to exercise it, it is only because they do not esteem it sufficiently. In New York City thousands of freeborn Caucasians surrender it in order to avoid jury duty; in the South thousands of Negroes surrender it in order to avoid having their homes burned and their heads broken. The two motives are fundamentally identical; in each case the potential voter values his peace and security more than he values the boon for which the Fathers bled. He certainly has a right to choose.

The Genius of Democracy*

BY ROBERT M. MacIVER

[1939]

THERE must be some universal appeal in the name of democracy, for even its destroyers proudly claim possession of its soul. . . . Soviet spokesmen assert that they have now the most democratic constitution on earth, and Stalin himself declared that the 1937 elections in Russia were "the most democratic the world has seen." . . .

There are those who think that democracy means giving everyone equal authority, so that no man has any more power than another. This scheme would assure not the presence of democracy but the absence of government—in short, chaos, the "state of nature." There are others who think it is democratic to take a vote in order to decide the merits of plays or pictures or cigarettes or movie actresses, as though there were some necessary relation between merit and popularity. The New York World's Fair, in planning its art exhibition, announced that it would have a nation-wide system of selection committees and that quotas would be established for the various regions throughout the United States, and they called this system "the most complete application of democratic methods ever attempted in an exhibition of this kind." There are also those who think that to be democratic is to show to everyone a "hail-fellow-well-met" spirit. There are certain organizations calling themselves democratic—not unknown, for example, in New York City—where the idea of democracy is to share the spoils among the members and to distribute bread and coal to the deserving democratic poor. There are other organizations which propose to establish democracy by handing things out all round. In the first democracy that history clearly reveals, the idea got abroad —it was in ancient Athens—that there was something peculiarly

---

* From *The Southern Review*, Vol. V (1939–40). Reprinted by permission of The Louisiana State University Press, University, Louisiana. Robert M. MacIver (b. 1882), of Scottish origin, is now Professor Emeritus of Political Philosophy and Sociology at Columbia University.

democratic in the rotation of office, and they created so many offices that practically every citizen had his turn in one of them. To make the principle complete and give every man an equal chance they decided to use the lot instead of the ballot.

A more important confusion is that which equates democracy with the government of the many, opposed to the government of the few. The many in this sense never actually govern. They never do and never can decide the specific issues of policy that governments are always facing. Mr. Walter Lippmann once wrote a book (*The Phantom Public*) in which he told us that the ordinary man was quite disillusioned about democracy, because he couldn't possibly give attention to all the pressing questions of the day. Democracy was an "unattainable ideal" because the man in the street was unable to attend to banking problems one day and Brooklyn sewers the next and Manchurian railroads the day after—and so on. Of course if any one entertains such an illusion about the nature of democracy he certainly ought to be disillusioned. No serious political thinker has ever put forward such a theory, with the possible exception of Jean Jacques Rousseau. Democracy can never be government by the people in that sense. They can broadly decide the general direction of governmental policy and little more.

Nor can we say that democracy is any system under which the majority of the people support the existing government. Apart from the mere technical point that it is sometimes difficult to say whether a democratically elected government still holds a majority or not—though no doubt Mr. Gallup is of great service to us here—there is the glaring fact that a dictatorship may have the support of a majority, and any definition that will not enable us to distinguish democracy from dictatorship is worse than useless. Assuredly that is not what any intelligent defender of democracy means. We must carefully avoid the definition of democracy as simply majority-rule. Democracy does involve one form of majority-rule, a form in which there is no fixed majority entrenched against the processes and tides of free opinion that could reduce it again to a minority. But a majority-system that silences all opposition and censors all contrary opinion is emphatically not to be named a democracy.

Still less can democracy be defined as mass-rule. Here I am using the term "mass" or "masses" in the sense given to it by the

Spanish writer Ortega y Gasset in his famous alarmist book, *The Revolt of the Masses*. For him the mass is the average man as a multitude and he declares that the mass in this sense is in our days triumphing over all leadership, over all distinction, creating a "hyperdemocracy." I call the book alarmist because it dogmatically presents as a universal modern menace a phenomenon that is far from universal and that is not particularly modern. It is not true that the average man normally hates distinction. He generally applauds it where he understands it, as is very evident in the field of sport. And if leadership in the field of politics is often commonplace, a charge to which the party-system of the United States has been peculiarly liable, realistic investigation shows that this condition is due more to the manipulation and machination of wire-pulling interest-groups than to the instincts of the average man. There is indeed an element of truth in the position of Senor y Gasset. It is that in times of stress and crisis the mass tends to coalesce into one or more dynamic movements, responsive to the orator or the demagogue who understands their mentality, though he may understand nothing else. Such movements can easily triumph in such times, and it is true that democracy affords the free and open forum in which they are bred. But so long as democracy endures, such movements sink as easily as they rise. It is only when democracy falls that they gain the character which Senor y Gasset attributes to them. "The mass," he says, "crushes beneath it everything that is different, everything that is excellent, individual, qualified, and select." The description has more obvious application to the dictatorial than to the democratic spirit.

There is one further confusion about the nature of democracy which we must seek to dispel before we turn to its positive character. Democracy first expressed itself in a certain type of representative system, a parliamentary system, and on the whole it is still associated with that system. But it is quite possible to conceive of democracy as existing without parliamentary institutions in the traditional sense; that is, apart from a central assembly composed of the elected representatives of the people, an assembly which debates in public, by majority vote, and constitutes the decisive and central organ of government. Historically the growth of democracy was the growth of parliamentary institutions, and it remains true that parliamentary institutions are impossible without democracy, without the free expression of opinion as the basis of national policy. But we must not assume that the free play of public opinion must register itself in parliamentary forms. Historical evolution may reveal an endless train of yet undreamed-of modes of government, adaptations to changing needs and changing demands. Democracy is on the whole a recent development. Parliamentary institutions arose when the problems of government were simpler than they are today, when public opinion was more homogeneous, less diversified by specialized corporate interests, when representation of localities or areas had a meaning that now it has for the most part lost, when agriculture was the predominant occupation of men and the relation to the land everywhere the paramount relation. All that is changed. Already, in every democracy, important activities of regulation are outside the direct control of parliaments. Everywhere the necessities of administration have created boards and commissions, controls and corporate functions, devoted to fundamental national tasks. If this process continues, parliaments and congresses may cease to be the main centers of national life. But if freedom continues, democracy will still prevail. Still the free tides of opinion will determine who shall govern, who shall be entrusted with power. The mechanism of democracy must always change if conditions change and the principle of liberty abides.

If then the institutions of democracy are subject to change and must be forever readapted to changing conditions and changing needs, must we give up the attempt to discover the political form of democracy and seek instead to identify it by its spirit? I would rather not resort to that refuge. It is too inconclusive. It is also dangerous, since even its enemies may and indeed do claim for themselves whatever we assign as the spiritual quality of democracy. I must still define it by its form or structure, though realizing that only a congenial spirit, only an appropriate set of attitudes, can sustain that structure. I believe the problem is solved by the distinction between the form of *government* and the form of the *state* itself. No form of government is permanent but there are abiding forms of state. Democracy is such a form, and wherever it has existed in the past or exists in the present, it can be identified by two simple criteria. By means of these we can tell whether democracy prevails or whether instead we are confronting a political system which should be called by some other name, such as that of dictatorship.

The two are as follows. (1) *Democracy puts into effect the distinction between the state and the community.* Among other things this implies the existence of constitutional guarantees and civil rights which the government is not empowered to abrogate. (2) *Democracy depends on the free operation of conflicting opinions.* Among other things this implies a system under which any major trend or change of public opinion can constitutionally register itself in the determination both of the composition and of the policies of government. Let us take in turn each of these criteria.

By a community I mean an inclusive area of social interaction within which men share the basic conditions of a common life, whether on the scale of a village or a city, a tribe or a nation. It is thus a relatively definite area of society, and the boundaries of a state, the political organization, may or may not coincide with some such area. But whether they do or do not the state and the community, or more broadly, the state and society, must be distinguished.... Because democracy in effect affirms this distinction, its foundations, however weak the superstructure, are sunk into the rock of reality. Because the totalitarian state in effect denies this distinction, its foundations rest on shifting sand. No might, no flourish of doctrine, no ruthlessness, can ever destroy the difference, can ever reduce society to the proportions of the state. For the state is men organized under government and no body of men, not even the most totalitarian, make the total surrender of themselves, of all their living and thinking and believing and loving and fearing, to the power-control of government. They could not if they tried. They are the creatures of customs and traditions and of morals and creeds, of hidden loyalties, of daily habits, that are not controlled or controllable from any mere center of power. We live in a social matrix that is immensely more rich and subtle than the rigid delineaments of the state. That matrix sustains our daily life whereas the state is aloof and impersonal, a majestic name we reverence or an ominous thing we fear. Behind the majesty or the dread there is only a group of men clinging to power, limited in their visions, in their sympathies, in their understanding. The state can do only what government can regulate, what the ruler, a mere man or assembly of men, can effectively decree. And when in arrogant pretension or insufferable narrow-mindedness the ruler decrees that all men shall think as he thinks and shall value only what he values, he passes beyond the limits of his power. For a time the credulities or the passions or the necessities or the despairs of the masses may lead them to support his inordinate claims. He may be the instrument of destiny, even the temporary Messiah of his people. But his will is only the will of a man, placed in power by the conjuncture of events, a man subject to sickness and to age, to the delusions of pride and to the corruptions of power. His will may rule the state but no man and no assembly of men can comprehend or control the creative forces of society. That the total being of a community should be shaped and measured by this totalitarian will is as absurd a pretension as the old belief that the sun goes round the earth.

The state is a particular type of social organization and in so far as it has intelligible meaning or function it is an agency of the community it regulates. Under all conditions it is a logical confusion to identify the state with the community, with the people, the nation, the country. The people engage in myriad activities, enter into myriad relationships, that by no stretch of language can be called political. The people display myriad differences of opinion, thought, morals, creed, and culture. The government of the state may formally suppress them, but they are still there, no longer in the state, the political system, but in the community, the social system. Unfortunately language abets the confusion of thought. The same word—"United States," "England," "Germany"—denotes both the state and the nation-community. We say indifferently, "the United States makes a treaty" and "the United States is recovering from a depression." The first sentence refers to the state, the second to the country. We speak of the "national" debt—it is the debt of the state, not of the country; it is in fact owed to the country. When we say that "Germany overthrew the Weimar Republic," we mean that the people, or a part of the people, overthrew the state, we do not mean that the state overthrew itself. As soon as we begin to think about it we perceive that the state and the community are two different things, that the state is not the community but the political organization of the community. The customs of the people may conflict with the laws of the state. Men and women, as social beings, are not merely citizens of states. They act in other relationships. Their thoughts, their strivings, their fears and hopes, their beliefs, their affections and interests, their family life, lie largely outside the scheme of government altogether. In war or in grave crisis the state commandeers the community, demanding that the citizens forget their other relationships, their other interests, but the cost is always heavy. Only at an immense temporary sacrifice does the state even approach the universal partnership that orators such as Edmund Burke have called it.

Now what democracy does is to establish through constitutional forms the principle that the community is more inclusive than, greater than, the state. In many older forms of state, in ancient empires, the distinction was implicit. The scheme of daily life, the customs of the people, remained almost untouched by government except for incursions by the tax-gatherer and the occasional disruptions of war. But only in democracy is the distinction made the foundation of a political system. In effect democracy asserts that the state is one form of organization of the community, for certain ends of the community. Not for all ends, since that would destroy the right to be different and therefore the possibility of democracy. The ends of the state must be somewhere limited if opinion remains free, if government is to be an agency of the people instead of the principal of which the people are an agent. Under democracy the cultural life of the community is in general withdrawn from the direct control of the state. For if culture is coordinated then divergent opinions and creeds are suppressed, free thought is suppressed, and democracy cannot exist. Democracy so understood is not a specific form of government attached to a specific historical set of institutions, it is a mode of government corresponding to a set of attitudes. The forms may change, must change with the conditions if the creative processes of democracy endure.

Our second criterion, the constitutional right of opinion to determine policy, is, as we have just suggested, a corollary of the first. It is the way in which the distinction between the state and the community is carried into effect. The state regulates the common interest or what is conceived to be the common interest. The community nourishes many interests that are not common to all the citizens but only at most to particular groups. The democratic state is a limited state in that it cannot, without destroying itself, suppress the freedom of opinion, with the possible exception of such opinion as advocates the abolition of free opinion. It is limited in that it cannot entertain policies abrogating the right of assembly or of association, policies preventing religious or other cultural groups from pursuing their particular principles or tenets after their own manner, provided they do not assault the peace and order of the community. . . .

The two criteria we have offered provide the sufficient and conclusive ground for distinguishing democracy from all other forms of state. In so far as these principles prevail a political system is democratic. No political system other than democracy is founded on these principles. A dictatorship may rest on a majority will, but if so it prevents all minority wills from attaining expression. Hence it is not the number who support a government that determines whether that state is a democracy or a dictatorship. Quite possibly a greater proportion of the citizens support or have supported the Nazi or the Soviet government than in this country favors the present administration at Washington. Again, in the earlier forms of dynastic state the people on the whole, whether through conviction, indoctrination, superstition, fear, or inertia, acquiesced in, approved, or even venerated the governing power, but the opinion of the people was not creative, could not constitutionally translate itself into policy. Only on the grounds we have mentioned can we adequately define democracy and set it properly apart from other forms of the state. . . .

In the modern world there is no way to save government from the people or to save the people from itself. It is idle to ask for a government of the best men, as distinct from a democratic government, for who will elect the best and if by some strange chance they should elect themselves how long would they remain the best? It is idle to seek a government of laws and not of men, for a government of laws will turn into a government of lawyers, who happen also to be men. Every alternative to democracy is subject to a charge more fatal than any that can be laid against it, the irremediable defect of irresponsible power. It is on that ground that many outstanding political thinkers, men who, like John Stuart Mill, were very conscious of its weaknesses, men who, like most of the Fathers of the Constitution, had no great confidence in the people, nevertheless have championed its cause.

Possibly no discovery of the physical sciences, however world-shaking it may be, has been of more profound importance to mankind than the discovery that power could be made effectively responsible. This was no easy achievement but the painful task of centuries. Before it could be realized the discovery had to break through age-old entrenchments of established interests, guarded by traditions, by ceremonies, by taboos, by magic, by religion, by dire penalties, by all the means, physical, economic, spiritual, that power itself can dispose. At best it has been a partial and a precarious achievement, but we should not minimize on that account its immense importance. Democracy is the generic name for that achievement, and its significance is understood only when we contemplate the effects of irresponsible power, not merely on those who are subject to it but above all on those who possess it. Who that has lived many years on this earth can have failed to observe how even a modicum of irresponsible power perverts the intelligence and hardens the sensibilities, how the jack-in-office struts in pompous undiscerning pride, how the bureaucrat loses touch with humanity, how the petty boss, when no superior watches, becomes a wretched bully? In every sphere of human activity, in the factory, in the trade-union, in the home, in the church, in the barracks, in the prison, even in the seats of learning, let power be uncontrolled and it will work the same effects. With some of his bitterest words Shakespeare characterized this phenomenon:

> Man, proud man,
> Drest in a little brief authority,
> Most ignorant of what he's most assured,
> His glassy essence, like an angry ape,
> Plays such fantastic tricks before high heaven
> As make the angels weep.

And if uncontrolled power works these effects on the lower level of every-day affairs, how much more evil it can be when it is set on high and calls for veneration! The intolerable experience of such exercise of power drove the politically-minded peoples to seek a safeguard against it—and they all had experience in plenty. The safeguard was democracy, and it is this hard-won safeguard that the dictatorships now ridicule and trample on. They laud and magnify the irresponsible power which corrupts the best and turns the worst into loathsome brutes.

In the development of this epoch-making discovery, that power can be made responsible, various peoples have played a part. The Greeks made the first great contribution, the Romans added something, the medieval cities began to explore it anew, the English rediscovered it on a national scale and broadened its foundations through the centuries. Here in conclusion I would say a word about the contribution of America. For at this juncture in world affairs the United States has taken on a new and more decisive role in the drama of democracy.

Although many of the pioneers in the making of America came in the quest of religious freedom, social emancipation, or economic opportunity, America contributed little to the practice or principle of democracy until quite modern times. In colonial days dogma had too strong a hold. Those who had found religious freedom for themselves were generally ready enough to deny it to others. Except in certain local arrangements there was relatively little democracy. Except for certain heretics, such as Roger Williams, there was relatively little democratic theory. Indeed there was very little political thinking at all that had any independence. Even in the crisis of the Revolution the eighteenth-century thinkers went back for their philosophical inspiration to the English thinkers of the seventeenth century. What was most congenial to them was the common-sense liberalism of John Locke, while the radicalism of Tom Paine had only an ephemeral flare of popularity and the democratic fervor of Rousseau had practically no appeal at all. The leaders of the Revolution accepted the principle that the people were the locus of sovereignty, but the people were conceived of in the Lockian sense. They were the substantial solid folk as distinct from the rabble. The appeal to the people was not the appeal to the whole people. Nor was it only the socially conservative, like Alexander Hamilton, John Adams, and Ames, who distrusted "the imprudence of democracy." There are evidences of the same distrust in Samuel Adams, even in Thomas Jefferson. The leaders of that age were in the dilemma that they must build the revolutionary state on the foundation of democracy and yet they were fearful of the foundation. So in the various states as well as in the Union they restrained the operation of majority voting by constitutional enactment—a process that particular states have since carried much further—and they further legislated property qualifications for voters and still more stringent ones for electoral candidates.

But before another fifty years had passed a new spirit was beginning to pervade the growing republic. It had cast off the intellectual dependence of colonial days. It had begun to create a distinctive social order in which property qualifications disappeared along with primogeniture, in which European traditions of rank and class were set at naught, in which a different philosophy emerged, signalized by the robust individualism of Whitman and Emerson and by the optimistic faith of Lincoln in the common people. In this vast movement the influence of the ever-expanding West predominated, fed by the vision of free men and by the presence of free land. Thither went those who had discarded tradition or who had none to discard, the dispossessed and those who had no possessions. There was generated the spiritual individualism which nursed the characteristic democracy of America. Individualism is not always the friend of democracy. . . . But in this age of free land and rich exploitable resources, individualism was tolerant of the liberties of the common man, even though it sustained also the boss, the spoils-seeker, and the robber-baron. The common man emerged from social subservience to an extent nowhere else attained in the civilized world. American democracy thus became a new thing, no longer dependent on European principles but developing along its own lines and growing fully conscious of its new-world quality.

Such in briefest outline has been the character of American democracy, reflecting the spirit of the individual's trust in himself, unhierarchical, tolerant of differences, lacking in class consciousness, and distrustful of the repressive powers of government. The picture is complicated by other factors, by the tendency to reverence the Constitution as a final political revelation, by the more powerful operation of the principle of judicial review with its influence in the direction of socio-economic conservatism, by the consolidation of economic power in vast corporate empires, by the presence of a large racially separate population not in effect accepted into the framework of the democratic system, and by the preponderant growth of an industrial urbanized population whose conditions of life and whose problems of security and employment are utterly remote from those that bred the American doctrine. It is not difficult to imagine that from some alignment of these factors might come a serious menace to the American democratic tradition. The frontier has receded into the Western ocean. Free land belongs to the past. Economic individualism forlornly fights a rear guard action against the billion-dollar corporation and centralized finance and organized labor. . . . New times have brought new attitudes and new needs. Democracy too must find a new voice, a new conviction. . . .

From Individualism to Mass Democracy*

BY EDWARD H. CARR

[1951]

MODERN democracy, as it grew up and spread from its focus in western Europe over the past three centuries, rested on three main propositions: first, that the individual conscience is the ultimate source of decisions about what is right and wrong; second, that there exists between different individuals a fundamental harmony of interests strong enough to enable them to live peacefully together in society; third, that where action has to be taken in the name of society, rational discussion between individuals is the best method of reaching a decision on that action. Modern democracy is, in virtue of its origins, individualist, optimistic and rational. The three main propositions on which it is based have all been seriously challenged in the contemporary world.

In the first place, the individualist conception of democracy rests on a belief in the inherent rights of individuals based on natural law. According to this conception, the function of democratic government is not to create or innovate, but to interpret and apply rights which already exist. This accounts for the importance attached in the democratic tradition to the rights of minorities within the citizen body. Decision by majority vote might be a necessary and convenient device. But individuals belonging to the minority had the same inherent rights as those belonging to the majority. Insistence on the rule of law, preferably inscribed in a written and permanent constitution, was an important part of the individualist tradition of democracy. The individual enjoyed certain indefeasible rights against the society of which he was a mem-

---

* From chapter IV of *The New Society*. Reprinted by permission of Macmillan & Co., Ltd. Edward H. Carr (b. 1892), at one time in the British foreign service and for many years Professor of International Relations at the University of Wales, is a noted authority on the theory of international politics, the history of political ideas and the history of Russia.

ber; these rights were often regarded as deriving from a real or hypothetical "social contract" which formed the title-deeds of society. Just as the individualist tradition in *laissez-faire* economics was hostile to all forms of combination, so the individualist tradition in politics was inimical to the idea of political parties. Both in Athenian democracy and in eighteenth-century Britain, parties were regarded with mistrust and denounced as "factions."

The French revolution with its announcement of the sovereignty of the people made the first serious assault on this view of democracy. The individualism of Locke's "natural law" was replaced by the collectivism of Rousseau's "general will." Both Pericles and Locke had thought in terms of a small and select society of privileged citizens. Rousseau for the first time thought in terms of the sovereignty of the whole people, and faced the issue of mass democracy. He did so reluctantly; for he himself preferred the tiny community where direct democracy, without representation or delegation of powers, was still possible. But he recognized that the large nation had come to stay, and held that in such conditions the people could be sovereign only if it imposed on itself the discipline of a "general will." The practical conclusion drawn from this doctrine, not by Rousseau himself, but by the Jacobins, was the foundation of a single political party to embody the general will. Its logical conclusions were still more far-reaching. The individual, far from enjoying rights against society assured to him by natural law, had no appeal against the deliverances of the general will. The general will was the repository of virtue and justice, the state its instrument for putting them into effect. The individual who dissented from the general will cut himself off from the community and was a self-proclaimed traitor to it. Rousseau's doctrine led directly to the Jacobin practice of revolutionary terror. It would be idle to embark on a theoretical discussion of the rival merits of the two conceptions of democracy. Individualism is an oligarchic doctrine—the doctrine of the select and enterprising few who refuse to be merged in the mass. The function of natural law in modern history, though it is susceptible of other interpretations, has been to sanctify existing rights and to brand as immoral attempts to overthrow them. A conception based on individual rights rooted in natural law was a natural product of the oligarchic and conservative eighteenth century. It was equally natural that this conception should be challenged and overthrown in the ferment of a revolution that proclaimed the supremacy of popular sovereignty.

While, however, the beginnings of mass democracy can be discerned in the doctrines of Rousseau and in the practice of the French revolution, the problem in its modern form was a product of the nineteenth century. The Industrial revolution started its career under the banner of individual enterprise. Adam Smith was as straightforward an example as could be desired of eighteenth-century individualism. But presently the machine overtook the man, and the competitive advantages of mass production ushered in the age of standardization and larger and larger economic units. And with the mammoth trust and the mammoth trade union came the mammoth organ of opinion, the mammoth political party and, floating above them all, the mammoth state, narrowing still further the field of responsibility and action left to the individual and setting the stage for the new mass society. It was the English Utilitarians who, by rejecting natural law, turned their backs on the individualist tradition and, by postulating the greatest good and the greatest number as the supreme goal, laid the theoretical foundation of mass democracy in Britain; in practice, they were also the first radical reformers. Before long, thinkers began to explore some of the awkward potentialities of mass democracy. The danger of the oppression of minorities by the majority was the most obvious. This was discerned by Tocqueville in the United States in the 1830's and by J. S. Mill in England twenty-five years later. In our own time the danger has reappeared in a more insidious form. Soviet Russia has a form of government which describes itself as a democracy. It claims, not without some historical justification, to stem from the Jacobins who stemmed from Rousseau and the doctrine of the general will. The general will is an orthodoxy which purports to express the common opinion; the minority which dissents can legitimately be suppressed. But we are not concerned here with the abuses and excesses of the Soviet form of government. What troubles us is the question how far, in moving from the individualism of restrictive liberal democracy to the mass civilization of today, we have ourselves become involved in a conception of democracy which postulates a general will. The question is all around us today not only in the form of loyalty tests, avowed or secret, or committees on un-American activities, but also in the form of the closed shop and of increasingly rigid standards of party discipline. . . .

The second postulate of Locke's conception of society, the belief in a fundamental harmony of interests between individuals, equally failed to stand the test of time, and for much the same reason. Even more than natural law, the harmony of interests was essentially a conservative doctrine. If the interest of the individual rightly understood coincided with the interest of the whole society, it followed that any individual who assailed the existing order was acting against his own true interests and could be condemned not only as wicked, but as short-sighted and foolish. Some such argument was, for instance, often invoked against strikers who failed to recognize the common interest uniting them with their employers. The French revolution, an act of self-assertion by the third estate against the two senior estates of nobility and clergy, demonstrated—like any other violent upheaval—the hollowness of the harmony of interests; and the doctrine was soon also to be powerfully challenged on the theoretical plane.

The challenge came from two quarters. The Utilitarians, while not making a frontal attack on the doctrine, implicitly denied it when they asserted that the harmony of interests had to be created by remedial action before it would work. They saw that some of the worst existing inequalities would have to be reformed out of existence before it was possible to speak without irony of a society based on a harmony of interests; and they believed in increased education, and the true liberty of thought which would result from it, as a necessary preparation for establishing harmony. Then Marx and Engels in the *Communist Manifesto* took the class struggle and made out of it a theory of history which, partial though it was, stood nearer to current reality than the theory of the harmony of interests had ever done. . . . The substitution of a planned economy for *laissez-faire* capitalism brought about a radical transformation in the attitude towards the state. The functions of the state were no longer merely supervisory, but creative and remedial. It was no longer an organ whose weakness was its virtue and whose activities should be restricted to a minimum in the interests of freedom. It was an organ which one sought to capture and control for the carrying out of necessary reforms; and, having captured it, one sought to make it as powerful and effective as possible in order to carry them out. The twentieth century has not only replaced individualist democracy by mass democracy, but has substituted the cult of the strong remedial state for the doctrine of the natural harmony of interests.

The third main characteristic of Locke's conception of society—a characteristic which helped to give the eighteenth century its nicknames of the Age of Reason or the Age of Enlightenment—was its faith in rational discussion as a guide to political action. This faith provided the most popular nineteenth-century justification of the rule of the majority as the basis of democracy. Since men were on the whole rational, and since the right answer to any given issue could be discovered by reason, one was more likely, in the case of dispute, to find right judgment on the side of the majority than on the side of the minority. Like other eighteenth-century conceptions, the doctrine of reason in politics was the doctrine of a ruling oligarchy. The rational approach to politics, which encouraged leisurely argument and eschewed passion, was eminently the approach of a well-to-do, leisured and cultured class. Its efficacy could be most clearly and certainly guaranteed when the citizen body consisted of a relatively small number of educated persons who could be trusted to reason intelligently and dispassionately on controversial issues submitted to them. The prominent rôle assigned to reason in the original democratic scheme provides perhaps the most convincing explanation why democracy has hitherto always seemed to flourish best with a restrictive franchise. Much has been written in recent years of the decline of reason, and of respect for reason, in human affairs, when sometimes what has really happened has been the abandonment of the highly simplified eighteenth-century view of reason in favour of a subtler and more sophisticated analysis. But it is none the less true that the epoch-making changes in our attitude towards reason provide a key to some of the profoundest problems of contemporary democracy.

First of all, the notion that men of intelligence and good will were likely by process of rational discussion to reach a correct opinion on controversial political questions could be valid only in an age when such questions were comparatively few and simple enough to be accessible to the educated layman. It implicitly denied that any specialized knowledge was required to solve political problems. This hypothesis was perhaps tenable so long as the state was not required to intervene in economic issues, and the questions on which decisions had to be taken turned on matters of practical detail or general political principles. In the first half of the twentieth century these conditions had everywhere ceased to exist. . . .

At this initial stage of the argument reason itself is not dethroned from its supreme rôle in the decision of political issues. The citizen is merely asked to surrender his right of decision to the superior reason of the expert. At the second stage of the argument reason itself is used to dethrone reason. The social psychologist, employing rational methods of investigation, discovers that men in the mass are often most effectively moved by non-rational emotions such as admiration, envy, hatred, and can be most effectively reached not by rational argument, but by emotional appeals to eye and ear, or by sheer repetition. Propaganda is as essential a function of mass democracy as advertising of mass production. The political organizer takes a leaf out of the book of the commercial advertiser and sells the leader or the candidate to the voter by the same methods used to sell patent medicines or refrigerators. The appeal is no longer to the reason of the citizen, but to his gullibility. A more recent phenomenon has been the emergence of what Max Weber called the "charismatic leader" as the expression of the general will. The retreat from individualism seemed to issue at last—and not alone in the so-called totalitarian countries—in the exaltation of a single individual leader who personified and resumed within himself the qualities and aspirations of the "little man," of the ordinary individual lost and bewildered in the new mass society. But the principal qualification of the leader is no longer his capacity to reason correctly on political or economic issues, or even his capacity to choose the best experts to reason for him, but a good public face, a convincing voice, a sympathetic fireside manner on the radio; and these qualities are deliberately built up for him by his publicity agents. In this picture of the techniques of contemporary democracy, the party headquarters, the directing brain at the centre, still operates rationally, but uses irrational rather than rational means to achieve its ends—means which are, moreover, not merely irrational but largely irrelevant to the purposes to be pursued or to the decisions to be taken.

The third stage of the argument reaches deeper levels. . . .

Marx played . . . a far more important part in what has been called "the flight from reason" than by the mere exaltation of the collective over the individual. By his vigorous assertion that "being determines consciousness, not consciousness being," that thinking is conditioned by the social environment of the thinker, and that ideas are the superstructure of a totality whose foundation is formed by the material conditions of life, Marx presented a clear challenge to what had hitherto been regarded as the sovereign or autonomous human reason. The actors who played significant parts in the historical drama were playing parts already written for them: this indeed was what made them significant. The function of individual reason was to identify itself with the universal reason which determined the course of history and to make itself the agent and executor of this universal reason. Some such view is indeed involved in any attempt to trace back historical events to underlying social causes; and Marx—and still more Engels—hedged a little in later years about the rôle of the individual in history. But the extraordinary vigour and conviction with which he drove home his main argument, and the political theory which he founded on it, give him a leading place among those nineteenth-century thinkers who shattered the comfortable belief of the Age of Enlightenment in the decisive power of individual reason in shaping the course of history.

Marx's keenest polemics were those directed to prove the "conditioned" character of the thinking of his opponents and particularly of the capitalist ruling class of the most advanced countries of his day. If they thought as they did it was because, as members of a class, "being" determined their "consciousness," and their ideas necessarily lacked any independent objectivity and validity. . . . Marx's writings gave a powerful impetus to all forms of relativism. . . .

Another thinker of the later nineteenth century also helped to mould the climate of political opinion. Like Darwin, Freud was a scientist without pretensions to be a philosopher or, still less, a political thinker. But in the flight from reason at the end of the nineteenth century, he played the same popular rôle as Darwin had played a generation earlier in the philosophy of *laissez-faire*. Freud demonstrated that the fundamental attitudes of human beings in action and thought are largely determined at levels beneath that of consciousness, and that the supposedly rational explanations of those attitudes which we offer to ourselves and others are artificial and erroneous "rationalizations" of processes which we have failed to understand. Reason is given to us, Freud seems to say, not to direct our thought and action, but to camouflage the hidden forces which do direct it. This is a still more devastating version of the Marxist thesis of substructure and superstructure. The substructure of reality resides in the unconscious: what appears above the surface is no more than the reflexion, seen in a distorting ideological mirror, of what goes on underneath. The political conclusion from all this—Freud himself drew none—is that any attempt to appeal to the reason of the ordinary man is waste of time, or is useful merely as camouflage to conceal the real nature of the process of persuasion; the appeal must be made to those subconscious strata which are decisive for thought and action. The debunking of ideology undertaken by the political science of Marx is repeated in a far more drastic and far-reaching way by the psychological science of Freud and his successors.

By the middle of the nineteenth century, therefore, the propositions of Locke on which the theory of liberal democracy were founded had all been subjected to fundamental attack, and the attack broadened and deepened as the century went on. Individualism began to give way to collectivism both in economic organization and in the forms and practice of mass democracy: the age of mass civilization had begun. The alleged harmony of interests between individuals was replaced by the naked struggle between powerful classes and organized interest groups. The belief in the settlement of issues by rational discussion was undermined, first, by recognition of the complex and technical character of the issues involved, later and more seriously, by recognition that rational arguments were merely the conditioned reflexion of the class interests of those who put them forward, and, last and most seriously of all, by the discovery that the democratic voter, like other human beings, is most effectively reached not by arguments directed to his reason, but by appeals directed to his irrational, subconscious prejudices. The picture of democracy which emerged from these criticisms was the picture of an arena where powerful interest-groups struggled for the mastery. The leaders themselves were often the spokesmen and instruments of historical processes which they did not fully understand; their followers consisted of voters recruited and marshalled for purposes of which they were wholly unconscious by all the subtle techniques of modern psychological science and modern commercial advertising.

The picture is overdrawn. But we shall not begin to understand the problems of mass democracy unless we recognize the serious elements of truth in it, unless we recognize how far we have moved away from the conceptions and from the conditions out of which the democratic tradition was born. From the conception of democracy as a select society of free individuals, enjoying equal rights and periodically electing to manage the affairs of the society, a small number of their peers, who deliberate together and decide by rational argument on the course to pursue (the assumption being that the course which appeals to the majority is likely to be the most rational), we have passed to the current reality of mass democracy. The typical mass democracy of today is a vast society of individuals, stratified by widely different social and economic backgrounds into a series of groups or classes, enjoying equal political rights the exercise of which is organized through two or more closely integrated political machines called parties. Between the parties and individual citizens stand an indeterminate number of entities variously known as unions, associations, lobbies or pressure-groups devoted to the promotion of some economic interest, or of some social or humanitarian cause in which keen critics usually detect a latent and perhaps unconscious interest. At the first stage of the democratic process, these associations and groups form a sort of exchange and mart where votes are traded for support of particular policies; the more votes such a group controls the better its chance of having its views incorporated in the party platform. At the second stage, when these bargains have been made, the party as a united entity "goes to the country" and endeavours by every form of political propaganda to win the support of the unattached voter. At the third stage, when the election has been decided, the parties once more dispute or bargain together, in the light of the votes cast, on the policies to be put into effect; the details of procedure at this third stage differ considerably in different democratic countries in accordance with varying constitutional requirements and party structures. What is important to note is that the first and third stages are fierce matters of bargaining. At the second stage, where the mass persuasion of the electorate is at issue, the methods employed now commonly approximate more and more closely to those of commercial advertisers, who, on the advice of modern psychologists, find the appeal to fear, envy or self-aggrandizement more effective than the appeal to reason. Certainly in the United States, where contemporary large-scale democracy has worked most successfully and where the strongest confidence is felt in its survival, experienced practitioners of politics would give little encouragement to the idea that rational argument exercises a major influence on the democratic process. We have returned to a barely disguised struggle of interest-groups in which the arguments used are for the most part no more than a rationalization of the interests concerned, and the rôle of persuasion is played by carefully calculated appeals to the irrational subconscious.

This discussion is intended to show not that mass democracy is more corrupt or less efficient than other forms of government (this I do not believe), but that mass democracy is a new phenomenon—a creation of the last half-century—which it is inappropriate and misleading to consider in terms of the philosophy of Locke or of the liberal democracy of the nineteenth century. It is new, because the new democratic society consists no longer of a homogeneous closed society of equal and economically secure individuals mutually recognizing one another's rights, but of ill co-ordinated, highly stratified masses of people of whom a large majority are primarily occupied with the daily struggle for existence. It is new, because the new democratic state can no longer be content to hold the ring in the strife of private economic interests, but must enter the arena at every moment and take the initiative in urgent issues of economic policy which affect the daily life of all the citizens, and especially of the least secure. It is new, because the old rationalist assumptions of Locke and of liberal democracy have broken down under the weight both of changed material conditions and of new scientific insights and inventions, and the leaders of the new democracy are concerned no longer primarily with the reflexion of opinion, but with the moulding and manipulation of opinion. To speak today of the defence of democracy as if we were defending something which we knew and had possessed for many decades or many centuries is self-deception and sham.

It is no answer to point to institutions that have survived from earlier forms of democracy. The survival of kingship in Great Britain does not prove that the British system of government is a monarchy; and democratic institutions survive in many countries today—some survived even in Hitler's Germany—which have little or no claim to be called democracies. The criterion must be sought not in the survival of traditional institutions, but in the question where power resides and how it is exercised. In this respect democracy is a matter of degree. Some countries today are more democratic than others. But none is perhaps very democratic, if any high standard of democracy is applied. Mass democracy is a difficult and hitherto largely uncharted territory; and we should be nearer the mark, and should have a far more convincing slogan, if we spoke of the need, not to defend democracy, but to create it.

. . . I discussed two of the basic problems which confront the new society—the problem of a planned economy and the problem of the right deployment and use of our human resources. These problems are basic in the sense that their solution is a condition of survival. The old methods of organizing production have collapsed, and society cannot exist without bringing new ones into operation. But those problems might conceivably be solved—are even, perhaps, in danger of being solved—by other than democratic means: here the task of mass democracy is to meet known and recognized needs by methods that are compatible with democracy, and to do it in time. The central problem which I have been discussing today touches the essence of democracy itself. Large-scale political organizations show many of the characteristics of large-scale economic organization, and have followed the same path of development. Mass democracy has, through its very nature, thrown up on all sides specialized groups of leaders—what are sometimes called élites. Everywhere, in government, in political parties, in trade unions, in co-operatives, these indispensable élites have taken shape with startling rapidity over the last thirty years. Everywhere the rift has widened between leaders and rank and file.

The rift takes two forms. In the first place, the interests of the leaders are no longer fully identical with those of the rank and file, since they include the special interest of the leaders in maintaining their own leadership—an interest which is no doubt rationalized, but not always justly, as constituting an interest of the whole group. The leaders, instead of remaining mere delegates of their equals, tend in virtue of their functions to become a separate professional, and then a separate social, group, forming the nucleus of a new ruling class or, more insidiously still, being absorbed into the old ruling class. Secondly, and most important of all, there is an ever-increasing gap between the terms in which an issue is debated and solved among leaders and the terms in which the same issue is presented to the rank and file. Nobody supposes that the arguments which the leaders and managers of a political party or a trade union use among themselves in private conclave are the same as those which they present to a meeting of their members; and the methods of persuasion used from the public platform or over the radio will diverge more widely still. When the decision of substance has been taken by the leaders, whether of government, of party or of union, a further decision is often required on the best method of selling the decision. Broadly speaking, the rôle of reason varies inversely with the number of those to whom the argument is addressed. The decision of the leaders may be taken on rational grounds. But the motivation of the decision to the rank and file of the party or union, and still more to the general public, will contain a larger element of the irrational the larger the audience becomes. The spectacle of an efficient élite maintaining its authority and asserting its will over the mass by the rationally calculated use of irrational methods of persuasion is the most disturbing nightmare of mass democracy.

The problem defies any rough-and-ready answer. It was implicit in Lincoln's formula of government "of the people" (meaning, I take it, belonging to the people in the sense of popular sovereignty), "by the people" (implying, I think, direct participation in the business of government) and "for the people" (requiring an identity of interests between governors and governed only obtainable when such participation occurs). It was implicit in Lenin's much-derided demand that every cook should learn to govern and that every worker should take his turn at the work of administration. The building of nineteenth-century democracy was long and arduous. The building of the new mass democracy will be no easier. . . .

. . . I have no faith in a flight into the irrational or in an exaltation of irrational values. Reason may be an imperfect instrument; and we can no longer take the simple view of its character and functions which satisfied the eighteenth and nineteenth centuries. But it is none the less in a widening and deepening of the power of reason that we must place our hope. Mass democracy calls just as much as individualist democracy for an educated society as well as for responsible and courageous leaders; for it is only thus that the gap between leaders and masses, which is the major threat to mass democracy, can be bridged.

V

THE DIFFUSION OF POWER

The Federalist No. 51*

BY JAMES MADISON

[1788]

To the People of the State of New York:

To what expedient, then, shall we finally resort, for maintaining in practice the necessary partition of power among the several departments, as laid down in the Constitution? The only answer that can be given is, that as all these exterior provisions are found to be inadequate, the defect must be supplied, by so contriving the interior structure of the government as that its several constituent parts may, by their mutual relations, be the means of keeping each other in their proper places. Without presuming to undertake a full development of this important idea, I will hazard a few general observations, which may perhaps place it in a clearer light, and enable us to form a more correct judgment of the principles and structure of the government planned by the convention.

In order to lay a due foundation for that separate and distinct exercise of the different powers of government, which to a certain extent is admitted on all hands to be essential to the preservation of liberty, it is evident that each department should have a will of its own; and consequently should be so constituted that the members of each should have as little agency as possible in the appointment of the members of the others. Were this principle rigorously adhered to, it would require that all the appointments for the supreme executive, legislative, and judiciary magistracies should be drawn from the same fountain of authority, the people, through channels having no communication whatever with one another. Perhaps such a plan of constructing the several departments would be less difficult in practice than it may in contemplation appear. Some difficulties, however, and some additional expense would attend the execution of it. Some deviations, there-

---

* James Madison (1751–1836) was the fourth President of the United States. He served from 1809 to 1817. He is the co-author, with Alexander Hamilton and John Jay, of *The Federalist*, a series of articles urging the ratification of the U. S. Constitution.

fore, from the principle must be admitted. In the constitution of the judiciary department in particular, it might be inexpedient to insist rigorously on the principle: first, because peculiar qualifications being essential in the members, the primary consideration ought to be to select that mode of choice which best secures these qualifications; secondly, because the permanent tenure by which the appointments are held in that department, must soon destroy all sense of dependence on the authority conferring them.

It is equally evident, that the members of each department should be as little dependent as possible on those of the others, for the emoluments annexed to their offices. Were the executive magistrate, or the judges, not independent of the legislature in this particular, their independence in every other would be merely nominal.

But the great security against a gradual concentration of the several powers in the same department, consists in giving to those who administer each department the necessary constitutional means and personal motives to resist encroachments of the others. The provision for defence must in this, as in all other cases, be made commensurate to the danger of attack. Ambition must be made to counteract ambition. The interest of the man must be connected with the constitutional rights of the place. It may be a reflection on human nature, that such devices should be necessary to control the abuses of government. But what is government itself, but the greatest of all reflections on human nature? If men were angels, no government would be necessary. If angels were to govern men, neither external nor internal controls on government would be necessary. In framing a government which is to be administered by men over men, the great difficulty lies in this: you must first enable the government to control the governed; and in the next place oblige it to control itself. A dependence on the people is, no doubt, the primary control on the government; but experience has taught mankind the necessity of auxiliary precautions.

This policy of supplying, by opposite and rival interests, the defect of better motives, might be traced through the whole system of human affairs, private as well as public. We see it particularly displayed in all the subordinate distributions of power, where the constant aim is to divide and arrange the several offices in such a manner as that each may be a check on the other—that the private interest of every individual may be a sentinel over the public rights. These inventions of prudence cannot be less requisite in the distribution of the supreme powers of the State.

But it is not possible to give to each department an equal power of self-defence. In republican government, the legislative authority necessarily predominates. The remedy for this inconveniency is to divide the legislature into different branches; and to render them, by different modes of election and different principles of action, as little connected with each other as the nature of their common functions and their common dependence on the society will admit. It may even be necessary to guard against dangerous encroachments by still further precautions. As the weight of the legislative authority requires that it should be thus divided, the weakness of the executive may require, on the other hand, that it should be fortified. An absolute negative on the legislature appears, at first view, to be the natural defence with which the executive magistrate should be armed. But perhaps it would be neither altogether safe nor alone sufficient. On ordinary occasions it might not be exerted with the requisite firmness, and on extraordinary occasions it might be perfidiously abused. May not this defect of an absolute negative be supplied by some qualified connection between this weaker department and the weaker branch of the stronger department, by which the latter may be led to support the constitutional rights of the former, without being too much detached from the rights of its own department?

If the principles on which these observations are founded be just, as I persuade myself they are, and they be applied as a criterion to the several State constitutions, and to the federal Constitution, it will be found that if the latter does not perfectly correspond with them, the former are infinitely less able to bear such a test.

There are, moreover, two considerations particularly applicable to the federal system of America, which place that system in a very interesting point of view.

First. In a single republic, all the power surrendered by the people is submitted to the administration of a single government; and the usurpations are guarded against by a division of the government into distinct and separate departments. In the compound republic of America, the power surrendered by the people is first divided between two distinct governments, and then the portion allotted to each subdivided among distinct and separate departments. Hence a double security arises to the rights of the people. The different governments will control each other, at the same time that each will be controlled by itself.

Second. It is of great importance in a republic not only to guard the society against the oppression of its rulers, but to guard one part of the society against the injustice of the other part. Different interests necessarily exist in different classes of citizens. If a majority be united by a common interest, the rights of the minority will be insecure. There are but two methods of providing against this evil: the one by creating a will in the community independent of the majority—that is, of the society itself; the other, by comprehending in the society so many separate descriptions of citizens as will render an unjust combination of a majority of the whole very improbable, if not impracticable. The first method prevails in all governments possessing an hereditary or self-appointed authority. This, at best, is but a precarious security; because a power independent of the society may as well espouse the unjust views of the major, as the rightful interests of the minor party, and may possibly be turned against both parties. The second method will be exemplified in the federal republic of the United States. Whilst all authority in it will be derived from and dependent on the society, the society itself will be broken into so many parts, interests and classes of citizens, that the rights of individuals, or of the minority, will be in little danger from interested combinations of the majority. In a free government the security for civil rights must be the same as that for religious rights. It consists in the one case in the multiplicity of interests, and in the other in the multiplicity of sects. The degree of security in both cases will depend on the number of interests and sects; and this may be presumed to depend on the extent of country and number of people comprehended under the same government. This view of the subject must particularly recommend a proper federal system to all the sincere and considerate friends of republican government, since it shows that in exact proportion as the territory of the Union may be formed into more circumscribed Confederacies, or States, oppressive combinations of a majority will be facilitated; the best security, under the republican forms, for the rights of every class of citizens, will be diminished; and consequently the stability and independence of some member of the government, the only other security, must be proportionally increased. Justice is the end of government. It is the end of civil society. It ever has been and ever will be pursued until it be obtained, or until liberty be lost in the pursuit. In a society under the forms of which the stronger faction can readily unite and oppress the weaker, anarchy may as truly be said to reign as in a state of nature, where the weaker individual is not secured against the violence of the stronger; and as, in the latter state, even the stronger individuals are prompted, by the uncertainty of their condition, to submit to a government which may protect the weak as well as themselves; so, in the former state, will the more powerful factions or parties be gradually induced, by a like motive, to wish for a government which will protect all parties, the weaker as well as the more powerful. It can be little doubted that if the State of Rhode Island was separated from the Confederacy and left to itself, the insecurity of rights under the popular form of government within such narrow limits would be displayed by such reiterated oppressions of factious majorities that some power altogether independent of the people would soon be called for by the voice of the very factions whose misrule had proved the necessity of it. In the extended republic of the United States, and among the great variety of interests, parties, and sects which it embraces, a coalition of a majority of the whole society could seldom take place on any other principles than those of justice and the general good; whilst there being thus less danger to a minor from the will of a major party, there must be less pretext, also, to provide for the security of the former, by introducing into the government a will not dependent on the latter, or, in other words, a will independent of the society itself. It is no less certain than it is important, notwithstanding the contrary opinions which have been entertained, that the larger the society, provided it lie within a practical sphere, the more duly capable it will be of self-government. And happily for the *republican cause,* the practicable sphere may be carried to a very great extent, by a judicious modification and mixture of the *federal principle.* PUBLIUS

Power and Responsibility*

A Comparison of the British and American Constitutions

BY WALTER BAGEHOT

[1867]

THE efficient secret of the English Constitution may be described as the close union, the nearly complete fusion, of the executive and legislative powers. No doubt by the traditional theory, as it exists in all the books, the goodness of our constitution consists in the entire separation of the legislative and executive authorities, but in truth its merits consists in their singular approximation. The connecting link is *the cabinet*. By that new word we mean a committee of the legislative body selected to be the executive body. The legislature has many committees, but this is its greatest. It chooses for this, its main committee, the men in whom it has most confidence.... As a rule, the nominal prime minister is chosen by the legislature, and the real prime minister for most purposes—the leader of the House of Commons—almost without exception is so. There is nearly always some one man plainly selected by the voice of the predominant party in the predominant house of the legislature to head that party, and consequently to rule the nation. We have in England an elective first magistrate as truly as the Americans have an elective first magistrate.... Nevertheless, our first magistrate differs from the American. He is not elected directly by the people, he is elected by the representatives of the people. He is an example of "double election." The legislature chosen, in name, to make laws, in fact finds its principal business in making and in keeping an executive....

But a cabinet, though it is a committee of the legislative assembly, is a committee with a power which no assembly would—unless

---

* From *The English Constitution*. Walter Bagehot (1826-1877) was for many years editor of the noted British magazine *The Economist* and wrote widely on political and economic subjects.

for historical accidents, and after happy experience—have been persuaded to entrust to any committee. It is a committee which can dissolve the assembly which appointed it; it is a committee with a suspensive veto—a committee with a power of appeal. Though appointed by one parliament, it can appeal if it chooses to the next. . . . The English system, therefore, is not an absorption of the executive power by the legislative power; it is a fusion of the two. Either the cabinet legislates and acts, or else it can dissolve. It is a creature, but it has the power of destroying its creators. It is an executive which can annihilate the legislature, as well as an executive which is the nominee of the legislature. It *was* made, but it *can* unmake; it was derivative in its origin, but it is destructive in its action.

This fusion of the legislative and executive functions may, to those who have not much considered it, seem but a dry and small matter to be the latent essence and effectual secret of the English Constitution; but we can only judge of its real importance by looking at a few of its principal effects, and contrasting it very shortly with its great competitor, which seems likely, unless care be taken, to outstrip it in the progress of the world. That competitor is the Presidential system. The characteristic of it is that the President is elected from the people by one process, and the House of Representatives by another. The independence of the legislative and executive powers is the specific quality of Presidential Government, just as their fusion and combination is the precise principle of Cabinet Government. . . .

The executive is crippled by not getting the laws it needs, and the legislature is spoiled by having to act without responsibility: the executive becomes unfit for its name since it cannot execute what it decides on; the legislature is demoralised by liberty, by taking decisions of which others (and not itself) will suffer the effects. . . .

In England, on a vital occasion, the cabinet can compel legislation by the threat of resignation, and the threat of dissolution; but neither of these can be used in a presidential state. There the legislature cannot be dissolved by the executive government; and it does not heed a resignation, for it has not to find the successor. Accordingly, when a difference of opinion arises, the legislature is forced to fight the executive, and the executive is forced to fight the legislative; and so very likely they contend to the conclusion of their respective terms.* There is, indeed, one condition of things in which this description, though still approximately true, is, nevertheless, not exactly true; and that is, when there is nothing to fight about. . . .

Nor is this the worst. Cabinet government educates the nation; the presidential does not educate it, and may corrupt it. It has been said that England invented the phrase, "Her Majesty's Opposition;" that it was the first government which made a criticism of administration as much a part of the polity as administration itself. This critical opposition is the consequence of cabinet government. The great scene of debate, the great engine of popular instruction and political controversy, is the legislative assembly. A speech there by an eminent statesman, a party movement by a great political combination, are the best means yet known for arousing, enlivening, and teaching a people. The cabinet system ensures such debates, for it makes them the means by which statesmen advertise themselves for future and confirm themselves in present governments. . . . The nation is forced to hear two sides—all the sides, perhaps, of that which most concerns it. And it likes to hear—it is eager to know. Human nature despises long arguments which come to nothing—heavy speeches which precede no motion—abstract disquisitions which leave visible things where they were. But all men heed great results, and a change of government is a great result. It has a hundred ramifications; it runs through society; it gives hope to many, and it takes away hope from many. It is one of those marked events which, by its magnitude and its melodrama, impress men even too much. And debates which have this catastrophe at the end of them—or may so have it—are sure to be listened to, and sure to sink deep into the national mind.

Travellers even in the Northern States of America, the greatest and best of presidential countries, have noticed that the nation was "not specially addicted to politics;" that they have not a public opinion finished and chastened as that of the English has been finished and chastened. A great many hasty writers have charged

---

* I leave this passage to stand as it was written, just after the assassination of Mr. Lincoln, and when every one said Mr. Johnson would be very hostile to the South.

this defect on the "Yankee race," on the Anglo-American character; but English people, if they had no motive to attend to politics certainly would not attend to politics. At present there is *business* in their attention. They assist at the determining crisis; they assist or help it. . . . But under a presidential government a nation has, except at the electing moment, no influence; it has not the ballot-box before it; its virtue is gone, and it must wait till its instant of despotism again returns. It is not incited to form an opinion like a nation under a cabinet government; nor is it instructed like such a nation. There are doubtless debates in the legislature, but they are prologues without a play. There is nothing of a catastrophe about them; you cannot turn out the government. The prize of power is not in the gift of the legislature, and no one cares for the legislature. The executive, the great centre of power and place, sticks irremovable; you cannot change it in any event. The teaching apparatus which has educated our public mind, which prepares our resolutions, which shapes our opinions, does not exist. No presidential country needs to form daily, delicate opinions, or is helped in forming them. . . .

After saying that the division of the legislature and the executive in presidential governments weakens the legislative power, it may seem a contradiction to say that it also weakens the executive power. But it is not a contradiction. The division weakens the whole aggregate force of government—the entire imperial power; and therefore it weakens both its halves. The executive is weakened in a very plain way. In England a strong cabinet can obtain the concurrence of the legislature in all acts which facilitate its administration; it is itself, so to say, the legislature. But a president may be hampered by the parliament, and is likely to be hampered. The natural tendency of the members of every legislature is to make themselves conspicuous. They wish to gratify an ambition laudable or blamable; they wish to promote the measures they think best for the public welfare; they wish to make their *will* felt in great affairs. All these mixed motives urge them to oppose the executive. They are embodying the purposes of others if they aid; they are advancing their own opinions if they defeat; they are first if they vanquish; they are auxiliaries if they support.

Unless a member of the legislature be sure of something more than speech, unless he is incited by the hope of action, and chastened by the chance of responsibility, a first-rate man will not care to take the place, and will not do much if he does take it. To belong to a debating society adhering to an executive (and this is no inapt description of a congress under a presidential constitution) is not an object to stir a noble ambition, and is a position to encourage idleness. The members of a parliament excluded from office can never be comparable, much less equal, to those of a parliament not excluded from office. The presidential government by its nature, divides political life into two halves, an executive half and a legislative half; and, by so dividing it, makes neither half worth a man's having—worth his making it a continuous career—worthy to absorb, as cabinet government absorbs, his whole soul. The statesmen from whom a nation chooses under a presidential system are much inferior to those from whom it chooses under a cabinet system, while the selecting apparatus is also far less discerning.

All these differences are more important at critical periods, because government itself is more important. Informed public opinion, a respectable, able, and disciplined legislature, a well-chosen executive, a parliament and an administration not thwarting each other, but co-operative with each other, are of greater consequence when great affairs are in progress than when small affairs are in progress—when there is much to do than when there is little to do. But in addition to this, a parliamentary or cabinet constitution possesses an additional and special advantage in very dangerous times. It has what we may call a reserve of power fit for and needed by extreme exigencies.

The principle of popular government is that the supreme power, the determining efficacy in matters political, resides in the people—not necessarily or commonly in the whole people, in the numerical majority, but in a *chosen* people, a picked and selected people. It is so in England; it is so in all free countries. Under a cabinet constitution at a sudden emergency this people can choose a ruler for the occasion. It is quite possible and even likely that he would not be ruler *before* the occasion. The great qualities, the imperious will, the rapid energy, the eager nature fit for a great crisis are not required—are impediments—in common times. A Lord Liverpool is better in everyday politics than a Chatham—a Louis Philippe far better than a Napoleon. By the structure of the world we often want, at the sudden occurrence of a grave tempest, to change the helmsman—to replace the pilot of the calm by the pilot of the storm. . . .

But under a presidential government you can do nothing of the kind. The American government calls itself a government of the supreme people; but at a quick crisis, the time when a sovereign power is most needed, you cannot *find* the supreme people. You have got a Congress elected for one fixed period, going out perhaps by fixed instalments, which cannot be accelerated or retarded—you have a President chosen for a fixed period, and immovable during that period: all the arrangements are for *stated* times. There is no *elastic* element, everything is rigid, specified, dated. Come what may, you can quicken nothing and can retard nothing. You have bespoken your government in advance, and whether it suits you or not, whether it works well or works ill, whether it is what you want or not, by law you must keep it. In a country of complex foreign relations it would mostly happen that the first and most critical year of every war would be managed by a peace premier, and the first and most critical years of peace by a war premier. In each case the period of transition would be irrevocably governed by a man selected not for what he was to introduce, but what he was to change—for the policy he was to abandon, not for the policy he was to administer. . . .

Even in quiet times, government by a president is, for the several various reasons which have been stated, inferior to government by a cabinet; but the difficulty of quiet times is nothing as compared with the difficulty of unquiet times.

Hobbes told us long ago, and everybody now understands that there must be a supreme authority, a conclusive power, in every state on every point somewhere. The idea of government involves it—when that idea is properly understood. But there are two classes of governments. In one the supreme determining power is upon all points the same; in the other, that ultimate power is different upon different points—now resides in one part of the constitution, and now in another. The Americans thought that they were imitating the English in making their constitution upon the last principle—in having one ultimate authority for one sort of matter, and another for another sort. But in truth, the English constitution is the type of the opposite species; it has only one authority for all sorts of matters....

The Americans of 1787 thought they were copying the English Constitution, but they were contriving a contrast to it. Just as the American is the type of *composite* governments, in which the supreme power is divided between many bodies and functionaries, so the English is the type of *simple* constitutions, in which the ultimate power upon all questions is in the hands of the same persons.

The ultimate authority in the English Constitution is a newly-elected House of Commons. No matter whether the question upon which it decides be administrative or legislative; no matter whether it concerns high matters of the essential constitution or small matters of daily detail; no matter whether it be a question of making a war or continuing a war; no matter whether it be the imposing a tax or the issuing a paper currency; no matter whether it be a question relating to India, or Ireland, or London,—a new House of Commons can despotically and finally resolve.

The House of Commons may, as was explained, assent in minor matters to the revision of the House of Lords, and submit in matters about which it cares little to the suspensive veto of the House of Lords; but when sure of the popular assent, and when freshly elected, it is absolute,—it can rule as it likes and decide as it likes. And it can take the best security that it does not decide in vain. It can ensure that its decrees shall be executed, for it, and it alone, appoints the executive; it can inflict the most severe of all penalties on neglect, for it can remove the executive. It can choose, to effect its wishes, those who wish the same; and so its will is sure to be done. A stipulated majority of both Houses of the American Congress can overrule by stated enactment their executive; but the popular branch of our legislature can make and unmake ours.

The English constitution, in a word, is framed on the principle of choosing a single sovereign authority, and making it good: the American, upon the principle of having many sovereign authorities, and hoping that their multitude may atone for their inferiority. The Americans now extol their institutions, and so defraud themselves of their due praise. But if they had not a genius for politics; if they had not a moderation in action singularly curious where superficial speech is so violent; if they had not a regard for law, such as no great people have yet evinced, and infinitely surpassing ours,—the multiplicity of authorities in the American Constitution would long ago have brought it to a bad end. Sensible shareholders, I have heard a shrewd attorney say, can work *any* deed of settlement; and so the men of Massachusetts could, I believe, work *any* constitution. But political philosophy must analyse political history; it must distinguish what is due to the excellence of the people, and what to the excellence of the laws; it must carefully calculate the exact effect of each part of the constitution, though thus it may destroy many an idol of the multitude, and detect the secret of utility where but few imagined it to lie.

A Government of Laws, Not of Men

The Supreme Court Fight of 1937

1. ADDRESS BY PRESIDENT FRANKLIN D. ROOSEVELT ON MARCH 9, 1937.

... Tonight, sitting at my desk in the White House, I make my first radio report to the people in my second term of office.

I am reminded of that evening in March 4 years ago, when I made my first radio report to you. We were then in the midst of the great banking crisis.

Soon after, with the authority of the Congress, we asked the Nation to turn over all of its privately held gold, dollar for dollar, to the Government of the United States.

Today's recovery proves how right that policy was.

But when, almost 2 years later, it came before the Supreme Court its constitutionality was upheld only by a 5-to-4 vote. The change of one vote would have thrown all the affairs of this great Nation back into helpless chaos. In effect, four Justices ruled that the right under a private contract to exact a pound of flesh was more sacred than the main objectives of the Constitution to establish an enduring Nation.

In 1933 you and I knew that we must never let our economic system get completely out of joint again—that we could not afford to take the risk of another great depression.

We also became convinced that the only way to avoid a repetition of those dark days was to have a government with power to prevent and to cure the abuses and the inequalities which had thrown that system out of joint.

We then began a program of remedying those abuses and inequalities—to give balance and stability to our economic system— to make it bombproof against the causes of 1929.

Today we are only part way through that program—and recovery is speeding up to a point where the dangers of 1929 are again becoming possible, not this week or month perhaps, but within a year or two.

National laws are needed to complete that program. Individual or local or State effort alone cannot protect us in 1937 any better than 10 years ago.

It will take time—and plenty of time—to work out our remedies administratively even after legislation is passed. To complete our program of protection in time, therefore, we cannot delay one moment in making certain that our National Government has power to carry through.

Four years ago action did not come until the eleventh hour. It was almost too late.

If we learned anything from the depression we will not allow ourselves to run around in new circles of futile discussion and debate, always postponing the day of decision.

The American people have learned from the depression. For in the last three national elections an overwhelming majority of them voted a mandate that the Congress and the President begin the task of providing that protection—not after long years of debate, but now.

The courts, however, have cast doubts on the ability of the elected Congress to protect us against catastrophe by meeting squarely our modern social and economic conditions.

We are at a crisis in our ability to proceed with that protection. It is a quiet crisis. There are no lines of depositors outside closed banks. But to the farsighted it is far-reaching in its possibilities of injury to America.

I want to talk with you very simply about the need for present action in this crisis—the need to meet the unanswered challenge of one-third of a nation ill-nourished, ill-clad, ill-housed.

Last Thursday I described the American form of government as a three-horse team provided by the Constitution to the American people so that their field might be plowed. The three horses are, of course, the three branches of government—the Congress, the executive, and the courts. Two of the horses are pulling in unison today; the third is not. Those who have intimated that the President of the United States is trying to drive that team overlook the simple fact that the President, as Chief Executive, is himself one of the three horses.

It is the American people themselves who are in the driver's seat.

It is the American people themselves who want the furrow plowed.

It is the American people themselves who expect the third horse to pull in unison with the other two.

I hope that you have reread the Constitution of the United States. Like the Bible, it ought to be read again and again.

It is an easy document to understand when you remember that it was called into being because the Articles of Confederation under which the Original Thirteen States tried to operate after the Revolution showed the need of a National Government with power enough to handle national problems. In its preamble the Constitution states that it was intended to form a more perfect Union and promote the general welfare; and the powers given to the Congress to carry out those purposes can be best described by saying that they were all the powers needed to meet each and every problem which then had a national character and which could not be met by merely local action.

But the framers went further. Having in mind that in succeeding generations many other problems then undreamed of would become national problems, they gave to the Congress the ample broad powers "to levy taxes . . . and provide for the common defense and general welfare of the United States."

That, my friends, is what I honestly believe to have been the clear and underlying purpose of the patriots who wrote a Federal Constitution to create a National Government with national power, intended as they said, "to form a more perfect union . . . for ourselves and our posterity."

For nearly 20 years there was no conflict between the Congress and the Court. Then, in 1803, Congress passed a statute which the Court said violated an express provision of the Constitution. The Court claimed the power to declare it unconstitutional and did so declare it. But a little later the Court itself admitted that it was an extra-ordinary power to exercise and through Mr. Justice Washington laid down this limitation upon it: "It is but a decent respect due to the wisdom, the integrity, and the patriotism of the legislative body, by which any law is passed, to presume in favor of its validity until its violation of the Constitution is proved beyond all reasonable doubt."

But since the rise of the modern movement for social and economic progress through legislation, the Court has more and more often and more and more boldly asserted a power to veto laws passed by the Congress and State legislatures in complete disregard of this original limitation.

In the last 4 years the sound rule of giving statutes the benefit of all reasonable doubt has been cast aside. The Court has been acting not as a judicial body, but as a policy-making body.

When the Congress has sought to stabilize national agriculture, to improve the conditions of labor, to safeguard business against unfair competition, to protect our national resources, and in many other ways to serve our clearly national needs, the majority of the Court has been assuming the power to pass on the wisdom of these acts of the Congress—and to approve or disapprove the public policy written into these laws.

That is not only my accusation. It is the accusation of most distinguished Justices of the present Supreme Court. I have not the time to quote to you all the language used by dissenting Justices in many of these cases. But in the case holding the Railroad Retirement Act unconstitutional, for instance, Chief Justice Hughes said in a dissenting opinion that the majority opinion was "a departure from sound principles," and placed "an unwarranted limitation upon the commerce clause." And three other Justices agreed with him.

In the case holding the A.A.A. unconstitutional, Justice Stone said of the majority opinion that it was a "tortured construction of the Constitution." And two other Justices agreed with him.

In the case holding the New York Minimum Wage Law unconstitutional, Justice Stone said that the majority were actually reading into the Constitution their own "personal economic predilections," and that if the legislatve power is not left free to choose the methods of solving the problems of poverty, subsistence, and health of large numbers in the community, then "government is to be rendered impotent." And two other Justices agreed with him.

In the face of these dissenting opinions, there is no basis for the claim made by some members of the Court that something in the Constitution has compelled them regretfully to thwart the will of the people.

In the face of such dissenting opinions, it is perfectly clear that as Chief Justice Hughes has said, "We are under a Constitution, but the Constitution is what the judges say it is."

The Court in addition to the proper use of its judicial functions has improperly set itself up as a third House of the Congress—a superlegislature, as one of the Justices has called it—reading into the Constitution words and implications which are not there, and which were never intended to be there.

We have, therefore, reached the point as a Nation where we must take action to save the Constitution from the Court and the Court from itself. We must find a way to take an appeal from the Supreme Court to the Constitution itself. We want a Supreme Court which will do justice under the Constitution—not over it. In our courts we want a government of laws and not of men.

I want—as all Americans want—an independent judiciary as proposed by the framers of the Constitution. That means a Supreme Court that will enforce the Constitution as written—that will refuse to amend the Constitution by the arbitrary exercise of judicial power—amendment by judicial say-so. It does not mean a judiciary so independent that it can deny the existence of facts universally recognized.

How, then, could we proceed to perform the mandate given us? It was said in last year's Democratic platform, "If these problems cannot be effectively solved within the Constitution, we shall seek such clarifying amendment as will assure the power to enact those laws, adequately to regulate commerce, protect public health and safety, and safeguard economic security." In other words, we said we would seek an amendment only if every other possible means by legislation were to fail.

When I commenced to review the situation with the problem squarely before me, I came by a process of elimination to the conclusion that short of amendments the only method which was clearly constitutional, and would at the same time carry out other much-needed reforms, was to infuse new blood into all our courts. We must have men worthy and equipped to carry out impartial justice. But at the same time we must have judges who will bring to the courts a present-day sense of the Constitution—judges who will retain in the courts the judicial functions of a court and reject the legislative powers which the courts have today assumed.

In 45 out of the 48 States of the Union, judges are chosen not for life but for a period of years. In many States judges must retire at the age of 70. Congress has provided financial security by offering life pensions at full pay for Federal judges on all courts who are willing to retire at 70. In the cases of Supreme Court Justices, that pension is $20,000 a year. But all Federal judges, once appointed, can, if they choose, hold office for life no matter how old they may get to be.

What is my proposal? It is simply this: Whenever a judge or justice of any Federal court has reached the age of 70 and does not avail himself of the opportunity to retire on a pension, a new member shall be appointed by the President then in office, with the approval, as required by the Constitution, of the Senate of the United States.

That plan has two chief purposes: By bringing into the judicial system a steady and continuing stream of new and younger blood, I hope, first, to make the administration of all Federal justice speedier and therefore less costly; secondly, to bring to the decision of social and economic problems younger men who have had personal experience and contact with modern facts and circumstances under which average men have to live and work. This plan will save our National Constitution from hardening of the judicial arteries.

The number of judges to be appointed would depend wholly on the decision of present judges now over 70 or those who would subsequently reach the age of 70.

If, for instance, any one of the six Justices of the Supreme Court now over the age of 70 should retire as provided under the plan, no additional place would be created. Consequently, although there never can be more than 15, there may be only 14, or 13, or 12, and there may be only 9.

There is nothing novel or radical about this idea. It seeks to maintain the Federal bench in full vigor. It has been discussed and approved by many persons of high authority ever since a similar proposal passed the House of Representatives in 1869. . . .

Those opposing this plan have sought to arouse prejudice and fear by crying that I am seeking to "pack" the Supreme Court and that a baneful precedent will be established.

What do they mean by the words "packing the Court"?

Let me answer this question with a bluntness that will end all honest misunderstanding of my purposes.

If by that phrase "packing the Court" it is charged that I wish to place on the bench spineless puppets who would disregard the law and would decide specific cases as I wished them to be decided, I make this answer: That no President fit for his office would appoint, and no Senate of honorable men fit for their office would confirm, that kind of appointees to the Supreme Court.

But if by that phrase the charge is made that I would appoint and the Senate would confirm Justices worthy to sit beside present members of the Court who understand those modern conditions; that I will appoint Justices who will not undertake to override the judgment of the Congress on legislative policy; that I will appoint Justices who will act as Justices and not as legislators—if the appointment of such Justices can be called "packing the Courts"—then I say that I, and with me the vast majority of the American people, favor doing just that thing—now.

Is it a dangerous precedent for the Congress to change the number of the Justices? The Congress has always had, and will have, that power. The number of Justices has been changed several times before—in the administrations of John Adams and Thomas Jefferson, both signers of the Declaration of Independence, Andrew Jackson, Abraham Lincoln, and Ulysses S. Grant.

I suggest only the addition of Justices to the bench in accordance with a clearly defined principle relating to a clearly defined age limit. Fundamentally, if in the future America cannot trust the Congress it elects to refrain from abuse of our constitutional usages, democracy will have failed far beyond the importance to it of any kind of precedent concerning the judiciary.

We think it so much in the public interest to maintain a vigorous judiciary that we encourage the retirement of elderly judges by offering them a life pension at full salary. Why then should we leave the fulfillment of this public policy to chance or make it dependent upon the desire or prejudice of any individual Justice?

It is the clear intention of our public policy to provide for a constant flow of new and younger blood into the judiciary. Normally, every President appoints a large number of district and circuit judges and a few members of the Supreme Court. Until my first term practically every President of the United States had appointed at least one member of the Supreme Court. President Taft appointed five members and named a Chief Justice; President Wilson three; President Harding four, including a Chief Justice; President Coolidge one; President Hoover three, including a Chief Justice.

Such a succession of appointments should have provided a court well balanced as to age. But chance and the disinclination of individuals to leave the Supreme Bench have now given us a Court in which five Justices will be over 75 years of age before next June and one over 70. Thus a sound public policy has been defeated.

I now propose that we establish by law an assurance against any such ill-balanced Court in the future. I propose that hereafter, when a judge reaches the age of 70, a new and younger judge shall be added to the Court automatically. In this way I propose to enforce a sound public policy by law instead of leaving the composition of our Federal courts, including the highest, to be determined by chance or the personal decision of individuals.

If such a law as I propose is regarded as establishing a new precedent, is it not a most desirable precedent?

Like all lawyers, like all Americans, I regret the necessity of this controversy. But the welfare of the United States, and indeed of the Constitution itself, is what we all must think about first. Our difficulty with the Court today rises not from the Court as an institution but from human beings within it. But we cannot yield our constitutional destiny to the personal judgment of a few men who, being fearful of the future, would deny us the necessary means of dealing with the present.

This plan of mine is no attack on the Court; it seeks to restore the Court to its rightful and historic place in our system of constitutional government and to have it resume its high task of building anew on the Constitution "a system of living law."

I have thus explained to you the reasons that lie behind our efforts to secure results by legislation within the Constitution. I hope that thereby the difficult process of constitutional amendment may be rendered unnecessary. . . .

Two groups oppose my plan on the ground that they favor a constitutional amendment. The first includes those who fundamentally object to social and economic legislation along modern lines. This is the same group who during the campaign last fall tried to block the mandate of the people.

Now they are making a last stand. And the strategy of that last stand is to suggest the time-consuming process of amendment in order to kill off by delay the legislation demanded by the mandate.

To them I say: I do not think you will be able long to fool the American people as to your purposes.

The other group is composed of those who honestly believe the amendment process is the best and who would be willing to support a reasonable amendment if they could agree on one.

To them I say: We cannot rely on an amendment as the immediate or only answer to our present difficulties. When the time comes for action, you will find that many of those who pretend to support you will sabotage any constructive amendment which is proposed. Look at these strange bedfellows of yours. When before have you found them really at your side in your fights for progress?

And remember one thing more. Even if an amendment were passed, and even if in the years to come it were to be ratified, its meaning would depend upon the kind of Justices who would be sitting on the Supreme Court bench. An amendment like the rest of the Constitution is what the Justices say it is rather than what its framers or you might hope it is.

This proposal of mine will not infringe in the slightest upon the civil or religious liberties so dear to every American.

My record as Governor and as President proves my devotion to those liberties. You who knew me can have no fear that I would tolerate the destruction by any branch of government of any part of our heritage of freedom.

The present attempt by those opposed to progress to play upon the fears of danger to personal liberty brings again to mind that crude and cruel strategy tried by the same opposition to frighten the workers of America in a pay-envelope propaganda against the social security law. The workers were not fooled by that propaganda then. The people of America will not be fooled by such propaganda now.

I am in favor of action through legislation—

First, because I believe that it can be passed at this session of the Congress.

Second, because it will provide a reinvigorated, liberal-minded judiciary necessary to furnish quicker and cheaper justice from bottom to top.

Third, because it will provide a series of Federal courts willing to enforce the Constitution as written, and unwilling to assert legislative powers by writing into it their own political and economic policies.

During the past half century the balance of power between the three great branches of the Federal Government has been tipped out of balance by the courts in direct contradiction of the high purposes of the framers of the Constitution. It is my purpose to restore that balance. You who know me will accept my solemn assurance that in a world in which democracy is under attack I seek to make American democracy succeed.

2. Report of the Committee on the Judiciary of the Senate, June 7, 1937.

The Committee on the Judiciary, to whom was referred the bill to reorganize the judicial branch of the Government, after full consideration, having unanimously amended the measure, hereby report the bill adversely with the recommendation that it do not pass. . . .

The committee recommends that the measure be rejected for the following primary reasons:

I. The bill does not accomplish any one of the objectives for which it was originally offered.

II. It applies force to the judiciary and in its initial and ultimate effect would undermine the independence of the courts.

III. It violates all precedents in the history of our Government and would in itself be a dangerous precedent for the future.

IV. The theory of the bill is in direct violation of the spirit of the American Constitution and its employment would permit alteration of the Constitution without the people's consent or approval; it undermines the protection our constitutional system gives to minorities and is subversive of the rights of individuals.

V. It tends to centralize the Federal district judiciary by the power of assigning judges from one district to another at will.

VI. It tends to expand political control over the judicial department by adding to the powers of the legislative and executive departments respecting the judiciary....

The President tells us in his address to the Nation of March 9 . . . :

> When the Congress has sought to stabilize national agriculture, to improve the conditions of labor, to safeguard business against unfair competition, to protect our national resources, and in many other ways, to serve our clearly national needs, the majority of the Court has been assuming the power to pass on the wisdom of these acts of the Congress and to approve or disapprove the public policy written into these laws. . . .
>
> We have, therefore, reached the point as a nation where we must take action to save the Constitution from the Court and the Court from itself. We must find a way to take an appeal from the Supreme Court to the Constitution itself. We want a Supreme Court which will do justice under the Constitution—not over it. In our courts we want a government of laws and not of men.

These words constitute a charge that the Supreme Court has exceeded the boundaries of its jurisdiction and invaded the field reserved by the Constitution to the legislative branch of the Government. At best the accusation is opinion only. It is not the conclusion of judicial process.

Here is the frank acknowledgment that neither speed nor "new blood" in the judiciary is the object of this legislation, but a change in the decisions of the Court—a subordination of the views of the judges to the views of the executive and legislative, a change to be brought about by forcing certain judges off the bench or increasing their number.

Let us, for the purpose of the argument, grant that the Court has been wrong, wrong not only in that it has rendered mistaken opinions but wrong in the far more serious sense that it has substituted its will for the congressional will in the matter of legislation. May we nevertheless safely punish the Court?

Today it may be the Court which is charged with forgetting its constitutional duties. Tomorrow it may be the Congress. The next day it may be the Executive. If we yield to temptation now to lay the lash upon the Court, we are only teaching others how to apply it to ourselves and to the people when the occasion seems to warrant. Manifestly, if we may force the hand of the Court to secure our interpretation of the Constitution, then some succeeding Congress may repeat the process to secure another and a different interpretation and one which may not sound so pleasant in our ears as that for which we now contend.

There is a remedy for usurpation or other judicial wrongdoing. If this bill be supported by the toilers of this country upon the ground that they want a Court which will sustain legislation limiting hours and providing minimum wages, they must remember that the procedure employed in the bill could be used in another administration to lengthen hours and to decrease wages. If farmers want agricultural relief and favor this bill upon the ground that it gives them a Court which will sustain legislation in their favor, they must remember that the procedure employed might some day be used to deprive them of every vestige of a farm relief.

When members of the Court usurp legislative powers or attempt to exercise political power, they lay themselves open to the charge of having lapsed from that "good behavior" which determines the period of their official life. But, if you say, the process of impeachment is difficult and uncertain, the answer is, the people made it so when they framed the Constitution. It is not for us, the servants of the people, the instruments of the Constitution, to find a more easy way to do that which our masters made difficult.

But, if the fault of the judges is not so grievous as to warrant impeachment, if their offense is merely that they have grown old, and we feel, therefore, that there should be a "constant infusion of new blood," then obviously the way to achieve that result is by constitutional amendment fixing definite terms for the members of the judiciary or making mandatory their retirement at a given age. Such a provision would indeed provide for the constant infusion of new blood, not only now but at all times in the future. The plan before us is but a temporary expedient which operates once and then never again, leaving the Court as permanently expanded to become once more a court of old men, gradually year by year falling behind the times. . . .

Shall we now, after 150 years of loyalty to the constitutional ideal of an untrammeled judiciary, duty bound to protect the constitutional rights of the humblest citizen even against the Government itself, create the vicious precedent which must necessarily undermine our system? The only argument for the increase which survives analysis is that Congress should enlarge the Court so as to make the policies of this administration effective.

We are told that a reactionary oligarchy defies the will of the majority, that this is a bill to "unpack" the Court and give effect to the desires of the majority; that is to say, a bill to increase the number of Justices for the express purpose of neutralizing the views of some of the present members. In justification we are told, but without authority, by those who would rationalize this program, that Congress was given the power to determine the size of the Court so that the legislative branch would be able to impose its will upon the judiciary. This amounts to nothing more than the declaration that when the Court stands in the way of a legislative enactment, the Congress may reverse the ruling by enlarging the Court. When such a principle is adopted, our constitutional system is overthrown! . . .

This is the first time in the history of our country that a proposal to alter the decisions of the court by enlarging its personnel has been so boldly made. Let us meet it. Let us now set a salutary precedent that will never be violated. Let us, of the Seventy-fifth Congress, in words that will never be disregarded by any succeeding Congress, declare that we would rather have an independent Court, a fearless Court, a Court that will dare to announce its honest opinions in what it believes to be the defense of the liberties of the people, than a Court that, out of fear or sense of obligation to the appointing power, or factional passion, approves any measure we may enact. We are not the judges of the judges. We are not above the Constitution.

Even if every charge brought against the so-called "reactionary" members of this Court be true, it is far better that we await orderly but inevitable change of personnel than that we impatiently overwhelm them with new members. Exhibiting this restraint, thus demonstrating our faith in the American system, we shall set an example that will protect the independent American judiciary from attack as long as this Government stands.

It is essential to the continuance of our constitutional democracy that the judiciary be completely independent of both the executive and legislative branches of the Government, and we assert that independent courts are the last safeguard of the citizen, where his rights, reserved to him by the express and implied provisions of the Constitution, come in conflict with the power of governmental agencies. . . .

We declare for the continuance and perpetuation of government and rule by law, as distinguished from government and rule by men, and in this we are but reasserting the principles basic to the Constitution of the United States. The converse of this would lead to and in fact accomplish the destruction of our form of government, where the written Constitution with its history, its spirit, and its long line of judicial interpretation and construction, is looked to and relied upon by millions of our people. Reduction of the degree of the supremacy of law means an increasing enlargement of the degree of personal government.

Personal government, or government by an individual, means autocratic dominance, by whatever name it may be designated. Autocratic dominance was the very thing against which the American Colonies revolted, and to prevent which the Constitution was in every particular framed.

Courts and the judges thereof should be free from a subservient attitude of mind, and this must be true whether a question of constitutional construction or one of popular activity is involved. If the court of last resort is to be made to respond to a prevalent sentiment of a current hour, politically imposed, that Court must ultimately become subservient to the pressure of public opinion of the hour, which might at the moment embrace mob passion abhorrent to a more calm, lasting consideration. . . .

The whole bill prophesies and permits executive and legislative interferences with the independence of the Court, a prophecy and a permission which constitute an affront to the spirit of the Constitution.

The complete independence of the courts of justice is peculiarly essential in a limited Constitution. By a limited Constitution, I understand one which contains certain specified exceptions to the legislative authority; such, for instance, as that it shall pass no bills of attainder, no ex-post-facto laws, and the like. Limitations of this kind can be preserved in practice no other way than through the medium of courts of justice, whose duty it must be to declare all acts contrary to the manifest tenor of the Constitution void. Without this, all the reservations of particular rights or privileges would amount to nothing (The Federalist, vol. 2, p. 100, no. 78). . . .

Inconvenience and even delay in the enactment of legislation is not a heavy price to pay for our system. Constitutional democracy moves forward with certainty rather than with speed. The safety and the permanence of the progressive march of our civilization are far more important to us and to those who are to come after us than the enactment now of any particular law. The Constitution of the United States provides ample opportunity for the expression of popular will to bring about such reforms and changes as the people may deem essential to their present and future welfare. It is the people's charter of the powers granted those who govern them. . . .

Minority political groups, no less than religious and racial groups, have never failed, when forced to appeal to the Supreme Court of the United States, to find in its opinions the reassurance and protection of their constitutional rights. No finer or more durable philosophy of free government is to be found in all the writings and practices of great statesmen than may be found in the decisions of the Supreme Court when dealing with great problems of free government touching human rights. This would not have been possible without an independent judiciary.

No finer illustration of the vigilance of the Court in protecting human rights can be found than in a decision wherein was involved the rights of a Chinese person, wherein the Court said:

When we consider the nature and the theory of our institutions of government, the principles upon which they are supposed to rest, and review the history of their development, we are constrained to conclude that they do not mean to leave room for the play and action of purely personal and arbitrary power. . . . The fundamental rights to life, liberty, and the pursuit of happiness considered as individual possessions are secured by those maxims of constitutional law which are the monuments showing the victorious progress of the race in securing to men the blessings of civilization under the reign of just and equal laws, so that in the famous language of the Massachusetts Bill of Rights, the government of the Commonwealth "may be a government of laws and not of men." For the very idea that one many may be compelled to hold his life or the means of living or any material right essential to the enjoyment of life, at the mere will of another, seems to be intolerable in any country where freedom prevails, as being the essence of slavery itself. (*Yick Wo* v. *Hopkins*, 118 U.S. 356.) . . .

The Constitution of the United States, courageously construed and upheld through 150 years of history, has been the bulwark of human liberty. It was bequeathed to us in a great hour of human destiny by one of the greatest characters civilization has produced—George Washington. It is in our hands now to preserve or to destroy. If ever there was a time when the people of America should heed the words of the Father of Their Country this is the hour. Listen to his solemn warning from the Farewell Address:

It is important, likewise, that the habits of thinking, in a free country, should inspire caution in those intrusted with its administration, to confine themselves within their respective constitutional spheres, avoiding, in the exercises of the powers of one department, to encroach upon another. The spirit of encroachment tends to consolidate the powers of all the departments in one, and thus to create, whatever the form of government, a real despotism. A first estimate of that love of power, and proneness to abuse it, which predominates in the human heart, is sufficient to satisfy us of the truth of this position. The necessity of reciprocal checks in the exercise of political power, by dividing and distributing it into different depositories, and constituting each the guardian of the public weal, against invasions by the others, has been evinced by experiment, ancient and modern; some of them in our own country, and under our own eyes. To preserve them must be as necessary as to institute them. If, in the opinion of the people, the distribution or modification of the constitutional powers be, in any particular, wrong, let it be corrected by an amendment in the way which the Constitution designates. But let there be no change by usurpation; for though this, in one instance, may be the instrument of good, it is the customary weapon by which free governments are destroyed. The precedent must always greatly overbalance, in permanent evil, any partial or transient benefit which the use can, at any time, yield.

We recommend the rejection of this bill as a needless, futile, and utterly dangerous abandonment of constitutional principle.

It was presented to the Congress in a most intricate form and for reasons that obscured its real purpose.

It would not banish age from the bench nor abolish divided decisions.

It would not affect the power of any court to hold laws unconstitutional nor withdraw from any judge the authority to issue injunctions.

It would not reduce the expense of litigation nor speed the decision of cases.

It is a proposal without precedent and without justification.

It would subjugate the courts to the will of Congress and the President and thereby destroy the independence of the judiciary, the only certain shield of individual rights.

It contains the germ of a system of centralized administration of law that would enable an executive so minded to send his judges into every judicial district in the land to sit in judgment on controversies between the Government and the citizen.

It points the way to the evasion of the Constitution and establishes the method whereby the people may be deprived of their right to pass upon all amendments of the fundamental law.

It stands now before the country, acknowledged by its proponents as a plan to force judicial interpretation of the Constitution, a proposal that violates every sacred tradition of American democracy.

Under the form of the Constitution it seeks to do that which is unconstitutional.

Its ultimate operation would be to make this Government one of men rather than one of law, and its practical operation would be to make the Constitution what the executive or legislative branches of the Government choose to say it is—an interpretation to be changed with each change of administration.

It is a measure which should be so emphatically rejected that its parallel will never again be presented to the free representatives of the free people of America.

William H. King	Tom Connally
Frederick Van Nuys	Joseph C. O'Mahoney
Patrick McCarran	William E. Borah
Carl A. Hatch	Warren R. Austin
Edward R. Burke	Frederick Steiwer

Property Rights and National Security

Constitutional Conflict and the Seizure of the Steel Mills

[1952]

1. EXECUTIVE ORDER DIRECTING THE SECRETARY OF COMMERCE TO TAKE POSSESSION OF AND OPERATE THE PLANTS AND FACILITIES OF CERTAIN STEEL COMPANIES.

WHEREAS on December 16, 1950, I proclaimed the existence of a national emergency which requires that the military, naval, air, and civilian defenses of this country be strengthened as speedily as possible to the end that we may be able to repel any and all threats against our national security and to fulfill our responsibilities in the efforts being made throughout the United Nations and otherwise to bring about a lasting peace; and

Whereas American fighting men and fighting men of other nations of the United Nations are now engaged in deadly combat with the forces of aggression in Korea, and forces of the United States are stationed elsewhere overseas for the purpose of participating in the defense of the Atlantic Community against aggression; and

Whereas the weapons and other materials needed by our armed forces and by those joined with us in the defense of the free world are produced to a great extent in this country, and steel is an indispensable component of substantially all of such weapons and materials; and

Whereas steel is likewise indispensable to the carrying out of programs of the Atomic Energy Commission of vital importance to our defense efforts; and

Whereas a continuing and uninterrupted supply of steel is also indispensable to the maintenance of the economy of the United States, upon which our military strength depends; and

Whereas a controversy has arisen between certain companies in the United States producing and fabricating steel and the elements thereof and certain of their workers represented by the United Steelworkers of America, CIO, regarding terms and conditions of employment; and

Whereas the controversy has not been settled through the processes of collective bargaining or through the efforts of the Government, including those of the Wage Stabilization Board, to which the controversy was referred on December 22, 1951, pursuant to Executive Order No. 10233, and a strike has been called for 12:01 A.M., April 9, 1952; and

Whereas a work stoppage would immediately jeopardize and imperil our national defense and the defense of those joined with us in resisting aggression, and would add to the continuing danger of our soldiers, sailors, and airmen engaged in combat in the field; and

Whereas in order to assure the continued availability of steel and steel products during the existing emergency, it is necessary that the United States take possession of and operate the plants, facilities, and other property of the said companies as hereinafter provided:

Now, therefore, by virtue of the authority vested in me by the Constitution and laws of the United States, and as President of the United States and Commander in Chief of the armed forces of the United States, it is hereby ordered as follows:

1. The Secretary of Commerce is hereby authorized and directed to take possession of all or such of the plants, facilities, and other property of the companies named in the list attached hereto, or any part thereof, as he may deem necessary in the interests of national defense; and to operate or to arrange for the operation thereof and to do all things necessary for, or incidental to, such operation.

2. In carrying out this order the Secretary of Commerce may act through or with the aid of such public or private instrumentalities or persons as he may designate; and all Federal agencies shall cooperate with the Secretary of Commerce to the fullest extent possible in carrying out the purposes of this order.

3. The Secretary of Commerce shall determine and prescribe terms and conditions of employment under which the plants, facilities, and other properties possession of which is taken pursuant to this order shall be operated. The Secretary of Commerce shall recognize the rights of workers to bargain collectively through representatives of their own choosing and to engage in concerted activities for the purpose of collective bargaining, adjustment of grievances, or other mutual aid or protection, provided that such activities do not interfere with the operation of such plants, facilities, and other properties.

4. Except so far as the Secretary of Commerce shall otherwise provide from time to time, the managements of the plants, facilities, and other properties possession of which is taken pursuant to this order shall continue their functions, including the collection and disbursement of funds in the usual and ordinary course of business in the names of their respective companies and by means of any instrumentalities used by such companies.

5. Except so far as the Secretary of Commerce may otherwise direct, existing rights and obligations of such companies shall remain in full force and effect, and there may be made, in due course, payments of dividends on stock and of principal, interest, sinking funds, and all other distributions upon bonds, debentures, and other obligations, and expenditures may be made for other ordinary corporate or business purposes.

6. Whenever in the judgment of the Secretary of Commerce further possession and operation by him of any plant, facility, or other property is no longer necessary or expedient in the interest of national defense, and the Secretary has reason to believe that effective future operation is assured, he shall return the possession and control thereof at the time possession was taken under this order.

7. The Secretary of Commerce is authorized to prescribe and issue such regulations and orders not inconsistent herewith as he may deem necessary or desirable for carrying out the purposes of this order; and he may delegate and authorize subdelegation of such of his functions under this order as he may deem desirable.

HARRY S. TRUMAN.

The White House, April 8, 1952.

2. SUPREME COURT OF THE UNITED STATES, THE YOUNGSTOWN SHEET AND TUBE COMPANY, *et al.*, *v.* CHARLES SAWYER (343 U.S. 579).

MR. JUSTICE BLACK delivered the opinion of the Court.

We are asked to decide whether the President was acting within his constitutional power when he issued an order directing the Secretary of Commerce to take possession of and operate most of the Nation's steel mills. The mill owners argue that the President's order amounts to lawmaking, a legislative function which the Constitution has expressly confided to the Congress and not to the President. The Government's position is that the order was made on findings of the President that his action was necessary to avert a national catastrophe which would inevitably result from a stoppage of steel production, and that in meeting this grave emergency the President was acting within the aggregate of his constitutional powers as the Nation's Chief Executive and the Commander in Chief of the Armed Forces of the United States. The issue emerges here from the following series of events:

In the latter part of 1951, a dispute arose between the steel companies and their employees over terms and conditions that should be included in new collective bargaining agreements. Long-continued conferences failed to resolve the dispute. . . . The indispensability of steel as a component of substantially all weapons and other war materials led the President to believe that the proposed work stoppage would immediately jeopardize our national defense and that governmental seizure of the steel mills was necessary in order to assure the continued availability of steel. Reciting these considerations for his action, the President, a few hours before the strike was to begin, issued Executive Order 10340. . . . The order directed the Secretary of Commerce to take possession of most of the steel mills and keep them running. The Secretary immediately issued his own possessory orders, calling upon the presidents of the various seized companies to serve as operating managers for the United States. They were directed to carry on their activities in accordance with regulations and directions of the Secretary. The next morning the President sent a message to Congress reporting his action. . . .

The President's power, if any, to issue the order must stem

either from an act of Congress or from the Constitution itself. There is no statute that expressly authorizes the President to take possession of property as he did here. Nor is there any act of Congress to which our attention has been directed from which such a power can fairly be implied. Indeed, we do not understand the Government to rely on statutory authorization for this seizure. There are two statutes which do authorize the President to take both personal and real property under certain conditions. However, the Government admits that these conditions were not met and that the President's order was not rooted in either of the statutes. The Government refers to the seizure provisions of one of these statutes . . . as "much too cumbersome, involved, and time-consuming for the crisis which was at hand." . . .

It is clear that if the President had authority to issue the order he did, it must be found in some provisions of the Constitution. And it is not claimed that express constitutional language grants this power to the President. The contention is that presidential power should be implied from the aggregate of his powers under the Constitution. Particular reliance is placed on provisions in Article II which say that "the executive Power shall be vested in a President . . ."; that "he shall take Care that the Laws be faithfully executed"; and that he "shall be Commander in Chief of the Army and Navy of the United States." , . .

. . . We cannot with faithfulness to our constitutional system hold that the Commander in Chief of the Armed Forces has the ultimate power as such to take possession of private property in order to keep labor disputes from stopping production. This is a job for the Nation's lawmakers, not for its military authorities.

Nor can the seizure order be sustained because of the several constitutional provisions that grant executive power to the President. In the framework of our Constitution, the President's power to see that the laws are faithfully executed refutes the idea that he is to be a lawmaker. The Constitution limits his functions in the law-making process to the recommending of laws he thinks wise and the vetoing of laws he thinks bad. . . .

The President's order does not direct that a congressional policy be executed in a manner prescribed by Congress—it directs that a presidential policy be executed in a manner prescribed by the President. The preamble of the order itself, like that of many statutes, sets out reasons why the President believes certain policies should be adopted, proclaims these policies as rules of conduct to be followed, and again, like a statute, authorizes a government official to promulgate additional rules and regulations consistent with the policy proclaimed and needed to carry that policy into execution. The power of Congress to adopt such public policies as those proclaimed by the order is beyond question. It can authorize the taking of private property for public use. It can make laws regulating the relationships between employers and employees, prescribing rules designed to settle labor disputes, and fixing wages and working conditions in certain fields of our economy. The Constitution does not subject this law-making power of Congress to presidential or military supervision or control. . . .

The Founders of this Nation entrusted the lawmaking power to the Congress alone in both good and bad times. It would do no good to recall the historical events, the fears of power and the hopes for freedom that lay behind their choice. Such a review would but confirm our holding that this seizure order cannot stand.

The judgment of the District Court is *Affirmed*. . . .

MR. JUSTICE DOUGLAS, concurring.

There can be no doubt that the emergency which caused the President to seize these steel plants was one that bore heavily on the country. But the emergency did not create power; it merely marked an occasion when power should be exercised. And the fact that it was necessary that measures be taken to keep steel in production does not mean that the President, rather than the Congress, had the constitutional authority to act. The Congress, as well as the President, is trustee of the national welfare. The President can act more quickly than the Congress. The President with the armed services at his disposal can move with force as well as with speed. All executive power—from the reign of ancient kings to the rule of modern dictators—has the outward appearance of efficiency.

Legislative power, by contrast, is slower to exercise. There must be delay while the ponderous machinery of committees, hearings, and debates is put into motion. That takes time; and while the Congress slowly moves into action, the emergency may take its toll in wages, consumer goods, war production, the standard of living of the people, and perhaps even lives. Legislative action may indeed often be cumbersome, time-consuming, and apparently inefficient. But as Mr. Justice Brandeis stated in his dissent in *Myers v. United States*, 272 U.S. 52, 293:

"The doctrine of the separation of powers was adopted by the Convention of 1787, not to promote efficiency but to preclude the exercise of arbitrary power. The purpose was, not to avoid friction, but, by means of the inevitable friction incident to the distribution of the governmental powers among three departments, to save the people from autocracy."

We therefore cannot decide this case by determining which branch of government can deal most expeditiously with the present crisis. The answer must depend on the allocation of powers under the Constitution. . . .

The legislative nature of the action taken by the President seems to me to be clear. When the United States takes over an industrial plant to settle a labor controversy, it is condemning property. The seizure of the plant is a taking in the constitutional sense. *United States v. Pewee Coal Co.*, 341 U.S. 114. A permanent taking would amount to the nationalization of the industry. A temporary taking falls short of that goal. But though the seizure is only for a week or a month, the condemnation is complete and the United States must pay compensation for the temporary possession. *United States v. General Motors Corp.*, 323 U.S. 373; *United States v. Pewee Coal Co., supra.* . . .

The great office of President is not a weak and powerless one. The President represents the people and is their spokesman in domestic and foreign affairs. The office is respected more than any other in the land. It gives a position of leadership that is unique. The power to formulate policies and mould opinion inheres in the Presidency and conditions our national life. The impact of the man and the philosophy he represents may at times be thwarted by the Congress. Stalemates may occur when emergencies mount and the Nation suffers for lack of harmonious, reciprocal action between the White House and Capitol Hill. That is a risk inherent in our system of separation of powers. The tragedy of such stalemates might be avoided by allowing the President the use of some legislative authority. The Framers with memories of the tyrannies produced by a blending of executive and legislative power rejected that political arrangement. Some future generation may, however, deem it so urgent that the President have legislative authority that the Constitution will be amended. We could not sanction the seizures and condemnations of the steel plants in this case without reading Article II as giving the President not only the power to execute the laws but to make some. Such a step would most assuredly alter the pattern of the Constitution.

We pay a price for our system of checks and balances, for the distribution of power among the three branches of government. It is a price that today may seem exorbitant to many. Today a kindly President uses the seizure power to effect a wage increase and to keep the steel furnaces in production. Yet tomorrow another President might use the same power to prevent a wage increase, to curb trade unionists, to regiment labor as oppressively as industry thinks it has been regimented by this seizure.

MR. JUSTICE JACKSON, concurring in the judgment and opinion of the Court.

That comprehensive and undefined presidential powers hold both practical advantages and grave dangers for the country will impress anyone who has served as legal adviser to a President in time of transition and public anxiety. While an interval of detached reflection may temper teachings of that experience, they probably are a more realistic influence on my views than the conventional materials of judicial decision which seem unduly to accentuate doctrine and legal fiction. But as we approach the question of presidential power, we half overcome mental hazards by recognizing them. The opinions of judges, no less than executives and publicists, often suffer the infirmity of confusing the issue of a power's validity with the cause it is invoked to promote, of confounding the permanent executive office with its temporary occupant. The tendency is strong to emphasize transient results upon policies—such as wages or stabilization—and lose sight of enduring consequences upon the balanced power structure of our Republic.

A judge, like an executive adviser, may be surprised at the poverty of really useful and unambiguous authority applicable to concrete problems of executive power as they actually present themselves. Just what our forefathers did envision, or would have envisioned had they foreseen modern conditions, must be divined from materials almost as enigmatic as the dreams Joseph was called upon to interpret for Pharaoh. A century and a half of partisan debate and scholarly speculation yields no net result but only supplies more or less apt quotations from respected sources on each side of any question. They largely cancel each other. And court decisions are indecisive because of the judicial practice of dealing with the largest questions in the most narrow way.

The actual art of governing under our Constitution does not and cannot conform to judicial definitions of the power of any of its branches based on isolated clauses or even single Articles torn from context. While the Constitution diffuses power the better to secure liberty, it also contemplates that practice will integrate the dispersed powers into a workable government. It enjoins upon its branches separateness but interdependence, autonomy but reciprocity. Presidential powers are not fixed but fluctuate, depending upon their disjunction or conjunction with those of Congress. . . .

The Solicitor General seeks the power of seizure in three clauses of the Executive Article [of the Constitution], the first reading, "The executive Power shall be vested in a President of the United States of America." Lest I be thought to exaggerate, I quote the interpretation which his brief puts upon it: "In our view, this clause constitutes a grant of all the executive powers of which the Government is capable." If that be true, it is difficult to see why the forefathers bothered to add several specific items, including some trifling ones.

The example of such unlimited executive power that must have most impressed the forefathers was the prerogative exercised by George III, and the description of its evils in the Declaration of Independence leads me to doubt that they were creating their new Executive in his image. Continental European examples were no more appealing. And if we seek instruction from our own times, we can match it only from the executive powers in those governments we disparagingly describe as totalitarian. I cannot accept the view that this clause is a grant in bulk of all conceivable executive power but regard it as an allocation to the presidential office of the generic powers thereafter stated.

The clause on which the Government next relies is that "The President shall be Commander in Chief of the Army and Navy of the United States. . . ." These cryptic words have given rise to some of the most persistent controversies in our constitutional history. Of course, they imply something more than an empty title. But just what authority goes with the name has plagued Presidential advisers who would not waive or narrow it by nonassertion yet cannot say where it begins or ends. It undoubtedly puts the Nation's armed forces under Presidential command. Hence, this loose appelation is sometimes advanced as support for any Presidential action, internal or external, involving use of force, the idea being that it vests power to do anything, anywhere, that can be done with an army or navy.

That seems to be the logic of an argument tendered at our bar—that the President having, on his own responsibility, sent American troops abroad derives from that act "affirmative power" to seize the means of producing a supply of steel for them. To quote, "Perhaps the most forceful illustrations of the scope of Presidential power in this connection is the fact that American troops in Korea, whose safety and effectiveness are so directly involved here, were sent to the field by an exercise of the President's constitutional powers." Thus, it is said he has invested himself with "war powers."

I cannot foresee all that it might entail if the Court should indorse this argument. Nothing in our Constitution is plainer than that declaration of a war is entrusted only to Congress. Of course, a state of war may in fact exist without a formal declaration. But no doctrine that the Court could promulgate would seem to me more sinister and alarming than that a President whose conduct of foreign affairs is so largely uncontrolled, and often even is unknown, can vastly enlarge his mastery over the internal affairs of the country by his own commitment of the Nation's armed forces to some foreign venture. I do not, however, find it necessary or appropriate to consider the legal status of the Korean enterprise to discountenance argument based on it.

Assuming that we are in a war *de facto*, whether it is or is not a war *de jure*, does that empower the Commander-in-Chief to seize industries he thinks necessary to supply our army? The Constitution expressly places in Congress power "to raise and *support* Armies" and "to *provide* and *maintain* a Navy." (Emphasis supplied.) This certainly lays upon Congress primary responsibility for supplying the armed forces. Congress alone controls the raising of revenues and their appropriation and may determine in what manner and by what means they shall be spent for military and naval procurement. I suppose no one would doubt that Congress can take over war supply as a Government enterprise. On the other hand, if Congress sees fit to rely on free private enterprise collectively bargaining with free labor for support and maintenance of our armed forces can the Executive because of lawful disagreements incidental to that process, seize the facility for operation upon Government-imposed terms?

There are indications that the Constitution did not contemplate that the title Commander-in-Chief *of the Army and Navy* will constitute him also Commander-in-Chief of the country, its industries and its inhabitants. He has no monopoly of "war powers," whatever they are. While Congress cannot deprive the President of the command of the army and navy, only Congress can provide him an army or navy to command. . . .

That military powers of the Commander-in-Chief were not to supersede representative government of internal affairs seems obvious from the Constitution and from elementary American history. Time out of mind, and even now in many parts of the world, a military commander can seize private housing to shelter his troops. Not so, however, in the United States, for the Third Amendment says, "No Soldier shall, in time of peace be quartered in any house, without the consent of the Owner, nor in time of war, but in a manner to be prescribed by law." Thus, even in war time, his seizure of needed military housing must be authorized by Congress. . . .

The third clause in which the Solicitor General finds seizure powers is that "he shall take Care that the Laws be faithfully executed. . . ." That authority must be matched against words of the Fifth Amendment that "No person shall be . . . deprived of life, liberty or property, without due process of law. . . ." One gives a governmental authority that reaches so far as there is law, the other gives a private right that authority shall go no farther. These signify about all there is of the principle that ours is a government of laws, not of men, and that we submit ourselves to rulers only if under rules.

The Solicitor General lastly grounds support of the seizure upon nebulous, inherent powers never expressly granted but said to have accrued to the office from the customs and claims of preceding administrations. The plea is for a resulting power to deal with a crisis or an emergency according to the necessities of the case, the unarticulated assumption being that necessity knows no law....

The appeal, however, that we declare the existence of inherent powers *ex necessitate* to meet an emergency asks us to do what many think would be wise, although it is something the forefathers omitted. They knew what emergencies were, knew the pressures they engender for authoritative action, knew, too, how they afford a ready pretext for usurpation. We may also suspect that they suspected that emergency powers would tend to kindle emergencies. Aside from suspension of the privilege of the writ of habeas corpus in time of rebellion or invasion, when the public safety may require it, they made no express provision for exercise of extraordinary authority because of a crisis. I do not think we rightfully may so amend their work, and, if we could, I am not convinced it would be wise to do so, although many modern nations have forthrightly recognized that war and economic crises may upset the normal balance between liberty and authority....

In view of the ease, expedition and safety with which Congress can grant and has granted large emergency powers, certainly ample to embrace this crisis, I am quite unimpressed with the argument that we should affirm possession of them without statute. Such power either has no beginning or it has no end. If it exists, it need submit to no legal restraint. I am not alarmed that it would plunge us straightway into dictatorship, but it is at least a step in that wrong direction.

As to whether there is imperative necessity for such powers, it is relevant to note the gap that exists between the President's paper powers and his real powers. The Constitution does not disclose the measure of the actual controls wielded by the modern presidential office. That instrument must be understood as an Eighteenth-Century sketch of a government hoped for, not as a blueprint of the Government that is. Vast accretions of federal power, eroded from that reserved by the States, have magnified the scope of presidential activity. Subtle shifts take place in the centers of real power that do not show on the face of the Constitution.

Executive power has the advantage of concentration in a single head in whose choice the whole Nation has a part, making him the focus of public hopes and expectations. In drama, magnitude and finality his decisions so far over-shadow any others that almost alone he fills the public eye and ear. No other personality in public life can begin to compete with him in access to the public mind through modern methods of communications. By his prestige as head of state and his influence upon public opinion he exerts a leverage upon those who are supposed to check and balance his power which often cancels their effectiveness.

Moreover, rise of the party system has made a significant extraconstitutional supplement to real executive power. No appraisal of his necessities is realistic which overlooks that he heads a political system as well as a legal system. Party loyalties and interests, sometimes more binding than law, extend his effective control into branches of government other than his own and he often may win, as a political leader, what he cannot command under the Constitution. . . .

The essence of our free Government is "leave to live by no man's leave, underneath the law"—to be governed by those impersonal forces which we call law. Our Government is fashioned to fulfill this concept so far as humanly possible. . . . With all its defects, delays and inconveniences, men have discovered no technique for long preserving free government except that the Executive be under the law, and that the law be made by parliamentary deliberations.

Such institutions may be destined to pass away. But it is the duty of the Court to be last, not first, to give them up. . . .

MR. CHIEF JUSTICE VINSON, with whom MR. JUSTICE REED and MR. JUSTICE MINTON join, dissenting.

. . . Because of the transcending importance of the questions presented not only in this critical litigation but also to the powers the President and of future Presidents to act in time of crisis, we are compelled to register this dissent.

In passing upon the question of Presidential powers in this case, we must first consider the context in which those powers were exercised.

Those who suggest that this is a case involving extraordinary powers should be mindful that these are extraordinary times. A world not yet recovered from the devastation of World War II has been forced to face the threat of another and more terrifying global conflict. . . . In 1950, when the United Nations called upon member nations "to render every assistance" to repel aggression in Korea, the United States furnished its vigorous support. For almost two full years, our armed forces have been fighting in Korea, suffering casualties of over 108,000 men. Hostilities have not abated. The "determination of the United Nations to continue its action in Korea to meet the aggression" has been reaffirmed. Congressional support of the action in Korea has been manifested by provisions for increased military manpower and equipment and for economic stabilization. . . .

Our treaties represent not merely legal obligations but show congressional recognition that mutual security for the free world is the best security against the threat of aggression on a global scale. The need for mutual security is shown by the very size of the armed forces outside the free world. Defendant's brief informs us that the Soviet Union maintains the largest air force in the world and maintains ground forces much larger than those presently available to the United States and the countries joined with us in mutual security arrangements. Constant international tensions are cited to demonstrate how precarious is the peace.

Even this brief review of our responsibilities in the world community discloses the enormity of our undertaking. Success of these measures may, as has often been observed, dramatically influence the lives of many generations of the world's peoples yet unborn. Alert to our responsibilities, which coincide with our own self preservation through mutual security, Congress has enacted a large body of implementing legislation. As an illustration of the magnitude of the over-all program, Congress has appropriated $130 billion for our own defense and for military assistance to our allies since the June, 1950, attack in Korea. . . .

Congress also directed the President to build up our own defenses. Congress, recognizing the "grim fact . . . that the United States is now engaged in a struggle for survival" and that "it is imperative that we now take those necessary steps to make our strength equal to the peril of the hour," granted authority to draft men into the armed forces. As a result, we now have over 3,500,000 men in our armed forces....

One is not here called upon even to consider the possibility of executive seizure of a farm, a corner grocery store or even a single industrial plant. Such considerations arise only when one ignores the central fact of this case—that the Nation's entire basic steel production would have shut down completely if there had been no Government seizure. Even ignoring for the moment whatever confidential information the President may possess as "the Nation's organ for foreign affairs," the uncontroverted affidavits in this record amply support the finding that "a work stoppage would immediately jeopardize and imperil our national defense."

Plaintiffs do not remotely suggest any basis for rejecting the President's finding that *any* stoppage of steel production would immediately place the Nation in peril. Moreover, even self-generated doubts that *any* stoppage of steel production constitutes an emergency are of little comfort here. The Union and the plaintiffs bargained for 6 months with over 100 issues in dispute—issues not limited to wage demands but including the union shop and other matters of principle between the parties. At the time of seizure there was not, and there is not now, the slightest evidence to justify the belief that any strike will be of short duration. The Union and the steel companies may well engage in a lengthy struggle. Plaintiff's counsel tells us that "sooner or later" the mills will operate again. That may satisfy the steel companies and, perhaps, the Union. But our soldiers and our allies will hardly be cheered with the assurance that the ammunition upon which their lives depend will be forthcoming—"sooner or later," or, in other words, "too little and too late."

Accordingly, if the President has any power under the Constitution to meet a critical situation in the absence of express statutory authorization, there is no basis whatever for criticizing the exercise of such power in this case.

The steel mills were seized for a public use. The power of eminent domain, invoked in this case, is an essential attribute of sovereignty and has long been recognized as a power of the Federal Government. *Kohl* v. *United States,* 91 U.S. 367 (1876). Plaintiffs cannot complain that any provision in the Constitution prohibits the exercise of the power of eminent domain in this case. The Fifth Amendment provides: "nor shall private property be taken for public use, without just compensation." It is no bar to this seizure for, if the taking is not otherwise unlawful, plaintiffs are assured of receiving the required just compensation. *United States* v. *Pewee Coal Co.*, 341 U.S. 114 (1951).

Admitting that the Government could seize the mills, plaintiffs claim that the implied power of eminent domain can be exercised only under an Act of Congress; under no circumstances, they say, can that power be exercised by the President unless he can point to an express provision in enabling legislation. This was the view adopted by the District Judge when he granted the preliminary injunction. Without an answer, without hearing evidence, he determined the issue on the basis of his "fixed conclusion . . . that defendant's acts are illegal" because the President's only course in the face of an emergency is to present the matter to Congress and await the final passage of legislation which will enable the Government to cope with threatened disaster.

Under this view, the President is left powerless at the very moment when the need for action may be most pressing and when no one, other than he, is immediately capable of action. Under this view, he is left powerless because a power not expressly given to Congress is nevertheless found to rest exclusively with Congress.

Consideration of this view of executive impotence calls for further examination of the nature of the separation of powers under our tripartite system of Government. . . . The whole of the "executive Power" is vested in the President. Before entering office, the President swears that he "will faithfully execute the Office of President of the United States, and will to the best of [his] ability, preserve, protect, and defend the Constitution of the United States." Art. II, Sec. 1.

This comprehensive grant of the executive power to a single person was bestowed soon after the country had thrown the yoke of monarchy. Only by instilling initiative and vigor in all of the three departments of Government, declared Madison, could tyranny in any form be avoided. Hamilton added: "Energy in the Executive is a leading character in the definition of good government. It is essential to the protection of the community against foreign attack; it is not less essential to the steady administration of the laws; to the protection of property against those irregular and high-handed combinations which sometimes interrupt the ordinary course of justice; to the security of liberty against the enterprises and assaults of ambition, of faction, and of anarchy." It is thus apparent that the Presidency was deliberately fashioned as an office of power and independence. Of course, the Framers created no autocrat capable of arrogating any power unto himself at any time. But neither did they create an automaton impotent to exercise the powers of Government at a time when the survival of the Republic itself may be at stake.

In passing upon the grave constitutional question presented in this case, we must never forget, as Chief Justice Marshall admonished, that the Constitution is "intended to endure for ages to come, and, consequently, to be adapted to the various *crises* of human affairs," and that "[i]ts means are adequate to its ends." Cases do arise presenting questions which could not have been foreseen by the Framers. In such cases, the Constitution has been treated as a living document adaptable to new situations. But we are not called upon today to expand the Constitution to meet a new situation. For, in this case, we need only look to history and time-honored principles of constitutional law—principles that have been applied consistently by all branches of the Government throughout our history. It is those who assert the invalidity of the Executive Order who seek to amend the Constitution in this case.

A review of executive action demonstrates that our Presidents have on many occasions exhibited the leadership contemplated by the Framers when they made the President Commander in Chief, and imposed upon him the trust to "take Care that the Laws be faithfully executed." With or without explicit statutory authorization, Presidents have at such times dealt with national emergencies by acting promptly and resolutely to enforce legislative programs, at least to save those programs until Congress could act. Congress and the courts have responded to such executive initiative with consistent approval.

Our first President displayed at once the leadership contemplated by the Framers. When the national revenue laws were openly flouted in some sections of Pennsylvania, President Washington, without waiting for a call from the state government, summoned the militia and took decisive steps to secure the faithful execution of the laws. When international disputes engendered by the French revolution threatened to involve this country in war, and while congressional policy remained uncertain, Washington issued his Proclamation of Neutrality. Hamilton, whose defense of the Proclamation has endured the test of time, invoked the argument that the Executive has the duty to do that which will preserve peace until Congress acts and, in addition, pointed to the need for keeping the Nation informed of the requirements of existing laws and treaties as part of the faithful execution of the laws. . . .

Jefferson's initiative in the Louisiana Purchase, the Monroe Doctrine, and Jackson's removal of Government deposits from the Bank of the United States further serve to demonstrate by deed what the Framers described by word when they vested the whole of the executive power in the President.

Without declaration of war, President Lincoln took energetic action with the outbreak of the Civil War. He summoned troops and paid them out of the Treasury without appropriation therefor. He proclaimed a naval blockade of the Confederacy and seized ships violating that blockade. Congress, far from denying the validity of these acts, gave them express approval. The most striking action of President Lincoln was the Emancipation Proclamation, issued in aid of the successful prosecution of the Civil War, but wholly without statutory authority.

In an action furnishing a most apt precedent for this case, President Lincoln directed the seizure of rail and telegraph lines leading to Washington without statutory authority. Many months later, Congress recognized and confirmed the power of the President to seize railroads and telegraph lines and provided criminal penalties for interference with Government operation. This Act did not confer on the President any additional powers of seizure. Congress plainly rejected the view that the President's acts had been without legal sanction until ratified by the legislature. Sponsors of the bill declared that its purpose was only to confirm the power which the President already possessed. Opponents insisted a statute authorizing seizure was unnecessary and might even be construed as limiting existing Presidential powers. . . .

During World War I, President Wilson established a War Labor Board without awaiting specific direction by Congress. With William Howard Taft and Frank P. Walsh as co-chairmen, the Board had as its purpose the prevention of strikes and lockouts interfering with the production of goods needed to meet the emergency. Effectiveness of War Labor Board decision was accomplished by Presidential action, including seizure of industrial plants. Seizure of the Nation's railroads was also ordered by President Wilson.

Beginning with the Bank Holiday Proclamation and continuing through World War II, executive leadership and initiative were characteristic of President Franklin D. Roosevelt's administration. In 1939, upon the outbreak of war in Europe, the President proclaimed a limited national emergency for the purpose of strengthening our national defense. By May of 1941, the danger from the Axis belligerents having become clear, the President proclaimed "an unlimited national emergency" calling for mobilization of the Nation's defenses to repel aggression. The President took the initiative in strengthening our defenses by acquiring rights from the British Government to establish air bases in exchange for overage destroyers.

In 1941, President Roosevelt acted to protect Iceland from attack by Axis powers when British forces were withdrawn by sending our forces to occupy Iceland. Congress was informed of this action on the same day that our forces reached Iceland. The occupation of Iceland was but one of "at least 125 incidents" in our history in which Presidents, "without Congressional authorization, and in the absence of a declaration of war, [have] ordered the Armed Forces to take action or maintain positions abroad."

Some six months before Pearl Harbor, a dispute at a single aviation plant at Inglewood, California, interrupted a segment of the production of military aircraft. In spite of the comparative insignificance of this work stoppage to total defense production as contrasted with the complete paralysis now threatened by a shutdown of the entire basic steel industry, and even though our armed forces were not then engaged in combat, President Roosevelt ordered the seizure of the plant "pursuant to the powers vested in [him] by the Constitution and laws of the United States, as President of the United States of America and Commander in Chief of the Army and Navy of the United States." The Attorney General (Jackson) vigorously proclaimed that the President had the moral duty to keep this Nation's defense effort a "going concern." His ringing moral justification was coupled with a legal justification equally well stated:

> "The Presidential proclamation rests upon the aggregate of the Presidential powers derived from the Constitution itself and from statutes enacted by the Congress.
>
> "The Constitution lays upon the President the duty 'to take care that the laws be faithfully executed.' Among the laws which he is required to find means to execute are those which direct him to equip an enlarged army, to provide for a strengthened navy, to protect Government property, to protect those who are engaged in carrying out the business of the Government, and to carry out the provisions of the Lend-Lease Act. For the faithful execution of such laws the President has back of him not only each general law-enforcement power conferred by the various acts of Congress but the aggregate of all such laws plus that wide discretion as to method vested in him by the Constitution for the purpose of executing the laws.
>
> "The Constitution also places on the President the responsibility and vests in him the powers of Commander in Chief of the Army and of the Navy. These weapons for the protection of the continued existence of the Nation are placed in his sole command and the implication is clear that he should not allow them to become paralyzed by failure to obtain supplies for which Congress has appropriated the money and which it has directed the President to obtain."

At this time, Senator Connally proposed amending the Selective Service and Training Act to authorize the President to seize any plant where an interruption of production would unduly impede the defense effort. Proponents of the measure in no way implied that the legislation would add to the powers already possessed by the President and the amendment was opposed as unnecessary since the President already had the power. The amendment relating to plant seizures was not approved at that session of Congress.

Meanwhile, and also prior to Pearl Harbor, the President ordered the seizure of a shipbuilding company and an aircraft parts plant. Following the declaration of war, but prior to the Smith-Connally Act of 1943, five additional industrial concerns were seized to avert interruption of needed production. During the same period, the President directed seizure of the Nation's coal mines to remove an obstruction to the effective prosecution of the war. . . .

More recently, President Truman acted to repel aggression by employing our armed forces in Korea. Upon the intervention of the Chinese Communists, the President proclaimed the existence of an unlimited national emergency requiring the speedy build-up of our defense establishment. Congress responded by providing for increased manpower and weapons for our own armed forces, by increasing military aid under the Mutual Security Program and by enacting economic stabilization measures, as previously described.

This is but a cursory summary of executive leadership. But it amply demonstrates that Presidents have taken prompt action to enforce the laws and protect the country whether or not Congress happened to provide in advance for the particular method of execution. At the minimum, the executive actions reviewed herein sustain the action of the President in this case. And many of the cited examples of Presidential practice go far beyond the extent of power necessary to sustain the President's order to seize the steel mills. The fact that temporary executive seizures of industrial plants to meet an emergency have not been directly tested in this Court furnishes not the slightest suggestion that such actions have been illegal. Rather, the fact that Congress and the courts have consistently recognized and given their support to such executive action indicates that such a power of seizure has been accepted throughout our history.

History bears out the genius of the Founding Fathers, who created a Government subject to law but not left subject to inertia when vigor and initiative are required. ...

Whatever the extent of Presidential power on more tranquil occasions, and whatever the right of the President to execute legislative programs as he sees fit without reporting the mode of execution to Congress, the single Presidential purpose disclosed on this record is to faithfully execute the laws by acting in an emergency to maintain the status quo, thereby preventing collapse of the legislative programs until Congress could act. The President's action served the same purposes as a judicial stay entered to maintain the status quo in order to preserve the jurisdiction of a court. In his Message to Congress immediately following the seizure, the President explained the necessity of his action in executing the military procurement and anti-inflation legislative programs and expressed his desire to cooperate with any legislative proposals approving, regulating or rejecting the seizure of the steel mills. Consequently, there is no evidence whatever of any Presidential purpose to defy Congress or act in any way inconsistent with the legislative will. . . .

The diversity of views expressed in the six opinions of the majority, the lack of reference to authoritative precedent, the repeated reliance upon prior dissenting opinions, the complete disregard of the uncontroverted facts showing the gravity of the emergency and the temporary nature of the taking all serve to demonstrate how far afield one must go to affirm the order of the District Court.

The broad executive power granted by Article II to an officer on duty 365 days a year cannot, it is said, be invoked to avert disaster. Instead, the President must confine himself to sending a message to Congress recommending action. Under this messenger-boy concept of the Office, the President cannot even act to preserve legislative programs from destruction so that Congress will have something left to act upon. There is no judicial finding that the executive action was unwarranted because there was in fact no basis for the President's finding of the existence of an emergency for, under this view, the gravity of the emergency and the immediacy of the threatened disaster are considered irrelevant as a matter of law.

Seizure of plaintiffs' property is not a pleasant undertaking. Similarly unpleasant to a free country are the draft which disrupts the home and military procurement which causes economic dislocation and compels adoption of price controls, wage stabilization and allocation of materials. The President informed Congress that even a temporary Government operation of plaintiffs' properties was "thoroughly distasteful" to him, but was necessary to prevent immediate paralysis of the mobilization program. Presidents have been in the past, and any man worthy of the Office should be in the future, free to take at least interim action necessary to execute legislative programs essential to survival of the Nation. A sturdy judiciary should not be swayed by the unpleasantness or unpopularity of necessary executive action, but must independently determine for itself whether the President was acting, as required by the Constitution, "to take Care that the Laws be faithfully executed."

As the District Judge stated, this is no time for "timorous" judicial action. But neither is this a time for timorous executive action. Faced with the duty of executing the defense programs which Congress had enacted and the disastrous effects that any stoppage in steel production would have on those programs, the President acted to preserve those programs by seizing the steel mills. There is no question that the possession was other than temporary in character and subject to congressional direction—either approving, disapproving or regulating the manner in which the mills were to be administered and returned to the owners. The President immediately informed Congress of his action and clearly stated his intention to abide by the legislative will. No basis for claims of arbitrary action, unlimited powers or dictatorial usurpation of congressional power appears from the facts of this case. On the contrary, judicial, legislative and executive precedents throughout our history demonstrate that in this case the President acted in full conformity with his duties under the Constitution. Accordingly, we would reverse the order of the District Court.

VI

THE POLITICS OF DEMOCRACY

Representing the People*

BY EDMUND BURKE

I

CERTAINLY, gentlemen, it ought to be the happiness and glory of a representative, to live in the strictest union, the closest correspondence, and the most unreserved communication with his constituents. Their wishes ought to have great weight with him; their opinion high respect; their business unremitted attention. It is his duty to sacrifice his repose, his pleasures, his satisfactions, to theirs; and, above all, ever, and in all cases, to prefer their interest to his own. But, his unbiased opinion, his mature judgment, his enlightened conscience, he ought not to sacrifice to you; to any man, or to any set of men living. These he does not derive from your pleasure; no, nor from the law and the constitution. They are a trust from Providence, for the abuse of which he is deeply answerable. Your representative owes you, not his industry only, but his judgment; and he betrays, instead of serving you, if he sacrifices it to your opinion.

My worthy colleague says, his will ought to be subservient to yours. If that be all, the thing is innocent. If government were a matter of will upon any side, yours, without question, ought to be superior. But government and legislation are matters of reason and judgment, and not of inclination; and, what sort of reason is that, in which the determination precedes the discussion; in which one set of men deliberate, and another decide; and where those who form the conclusion are perhaps three hundred miles distant from those who hear the arguments?

To deliver an opinion, is the right of all men; that of constituents is a weighty and respectable opinion, which a representative ought always to rejoice to hear; and which he ought always most seriously to consider. But *authoritative* instructions; *mandates* issued, which the member is bound blindly and implicitly to obey, to vote, and to argue for, though contrary to the clearest conviction of his

---

* The first selection is taken from a speech to his constituents at Bristol, England, in 1774. The second is taken from Burke's pamphlet *Thoughts on the Cause of the Present Discontents*, 1770.

judgment and conscience; these are things utterly unknown to the laws of this land, and which arise from a fundamental mistake of the whole order and tenour of our constitution.

Parliament is not a *congress* of ambassadors from different and hostile interests; which interests each must maintain, as an agent and advocate, against other agents and advocates; but parliament is a *deliberative* assembly of *one* nation, with *one* interest, that of the whole; where, not local purposes, not local prejudices ought to guide, but the general good, resulting from the general reason of the whole. You chuse a member indeed; but when you have chosen him, he is not a member of Bristol, but he is a member of *parliament*. If the local constituent should have an interest, or should form an hasty opinion, evidently opposite to the real good of the rest of the community, the member for that place ought to be as far, as any other, from any endeavour to give it effect. I beg pardon for saying so much on this subject. I have been unwillingly drawn into it; but I shall ever use a respectful frankness of communication with you. Your faithful friend, your devoted servant, I shall be to the end of my life: a flatterer you do not wish for.

II

Every profession, not excepting the glorious one of a soldier, or the sacred one of a priest, is liable to its own particular vices; which, however, form no argument against those ways of life; nor are the vices themselves inevitable to every individual in those professions. Of such a nature are connexions in politicks; essentially necessary for the full performance of our public duty, accidentally liable to degenerate into faction. Commonwealths are made of families, free commonwealths of parties also; and we may as well affirm, that our natural regards and ties of blood tend inevitably to make men bad citizens, as that the bonds of our party weaken those by which we are held to our country. . . .

When bad men combine, the good must associate; else they will fall, one by one, an unpitied sacrifice in a contemptible struggle.

It is not enough in a situation of trust in the commonwealth, that a man means well to his country; it is not enough that in his single person he never did an evil act, but always voted according to his conscience, and even harangued against every design which he apprehended to be prejudicial to the interests of his country. This innoxious and ineffectual character, that seems formed upon a plan of apology and disculpation, falls miserably short of the mark of public duty. That duty demands and requires, that what is right should not only be made known, but made prevalent; that what is evil should not only be detected, but defeated. When the public man omits to put himself in a situation of doing his duty with effect, it is an omission that frustrates the purposes of his trust almost as much as if he had formally betrayed it. It is surely no very rational account of a man's life, that he has always acted right; but has taken special care to act in such a manner that his endeavours could not possibly be productive of any consequence.

I do not wonder that the behaviour of many parties should have made persons of tender and scrupulous virtue somewhat out of humour with all sorts of connexion in politicks. I admit that people frequently acquire in such confederacies a narrow, bigotted, and proscriptive spirit; that they are apt to sink the idea of the general good in this circumscribed and partial interest. But, where duty renders a critical situation a necessary one, it is our business to keep free from the evils attendant upon it; and not to fly from the situation itself. . . .

Party is a body of men united, for promoting by the joint endeavours the national interest, upon some particular principle in which they are all agreed. For my part, I find it impossible to conceive, that any one believes in his own politicks, or thinks them to be of any weight, who refuses to adopt the means of having them reduced into practice. It is the business of the speculative philosopher to mark the proper ends of Government. It is the business of the politician, who is the philosopher in action, to find out proper means towards those ends, and to employ them with effect. Therefore every honourable connexion will avow it as their first purpose, to pursue every just method to put the men who hold their opinions into such a condition as may enable them to carry their common plans into execution, with all the power and authority of the State. As this power is attached to certain situations, it is their duty to contend for these situations. Without a proscription of others, they are bound to give to their own party the preference in all things; and by no means, for private considerations, to accept any offers of power in which the whole body is not included; nor to suffer themselves to be led, or to be controuled, or to be over-balanced, in office or in council, by those who contradict the very fundamental principles on which their party is formed, and even those upon which every fair connexion must stand. Such a generous contention for power, on such manly and honourable maxims, will easily be distinguished from the mean and interested struggle for place and emolument. The very stile of such persons will serve to discriminate them from those numberless impostors, who have deluded the ignorant with professions incompatible with human practice, and have afterwards incensed them by practices below the level of vulgar rectitude.

The Federalist No. 10

BY JAMES MADISON

[1787]

To the People of the State of New York:

Among the numerous advantages promised by a well-constructed Union, none deserves to be more accurately developed than its tendency to break and control the violence of faction. The friend of popular governments never finds himself so much alarmed for their character and fate, as when he contemplates their propensity to this dangerous vice. He will not fail, therefore, to set a due value on any plan which, without violating the principles to which he is attached, provides a proper cure for it. The instability, injustice, and confusion introduced into the public councils, have, in truth, been the mortal diseases under which popular governments have everywhere perished; as they continue to be the favorite and fruitful topics from which the adversaries to liberty derive their most specious declamations. The valuable improvements made by the American constitutions on the popular models, both ancient and modern, cannot certainly be too much admired; but it would be an unwarrantable partiality, to contend that they have as effectually obviated the danger on this side, as was wished and expected. Complaints are everywhere heard from our most considerate and virtuous citizens, equally the friends of public and private faith, and of public and personal liberty, that our governments are too unstable, that the public good is disregarded in the conflicts of rival parties, and that measures are too often decided, not according to the rules of justice and the rights of the minor party, but by the superior force of an interested and overbearing majority. However anxiously we may wish that these complaints had no foundation, the evidence of known facts will not permit us to deny that they are in some degree true. It will be found, indeed, on a candid review of our situation, that some of the distresses under which we labor have been erroneously charged on the operation of our governments; but it will be found, at the same time, that other causes will not alone account for many of our heaviest misfortunes; and, particularly, for that prevailing and increasing distrust of public engagements, and alarm for private rights, which are echoed from one end of the continent to the other. These must be chiefly, if not wholly, effects of the unsteadiness and injustice with which a factious spirit has tainted our public administrations.

By a faction, I understand a number of citizens, whether amounting to a majority or minority of the whole, who are united and actuated by some common impulse of passion, or of interest, adverse to the rights of other citizens, or to the permanent and aggregate interests of the community.

There are two methods of curing the mischiefs of faction: the one, by removing its causes; the other, by controlling its effects.

There are again two methods of removing the causes of faction: the one, by destroying the liberty which is essential to its existence; the other, by giving to every citizen the same opinions, the same passions, and the same interests.

It could never be more truly said than of the first remedy, that it was worse than the disease. Liberty is to faction what air is to fire, an aliment without which it instantly expires. But it could not be less folly to abolish liberty, which is essential to political life, because it nourishes faction, than it would be to wish the annihilation of air, which is essential to animal life, because it imparts to fire its destructive agency.

The second expedient is as impracticable as the first would be unwise. As long as the reason of man continues fallible, and he is at liberty to exercise it, different opinions will be formed. As long as the connection subsists between his reason and his self-love, his opinions and his passions will have a reciprocal influence on each other; and the former will be objects to which the latter will attach themselves. The diversity in the faculties of men, from which the rights of property originate, is not less an insuperable obstacle to a uniformity of interests. The protection of these faculties is the first object of government. From the protection of different and unequal faculties of acquiring property, the possession of different degrees and kinds of property immediately results; and from the influence of these on the sentiments and views of the respective proprietors, ensues a division of the society into different interests and parties.

The latent causes of faction are thus sown in the nature of man; and we see them everywhere brought into different degrees of activity, according to the different circumstances of civil society. A zeal for different opinions concerning religion, concerning government, and many other points, as well of speculation as of practice; an attachment to different leaders ambitiously contending for pre-eminence and power; or to persons of other descriptions whose fortunes have been interesting to the human passions, have, in turn, divided mankind into parties, inflamed them with mutual animosity, and rendered them much more disposed to vex and oppress each other than to co-operate for their common good. So strong is this propensity of mankind to fall into mutual animosities, that where no substantial occasion presents itself, the most frivolous and fanciful distinctions have been sufficient to kindle their unfriendly passions and excite their most violent conflicts. But the most common and durable source of factions has been the various and unequal distribution of property. Those who hold and those who are without property have ever formed distinct interests in society. Those who are creditors, and those who are debtors, fall under a like discrimination. A landed interest, a manufacturing interest, a mercantile interest, a moneyed interest, with many lesser interests, grow up of necessity in civilized nations, and divide them into different classes, actuated by different sentiments and views. The regulation of these various and interfering interests forms the principal task of modern legislation, and involves the spirit of party and faction in the necessary and ordinary operations of the government.

No man is allowed to be a judge in his own cause, because his interest would certainly bias his judgment, and, not improbably, corrupt his integrity. With equal, nay with greater reason, a body of men are unfit to be both judges and parties at the same time; yet what are many of the most important acts of legislation, but so many judicial determinations, not indeed concerning the rights of single persons, but concerning the rights of large bodies of citizens? And what are the different classes of legislators but advocates and parties to the causes which they determine? Is a law proposed concerning private debts? It is a question to which the creditors are parties on one side and the debtors on the other. Justice ought to hold the balance between them. Yet the parties are, and must be, themselves the judges; and the most numerous party, or, in other words, the most powerful faction must be expected to prevail. Shall domestic manufactures be encouraged, and in what degree, by restrictions on foreign manufactures? are questions which would be differently decided by the landed and the manufacturing classes, and probably by neither with a sole regard to justice and the public good. The apportionment of taxes on the various descriptions of property is an act which seems to require the most exact impartiality; yet there is, perhaps, no legislative act in which greater opportunity and temptation are given to a predominant party to trample on the rules of justice. Every shilling with which they overburden the inferior number, is a shilling saved to their own pockets.

It is in vain to say that enlightened statesmen will be able to adjust these clashing interests, and render them all subservient to the public good. Enlightened statesmen will not always be at the helm. Nor, in many cases, can such an adjustment be made at all without taking into view indirect and remote considerations, which will rarely prevail over the immediate interest which one party may find in disregarding the rights of another or the good of the whole.

The inference to which we are brought is, that the *causes* of faction cannot be removed, and that relief is only to be sought in the means of controlling its *effects*.

If a faction consists of less than a majority, relief is supplied by the republican principle, which enables the majority to defeat its sinister views by regular vote. It may clog the administration, it may convulse the society; but it will be unable to execute and mask its violence under the forms of the Constitution. When a majority is included in a faction, the form of popular government, on the other hand, enables it to sacrifice to its ruling passion or interest both the public good and the rights of other citizens. To secure the public good and private rights against the danger of such a faction, and at the same time to preserve the spirit and the form of popular government, is then the great object to which our inquiries are directed. Let me add that it is the great desideratum by which this form of government can be rescued from the opprobrium under which it has so long labored, and be recommended to the esteem and adoption of mankind.

By what means is this object attainable? Evidently by one of two only. Either the existence of the same passion or interest in a majority at the same time must be prevented, or the majority, having such coexistent passion or interest, must be rendered, by their number and local situation, unable to concert and carry into effect schemes of oppression. If the impulse and the opportunity be suffered to coincide, we well know that neither moral nor religious motives can be relied on as an adequate control. They are not found to be such on the injustice and violence of individuals, and lose their efficacy in proportion to the number combined together, that is, in proportion as their efficacy becomes needful.

From this view of the subject it may be concluded that a pure democracy, by which I mean a society consisting of a small number of citizens, who assemble and administer the government in person, can admit of no cure for the mischiefs of faction. A common passion or interest will, in almost every case, be felt by a majority of the whole; a communication and concert result from the form of government itself; and there is nothing to check the inducements to sacrifice the weaker party or an obnoxious individual. Hence it is that such democracies have ever been spectacles of turbulence and contention; have ever been found incompatible with personal security or the rights of property; and have in general been as short in their lives as they have been violent in their deaths. Theoretic politicians, who have patronized this species of government, have erroneously supposed that by reducing mankind to a perfect equality in their political rights, they would, at the same time, be perfectly equalized and assimilated in their possessions, their opinions, and their passions.

A republic, by which I mean a government in which the scheme of representation takes place, opens a different prospect, and promises the cure for which we are seeking. Let us examine the points in which it varies from pure democracy, and we shall comprehend both the nature of the cure and the efficacy which it must derive from the Union.

The two great points of difference between a democracy and a republic are: first, the delegation of the government, in the latter, to a small number of citizens elected by the rest; secondly, the greater number of citizens, and greater sphere of country, over which the latter may be extended.

The effect of the first difference is, on the one hand, to refine and enlarge the public views, by passing them through the medium of a chosen body of citizens, whose wisdom may best discern the true interest of their country, and whose patriotism and love of justice will be least likely to sacrifice it to temporary or partial considerations. Under such a regulation, it may well happen that the public voice, pronounced by the representatives of the people, will be more consonant to the public good than if pronounced by the people themselves, convened for the purpose. On the other hand, the effect may be inverted. Men of factious tempers, of local prejudices, or of sinister designs, may, by intrigue, by corruption, or by other means, first obtain the suffrages, and then betray the interests, of the people. The question resulting is, whether small or extensive republics are more favorable to the election of proper guardians of the public weal; and it is clearly decided in favor of the latter by two obvious considerations:

In the first place, it is to be remarked that, however small the republic may be, the representatives must be raised to a certain number, in order to guard against the cabals of a few; and that, however large it may be, they must be limited to a certain number, in order to guard against the confusion of a multitude. Hence, the number of representatives in the two cases not being in proportion to that of the two constituents, and being proportionally greater in the small republic, it follows that, if the proportion of fit characters be not less in the large than in the small republic, the former will present a greater option, and consequently a greater probability of a fit choice.

In the next place, as each representative will be chosen by a greater number of citizens in the large than in the small republic, it will be more difficult for unworthy candidates to practise with success the vicious arts by which elections are too often carried; and the suffrages of the people being more free, will be more likely to centre in men who possess the most attractive merit and the most diffusive and established characters.

It must be confessed that in this, as in most other cases, there is a mean, on both sides of which inconveniences will be found to lie. By enlarging too much the number of electors, you render the representative too little acquainted with all their local circumstances and lesser interests; as by reducing it too much, you render him unduly attached to these, and too little fit to comprehend and pursue great and national objects. The federal Constitution forms a happy combination in this respect; the great and aggregate interests being referred to the national, the local and particular to the State legislatures.

The other point of difference is, the greater number of citizens and extent of territory which may be brought within the compass of republican than of democratic government; and it is this circumstance principally which renders factious combinations less to be dreaded in the former than in the latter. The smaller the society, the fewer probably will be the distinct parties and interests composing it; the fewer the distinct parties and interests, the more frequently will a majority be found of the same party; and the smaller the number of individuals composing a majority, and the smaller the compass within which they are placed, the more easily will they concert and execute their plans of oppression. Extend the sphere, and you take in a greater variety of parties and interests; you make it less probable that a majority of the whole will have a common motive to invade the rights of other citizens; or if such a common motive exists, it will be more difficult for all who feel it to discover their own strength, and to act in unison with each other. Besides other impediments, it may be remarked that, where there is a consciousness of unjust or dishonorable purposes, communication is always checked by distrust in proportion to the number whose concurrence is necessary.

Hence, it clearly appears, that the same advantage which a republic has over a democracy, in controlling the effects of faction, is enjoyed by a large over a small republic,—is enjoyed by the Union over the States composing it. Does the advantage consist in the substitution of representatives whose enlightened views and virtuous sentiments render them superior to local prejudices and to schemes of injustice? It will not be denied that the representation of the Union will be most likely to possess these requisite endowments. Does it consist in the greater security afforded by a greater variety of parties, against the event of any one party being able to outnumber and oppress the rest? In an equal degree does the increased variety of parties comprised within the Union, increase this security. Does it, in fine, consist in the greater obstacles opposed to the concert and accomplishment of the secret wishes of an unjust and interested majority? Here, again, the extent of the Union gives it the most palpable advantage.

The influence of factious leaders may kindle a flame within their particular States, but will be unable to spread a general conflagration through the other States. A religious sect may degenerate into a political faction in a part of the Confederacy; but the variety of sects dispersed over the entire face of it must secure the national councils against any danger from that source. A rage for paper money, for an abolition of debts, for an equal division of property, or for any other improper or wicked project, will be less apt to pervade the whole body of the Union than a particular member of it; in the same proportion as such a malady is more likely to taint a particular county or district, than an entire State.

In the extent and proper structure of the Union, therefore, we behold a republican remedy for the diseases most incident to republican government. And according to the degree of pleasure and pride we feel in being republicans, ought to be our zeal in cherishing the spirit and supporting the character of Federalists.

PUBLIUS

On Party*

BY JAMES FENIMORE COOPER

[1838]

It is commonly said that political parties are necessary to liberty. This is one of the mistaken opinions that have been inherited from those who, living under governments in which there is no true political liberty, have fancied that the struggles which are inseparable from their condition, must be common to the conditions of all others.

England, the country from which this people is derived, and, until the establishment of our own form of government, the freest nation of Christendom, enjoys no other liberty than that which has been obtained by the struggles of parties. Still retaining in the bosom of the state, a power in theory, which, if carried out in practice, would effectually overshadow all the other powers of the state, it may truly be necessary to hold such a force in check, by the combinations of political parties. But the condition of America, in no respect, resembles this. Here, the base of the government is the constituencies, and its balance is in the divided action of their representatives, checked as the latter are by frequent elections. As these constituencies are popular, the result is a free, or a popular government.

Under such a system, in which the fundamental laws are settled by a written compact, it is not easy to see what good can be done by parties, while it is easy to see that they may effect much harm. It is the object of this article, to point out a few of the more prominent evils that originate from such a source.

Party is known to encourage prejudice, and to lead men astray in the judgment of character. Thus it is we see one half the nation extolling those that the other half condemns, and condemning

---

* Reprinted from *The American Democrat* by James Fenimore Cooper, by permission of Alfred A. Knopf, Inc. Copyright 1931, 1956 by Alfred A. Knopf, Inc. Published by Vintage Books, Inc., 1956. James Fenimore Cooper (1789–1851), American novelist, is the author of the famous Leatherstocking Series.

those that the other half extols. Both cannot be right, and as passions, interests and prejudices are all enlisted on such occasions, it would be nearer the truth to say that both are wrong.

Party is an instrument of error, by pledging men to support its policy, instead of supporting the policy of the state. Thus we see party measures almost always in extremes, the resistance of opponents inducing the leaders to ask for more than is necessary.

Party leads to vicious, corrupt and unprofitable legislation, for the sole purpose of defeating party. Thus have we seen those territorial divisions and regulations which ought to be permanent, as well as other useful laws, altered, for no other end than to influence an election.

Party, has been a means of entirely destroying that local independence, which elsewhere has given rise to a representation that acts solely for the nation, and which, under other systems is called the country party, every legislator being virtually pledged to support one of two opinions; or, if a shade of opinion between them, a shade that is equally fettered, though the truth be with neither.

The discipline and organization of party, are expedients to defeat the intention of the institutions, by putting managers in the place of the people; it being of little avail that a majority elect, when the nomination rests in the hands of a few.

Party is the cause of so many corrupt and incompetent men's being preferred to power, as the elector, who, in his own person, is disposed to resist a bad nomination, yields to the influence and a dread of factions.

Party pledges the representative to the support of the executive, right or wrong, when the institutions intend that he shall be pledged only to justice, expediency and the right, under the restrictions of the constitution.

When party rules, the people do not rule, but merely such a portion of the people as can manage to get the control of party. The only method by which the people can completely control the country, is by electing representatives known to prize and understand the institutions; and, who, so far from being pledged to support an administration, are pledged to support nothing but the right, and whose characters are guarantees that this pledge will be respected.

The effect of party is always to supplant established power. In a monarchy it checks the king; in a democracy it controls the people.

Party, by feeding the passions and exciting personal interests, overshadows truth, justice, patriotism, and every other publick virtue, completely reversing the order of a democracy, by putting unworthy motives in the place of reason.

It is a very different thing to be a democrat, and to be a member of what is called a democratic party; for the first insists on his independence and an entire freedom of opinion, while the last is incompatible with either.

The great body of the nation has no real interest in party. Every local election should be absolutely independent of great party divisions, and until this be done, the intentions of the American institutions will never be carried out, in their excellence.

Party misleads the public mind as to the rights and duties of the citizen. An instance has recently occurred, in which a native born citizen of the United States of America, the descendant of generations of Americans, has become the object of systematic and com-combined persecution, because he published a constitutional opinion that conflicted with the interests and passions of party, although having no connection with party himself; very many of his bitterest assailants being foreigners, who have felt themselves authorized to pursue this extraordinary course, as the agents of party!

No freeman, who really loves liberty, and who has a just perception of its dignity, character, action and objects, will ever become a mere party man. He may have his preferences as to measures and men, may act in concert with those who think with himself, on occasions that require concert, but it will be his earnest endeavour to hold himself a free agent, and most of all to keep his mind untrammelled by the prejudices, frauds, and tyranny of factions.

The Price of Union*

BY HERBERT AGAR

[1950]

In 1788, when Alexander Hamilton, James Madison, and John Jay were struggling to persuade the New York convention to ratify the Constitution of the United States, young De Witt Clinton composed a prayer for the Opposition: "From the insolence of great men—from the tyranny of the rich—from the unfeeling rapacity of the excise-men and tax-gatherers—from the misery of despotism—from the expense of supporting standing armies, navies, placemen, sinecures, federal cities, senators, presidents, and a long train of etceteras, Good Lord deliver us." There speaks the deep American distaste for government, the belief that it is evil and that it must be kept weak.

The proposed constitution seemed weak enough to its friends—too weak for safety in the opinion of Hamilton. The thirteen states were left with all the powers which would normally concern "the lives, liberties and properties of the people," while the delegated powers of the Union were divided among the Executive, the Legislature, and the Judiciary in the belief that they would check each other and prevent rash or oppressive deeds. Yet behind this modest proposal Clinton saw threats of insolence, rapacity, misery, needless expense, and all the corruptions of history. There is little doubt that he spoke for the majority in New York; and many citizens would still agree with him. A constant factor in American history is the fear of Leviathan, of the encroaching state.

We can trace this fear from prerevolutionary days, and we can trace the forces which have nevertheless caused Leviathan to grow steadily more ponderous. War and industrial revolution promote strong government. Foreign dangers, business depressions, sectional or class strife—whenever these are acute the people look to

---

* Introduction and Chapter XXXV from *The Price of Union*. Copyright 1950 by Herbert Agar. Reprinted by permission of Houghton Mifflin Company. Herbert Agar (b. 1897) is an American historian who now resides in England. His latest book is *The Price of Power* (1957).

central power for help. Yet they have not abandoned hope that life might sometime be peaceful and that government might become frugal and unassuming, as Jefferson promised. And they have not abandoned the constitution which seemed to many of those who wrote it to err on the side of weakness, to put the liberties of the citizen before the safety of the commonwealth. Yet by unwritten means, out of simple self-protection, that constitution has been given strength and flexibility to meet the threats and disasters of a hundred and sixty years. The result is one of the most interesting forms of government the world has seen, and in the light of its problems one of the most successful.

The special problems of the American Government derive from geography, national character, and the nature both of a written constitution and of a federal empire. The government is cramped and confined by a seemingly rigid bond; yet it must adapt itself to a rate of change in economics, technology, and foreign relations which would have made all previous ages dizzy. In good times the government must abide by the theory that its limited sovereignty has been divided between the Union and the several states; yet when the bombs fall or the banks close or the breadlines grow by millions it must recapture the distributed sovereignty and act like a strong centralized nation. The government must regard the separation of its own powers, especially those of the Executive and the Legislature, as an essential and indeed a sacred part of the system; yet when the separation threatens deadlock and danger it must reassemble those powers informally and weld them into a working team. Finally, the government must accept the fact that in a country so huge, containing such diverse climates and economic interests and social habits and racial and religious backgrounds, most politics will be parochial, most politicians will have small horizons, seeking the good of the state or the district rather than of the Union; yet by diplomacy and compromise, never by force the government must water down the selfish demands of regions, races, classes, business associations, into a national policy which will alienate no major group and which will contain at least a small plum for everybody. This is the price of unity in a continent-wide federation. Decisions will therefore be slow, methods will be cumbersome, political parties will be illogical and inconsistent; but the people remain free, reasonably united, and as lightly burdened by the state as is consistent with safety.

It may be asked, if the inevitable problems are so acute and so contradictory why not change the form of government? The answer is that the American political system with all its absurdities is one of the few successes in a calamitous age. Step by step, it has learned to avoid many of the worst mistakes of empire, in a nation which would stretch from London to the Ural Mountains and from Sweden to the Sahara; it has learned to circumvent threats of secession (the mortal illness of federalism) before they appear; it has learned to evade class warfare (the mortal illness of liberty), and to the dismay of its critics it shows no sign of moving toward class parties. Once, in the midst of the long period of learning, the political system failed totally. The result was civil war. The system must always fail partially, since politics cannot rise above the mixed nature of man. "Government is a very rough business," said Sir George Cornewall Lewis to the young Gladstone; "you must be content with very unsatisfactory results." In a world condemned to such results the American political system deserves attention—especially that part of the system which combines compromise with energy, minority rights with government by a majority which may live thousands of miles away. This is the special province of the unwritten constitution.

The written constitution has been unofficially revised, without the change of a word or a comma, in several ways. First, the government which had been planned as a very loose federation grew steadily more centralized. Not even the most power-fearing statesmen could prevent this drift.

Second, the office of the presidency was captured by the emerging democracy—much to the surprise of the Fathers, who thought they had put it beyond the clutches of what they called the "mob." The President, thereafter, was the one man elected by all the voters, so when the country became a thorough democracy the President became the voice of the people. For the most part the members of the Senate and of the House must represent their own states and districts. It is not waywardness which makes them do this; it is the nature of the federal system. The representatives from Delaware do not and should not spend their time serving the interests of Idaho. Yet there may be occasions when the welfare of the national majority conflicts with that of Idaho. It then becomes the duty of the congressmen from Idaho to argue for their own region and to ask for compromise. But the President talks for the nation.

Third, when the President became the voice of the people it was important that his voice carry weight. Under the written constitution, as interpreted from the days of Madison (who came to the White House in 1809) to those of J. Q. Adams (who left it in 1829), not even the most popular President could impose a national policy. The Congress was "a scuffle of local interests," and could make only desultory policy. So the nation drifted: in and out of war, in and out of depression, in and out of sectional strife. Thus came the demand for the third unwritten change, which stabilized the others and saved the system from stagnation: the modern political parties.

These parties are unique. They cannot be compared to the parties of other nations. They serve a new purpose in a new way. Unforeseen and unwanted by the Fathers, they form the heart of the unwritten constitution and help the written one to work.

It is through the parties that the clashing interests of a continent find grounds for compromise; it is through the parties that majority rule is softened and minorities gain a suspensive veto; it is through the parties that the separation of powers within the federal government is diminished and the President is given strength (when he dares use it) to act as tribune of the people; it is through the parties that the dignity of the states is maintained, and the tendency for central power to grow from its own strength is to some extent resisted. It is through the parties, also, that many corruptions and vulgarities enter the national life, and many dubious habits—like that of ignoring issues which are too grave for compromise.

The parties are never static. They are as responsive to shifting conditions as the Constitution of Great Britain. Above all, they are not addicted to fixed ideas, to rigid principles. It is not their duty to be Left, or Right, or Center—but to be all three. It is not their function to defend political or economic doctrines, but to administer the doctrines chosen by the majority, with due regard regard to special interests and to the habits of the regions.

Third parties may preach causes and adhere to creeds. They may educate the public and thus mold history. But the two major parties have a more absorbing, a more subtle, and a more difficult task. The task is peculiar to a federal state covering so much land and containing so many people that its economic and cultural life cannot and should not be merged. Over such an area, where there is no unity of race, no immemorial tradition, no throne to revere, no ancient roots in the land, no single religion to color all minds alike—where there is only language in common, and faith, and the pride of the rights of man—the American party system helps to build freedom and union: the blessings of liberty, and the strength that comes from letting men, money, goods, and ideas move without hindrance.

Although the major parties do not stand for opposed philosophies and do not represent opposed classes, there are traditional differences which divide them. On the whole, it is roughly true that the ancestors of the Democratic Party are Thomas Jefferson and his friends, who made a political alliance in the seventeen-nineties and who came to power in 1801. And it is roughly true that the ancestors of the Republican Party are Alexander Hamilton and his friends, who formed the dominant group in the Cabinets of the first two Presidents (1789–1801).

Broadly speaking, and subject to many qualifications, the Jeffersonians drew their strength from the landed interests (the small farmers and in some cases the great plantation owners) and from the mechanics and manual workers of the towns, whereas the Hamiltonians drew their strength from the business and banking and commercial communities, from the middle-class workers who associated their interests with these communities, and from some of the richest owners of plantations. There are so many exceptions to the above statements that it would be misleading to take them literally; yet in a very general sense they tend to be true. . . . Yet the alignment is insignificant compared to the fundamental forces which make the party system.

Another tendency toward strict party alignment stems from the Civil War—a tendency which cuts across and confuses the alignment which has just been described. The Republican Party was the party of Union during the Civil War, the party of Northern victory. As a result the farmers of the Middle West and of New England, for whom the Union was a sacred cause, long tended to vote the Republican ticket. So did the Negroes, for whom Northern victory meant release from slavery. And so did the veterans of the Union armies, for whom the Republican Party meant generous pensions. And on the other side the states of the Confederacy formed the "solid South" wherein only the Democratic Party was respectable. It is the remnants of such wartime passions which explain the Georgia judge who said, "I shall die a penitent Christian, but meet my Maker as an impenitent Democrat."

Yet in recent years Franklin Roosevelt won the whole of the Middle West for the Democrats, and the whole of New England with the exception of Maine and Vermont. And in two elections he received most of the Negro votes. And in 1928 (under very exceptional circumstances, to be sure) Herbert Hoover, the Republican candidate, received the votes of Virginia, Tennessee, Texas, North Carolina, and Florida—all parts of the "solid South."

The meaning and the purpose of the national parties cannot be sought in traditional alignments or old animosities, and still less in economic or political creeds. They can only be understood in the light of the regional problem created by America's size, and of the constitutional problem created by the separation of powers.

The American political system sometimes fails lamentably. Although on the whole it promotes freedom and union, too many citizens have a sham freedom, and too often the Union is used to press colonial servitude upon entire regions. Yet we must remember, as Dennis Brogan says, that when the Constitution went into effect in 1789 "there was still a King of France and Navarre, a King of Spain and the Indies, a Venetian and a Dutch Republic, an Emperor in Pekin; a Pope-King ruled in Bologna, a Tsarina in Petersburg and a Shogun in Yedo, not yet Tokio and not yet the residence of the Divine Mikado."

The form of government which has weathered the wars, revolutions, and economic collapses ravaging the world during this century and a half must satisfy some deep need of man. And it must have resilience and adaptability. The unwritten constitution supplies the latter qualities.

During Grover Cleveland's first term in the White House, James Bryce published his remarkable book, *The American Commonwealth.* Surveying the party system from the English point of view, and with quiet surprise, he made the classic statement of the difference between the Republicans and the Democrats.

What are their principles [he wrote], their distinctive tenets, their tendencies? Which of them is for free trade, for civil-service reform, for a spirited foreign policy . . . for changes in the currency, for any other of the twenty issues which one hears discussed in the country as seriously involving its welfare? This is what a European is always asking of intelligent Republicans and intelligent Democrats. He is always asking because he never gets an answer. The replies leave him in deeper perplexity. After some months the truth begins to dawn on him. Neither party has any principles, any distinctive tenets. Both have traditions. Both claim to have tendencies. Both have certainly war cries, organizations, interests, enlisted in their support. But those interests are in the main the interests of getting or keeping the patronage of the government. Tenets and policies, points of political doctrine and points of political practice, have all but vanished. They have not been thrown away but have been stripped away by Time and the progress of events, fulfilling some policies, blotting out others. All has been lost, except office or the hope of it.

This is a true description of the parties as they were, and as they still are; but Bryce's explanation of how they came to be that way is misleading. He assumes that if the American parties were healthy they would resemble the parties of Great Britain. They would have "principles" and "tenets," and would thus be forced to take sides on all "the twenty issues that one hears discussed." And he assumes that "Time and the progress of events" have deprived the parties of their principles, leaving them with nothing but "office or the hope of it." But this is too short a view; Lord Bryce was confused by the brief history of the Republican Party, which possessed principles in 1856 and none in 1886. He thought this was a sign of failure and decay; but in fact it was a sign of health: 1856 had been the exception and the danger; 1886 was the reassuring norm.

The purpose—the important and healthy purpose—of an American party is to be exactly what Lord Bryce describes, and by implication deplores. The party is intended to be an organization for "getting or keeping the patronage of government." Instead of seeking "principles," or "distinctive tenets," which can only divide a federal union, the party is intended to seek bargains between the regions, the classes, and the other interest groups. It is intended to bring men and women of all beliefs, occupations, sections, racial backgrounds, into a combination for the pursuit of power. The combination is too various to possess firm convictions. The members may have nothing in common except a desire for office. Unless driven by a forceful President they tend to do as little as possible. They tend to provide some small favor for each noisy group, and to call that a policy. They tend to ignore any issue that rouses deep passion. And by so doing they strengthen the Union.

The decisive American experience—the warning against politics based on principles—took place between 1850 and 1860. A subtle and healing compromise had been effected in 1850; yet year by year, whether through fate or through human folly, it slowly disintegrated. The best men watched in anguish but could not halt the ruin. In the name of principles and distinctive tenets the Whig Party was ground to bits. A new party was born which met Lord Bryce's requirements. The Republicans knew exactly where they stood on the major issue and would not give an inch. Finally, the same "principles" broke the Democratic Party, and the Union of 1789 perished.

The lesson which America learned was useful: in a large federal nation, when a problem is passionately felt, and is discussed in terms of morals, each party may divide within itself, against itself. And if the parties divide, the nation may divide; for the parties, with their enjoyable pursuit of power, are a unifying influence. Wise men, therefore, may seek to dodge such problems as long as possible. And the easiest way to dodge them is for both parties to take both sides. This is normal American practice, whether the issue turns section against section, like "cheap money"; or town against country, like Prohibition; or class against class like the use of injunctions in labor disputes. It is a sign of health when the Democrats choose a "sound-money" candidate for the presidency and a "cheap-money" platform, as they did in 1868; or when they choose a "wet" Eastern candidate for the presidency and a "dry" Western candidate for the vice-presidency, as they did in 1924. It is a sign of health when the Republicans choose a "sound-money" platform but cheerfully repudiate it throughout the "cheap-money" states, as they did in 1868.

A federal nation is safe so long as the parties are undogmatic and contain members with many contradictory views. But when the people begin to divide according to reason, with all the voters in one party who believe one way, the federal structure is strained. We saw this in 1896, during the last great fight for "free silver." To be sure, there remained some "gold Democrats" and some "silver Republicans" in 1896; yet the campaign produced the sharpest alignment on principle since the Civil War. And the fierce sectional passions racked the nation. Luckily, the silver issue soon settled itself, and removed itself from politics, so the parties could relapse into their saving illogicality.

The faults of such irrational parties are obvious. Brains and energy are lavished, not on the search for truth, but on the search for bargains, for concessions which will soothe well-organized minorities, for excuses to justify delay and denial. . . . Every interest which is strong enough to make trouble must usually be satisfied before anything can be done. This means great caution in attempting new policies, so that a whole ungainly continent may keep in step. Obstruction, evasion, well-nigh intolerable slowness—these are the costs of America's federal union. And the endless bartering of minor favors . . . is also part of the price. And so is the absence of a clear purpose whenever the President is weak or self-effacing, since the sum of sectional and class interests is not equal to the national interest, and the exchange of favors between blocs or pressure groups does not make a policy.

Yet no matter how high one puts the price of federal union, it is small compared to the price which other continents have paid for disunion, and for the little national states in which parties of principle can live (or more often die) for their clearly defined causes. And the price is small compared to what America paid for her own years of disunion. The United States, of course, may some day attain such uniformity (or have it thrust upon her) that she will abandon her federal structure; but until that happens she will be governed by concurrent majorities, by vetoes and filibusters, by parties which take both sides of every dangerous question, which are held together by the amusements and rewards of office-seeking, and which can only win an election by bringing many incompatible groups to accept a token triumph in the name of unity, instead of demanding their full "rights" at the cost of a fight.

The world today might do worse than study the curious methods by which such assuagements are effected.

Tweedledum and Tweedledee*

BY FRANKLIN D. ROOSEVELT

[1941]

THE SYSTEM of party responsibility in America requires that one of its parties be the liberal party and the other the conservative party. This has been the division by which the major parties in American history have identified themselves whenever crises have developed which required definite choice of direction. . . .

One great difference which has characterized this division has been that the liberal party—no matter what its particular name was at the time—believed in the wisdom and efficacy of the will of the great majority of the people, as distinguished from the judgment of a small minority of either education or wealth.

The liberal group has always believed that control by a few—political control or economic control—if exercised for a long period of time, would be destructive of a sound representative democracy. For this reason, for example, it has always advocated the extension of the right of suffrage to as many people as possible, trusting the combined judgment of all the people in political matters rather than the judgment of a small minority.

The other great difference between the two parties has been this: the liberal party is a party which believes that, as new conditions and problems arise beyond the power of men and women to meet as individuals, it becomes the duty of the government itself to find new remedies with which to meet them. The liberal party insists that the government has the definite duty to use all its power and resources to meet new social problems with new social controls. . . . That theory of the role of government was expressed by Abraham Lincoln when he said "the legitimate object of government is to do for a community of people whatever they need to have done, but cannot do at all, or cannot do so well for themselves, in their separate or individual capacities."

---

* From the Introduction to Volume 7 of *The Public Papers and Addresses* of *Franklin Delano Roosevelt*, Macmillan, 1941. Reprinted by permission. Franklin Delano Roosevelt (1882–1945) was the thirty-second President of the United States. He served from 1933 to 1945.

The conservative party in government honestly and conscientiously believes the contrary. It believes that there is no necessity for the government to step in, even when new conditions and new problems arise. It believes that, in the long run, individual initiative and private philanthropy can take care of all situations. The text of allegiance to one or the other of these schools of political and economic thought cannot be based on a person's views with respect to one particular measure or policy or even several of them. The test is rather whether a person adheres to the broad general objectives of the particular party as expressed in its fundamental principles. . . .

It is a comparatively simple thing for a nation to determine by its votes whether it chooses the liberal or the conservative form of government. On the other hand, a nation can never intelligently determine its policy, if it has to go through the confusion of voting for candidates who pretend to be one thing but who act the other.

I have always believed, and I have frequently stated, that my own party can succeed at the polls only so long as it continues to be the party of militant liberalism. . . .

. . . I took an active part in some of the primary elections of 1938—in an effort to keep liberalism in the foreground in the councils of my own party, as well as in the legislative and executive branches of the government itself. My participation in these primary campaigns was slurringly referred to by those who were opposed to liberalism as a "purge." . . . Nothing could be further from the truth. I was not interested in personalities. Nor was I interested in particular measures because most of the liberal measures I had recommended had already passed. I was, however, primarily interested in seeing to it that the Democratic Party and the Republican Party should not be Tweedledum and Tweedledee to each other.

Republicans, Democrats: Who's Who?*

BY JAMES MacGREGOR BURNS

[1955]

NOT LONG AGO a London editor was trying to guide his readers through the wilderness of the American party system. There are four parties, he explained—liberal Republicans, conservative Republicans, conservative Democrats and liberal Democrats. The first three parties, he went on, combined to elect Mr. Eisenhower President, and the last three now combine to oppose him in Congress.

The story is pertinent to the news out of Washington. . . . Eisenhower solicits bipartisan support for his program of "progressive moderation." What does this mean? . . . What is a Republican, anyway, and what's a Democrat? . . .

A little more than ten years ago the most successful politician in American history, plagued like Mr. Eisenhower by hostile Senators in his own party, took the first tentative steps toward a party realignment. Franklin D. Roosevelt, according to his intimate adviser, Judge Samuel I. Rosenman, had decided in 1944, after thirty-five years in American politics, that the time was ripe for a party reorganization along liberal-conservative lines. Moreover, he said, he had been told by a mutual friend of his and Wendell Willkie's that Willkie, who had just lost the 1944 Republican nomination, was of like mind.

"We ought to have two real parties—one liberal and the other conservative," the President said to Rosenman. "As it is now, each party is split by dissenters." He added that party realignment

---

* From *The New York Times Magazine*, January 2, 1955. Reprinted by permission. James MacGregor Burns (b. 1918) is Professor of Political Science at Williams College and author of *The Lion and the Fox: A Political Biography of F. D. Roosevelt* (1956).

would take time, but that it could be accomplished. "From the liberals of both parties Willkie and I can form a new, really liberal party in America."

Roosevelt wrote to Willkie, but nothing came of the idea. Neither of them wished to pursue the matter during the 1944 election; Willkie died before the election, and Roosevelt a few months later.

This raises the question of what we mean by "Republican" and by "Democrat." It is fashionable to say that the terms are utterly meaningless. But this is not so. Party platforms and Presidential statements show that *most* Democrats stand for the increased use of government for the broader distribution of social welfare even if it means unbalanced budgets, big government and higher taxes, especially on the rich. They show that *most* Republicans would restrict government in order to give more scope to private initiative and investment, even if this means considerable inequality of income and even some temporary hardship for the mass of people.

This, of course, is a generalization, but a generalization that focuses on the crucial issue separating Democrats from Republicans—the extent to which government should be used to distribute social welfare.

Does this division involve foreign policy too? More and more it does. Today the parties are substantially agreed on some of the great policial and military problems such as recognition of Red China, European unity and opposition to Communist expansion. Where they differ is in the field of foreign economic policies—tariffs, the extent of United States economic aid, the size of Point Four.

If party realignment takes place, then the only logical form would be along the lines of domestic and international economic policy. What are the chances of such a realignment? Political scientists see underlying economic, social and political changes that may make party realignment increasingly possible as time passes. Some of these are:

Increasing urbanization. Fifty years ago most people lived in rural areas; today twice as many Americans live in urban areas as in rural. The population of suburban areas has also increased. Politically, this means more voters who will divide over the problems involved in the Government's relation to wages, prices, taxes and social benefits.

Less sectionalism. Five or six decades ago Americans tended to divide far more on a sectional basis than they do now. Parties were strongly rooted in certain areas and tended to embrace both the liberals and conservatives in those areas. As our politics have become increasingly "nationalized," political divisions have gradually divided voters on an economic and social, rather than on a geographical, basis.

Changes in the South. Economic and social developments are rapidly bringing the South more in step politically with the rest of the country. These are the growth of industry, the diversification and mechanization of farming, urbanization, the political organization of labor and of Negroes. While these changes vary widely from state to state, the South as a whole is moving toward the same horizontal political cleavages that cut across the rest of the population.

In the face of these developments, however, it is well to remind ourselves that somehow the American party system has survived deep-seated changes in the social and economic pattern and still has kept its essential form.

One of the elements ingrained in our system is party organization. Our parties are not centralized agencies that can be easily reformed or "purified" from the top. They are really vast holding companies for thousands of factional groups fighting for local, state and Congressional offices. The men who run the party at the base of the pyramid often have little interest in the fortunes of the party nationally; they are concerned with winning races for District Attorney, Mayor, Sheriff, State Representative, Governor, County Commissioner.

A second factor that will resist quick party change is the sheer weight of habit. Millions of Americans stick to their party year in year out regardless of its changing liberalism or conservatism nationally. For example, in Republican New Hampshire there are little pockets of voters who have cast their ballots for Democrats ever since Jackson's time; "mountain Republicans" in the South have been voting for their party steadily since the Civil War.

A third factor against easy party realignment is the tendency of *both* parties to move in the same direction in the face of a strong trend in public opinion. In the Eighteen Seventies and Eighteen Eighties both parties were essentially conservative parties. These days the Republicans pick national candidates—Willkie, Dewey, and Eisenhower—who take moderately liberal attitudes toward social and economic policies. The crucial changes in party development have often taken place within each of the major parties rather than between them. New England was once the spawning ground of the most conservative Republicans; now it sends to Congress some of the most liberal Republicans.

The seeming paradox of liberals and conservatives in the same party is actually the normal pattern in American parties. Early in this century the Democratic party was split into Bryanites and Eastern conservatives, the Republican party into Roosevelt progressives and Taft stand-patters.

In the Nineteen Twenties the Democrats embraced both Al Smith progressives and conservatives of the John W. Davis stripe and the Republicans had their Hardings and Coolidges but also their "sons of the wild jackass," such as William Borah and George W. Norris. Is it strange that today there are Aiken and Javits Republicans as well as Capehart and Bridges Republicans, that there are Walter George and Richard Russell Democrats as well as Paul Douglas and Averell Harriman Democrats?

Does all this mean that party realignment is out of the question? Many political scientists think not. What we have come to, they suggest, is one of history's crucial periods when the new forces pressing for social change are in a condition of precarious balance with the forces of inertia. The outcome will depend on two elements—the nature of party leadership and the attitude of the rank-and-file.

The leaders of neither party show much disposition to press at the moment for the kind of change that Roosevelt sought. President Eisenhower wants backing for his policy of "progressive moderation," but he has shown little flair for the type of creative political leadership that could reorganize the Republican party, slough off its reactionary wing and yet keep it "slightly right of center."

Democratic leaders seem no more eager for a liberal-conservative realignment. The recent "harmony" meeting of the Democratic National Committee indicates that they would prefer to keep their Southern wing as intact as possible, even if this policy jeopardizes the allegiance of Northern Negroes and other minority groups concerned about civil rights and social welfare.

The attitudes of leaders will change, however, if the attitudes of their followers change, and it is here that important shifts may take place. The disposition of millions of workers and their union leaders to continue to work and to enlarge their influence in the Democratic party holds tremendous implications for the strengthening of liberalism in that party.

The willingness of the executive and white-collar class in the burgeoning industry of the South to turn openly and without embarassment to the Republican party as a vehicle for conservatism in the long run will outweigh century-old emotional attitudes toward the hated Republicans of the North.

Ultimately, party realignment will turn on a rational calculation by intelligent Americans as to whether or not such a realignment on a liberal-conservative basis would benefit the general welfare. What is the case for and against such a realignment?

The essential case against ideological parties is that the present system blurs and softens the political antagonisms which divide Americans. If all people of one mind were in one party, and of the opposite mind in the other, the result, it is said, might be fanaticism and intransigeance. By compromising among diverse groups *within* parties, change can be brought about easily and quietly and without tension or open hostility.

Both parties should reflect the rich diversity of our group life. It is especially important, according to this argument, that the lovers of freedom be represented in both parties, so that party leaders will feel pressure from their own followers on the paramount question of civil liberties.

The opponents of a liberal-conservative party realignment grant that the present system leads to inefficiency, obstruction, and evasion. But this is the price they are willing to pay—the "price of union," Herbert Agar has called it—to maintain underlying unity in a continental nation of many diverse sections, minority groups, national backgrounds, and political attitudes. The Civil War is cited as an example of what follows when parties divide irretrievably over some crucial issue.

What is the other side of the case? How do supporters of party realignment answer these arguments?

They agree that political harmony must be preserved. They fear, however, that "me-too" parties that stand for little will disrupt American politics. If the average American comes to believe that parties will not take relatively definite and clear-cut positions on major issues, he may suspect that democratic government evades issues instead of facing and solving them. Extremist leaders and parties may arise to win the votes of such discontented people, especially in a time of economic crisis, war or long drawn-out "cold war."

Ideological parties need not be sharply divided parties, according to this view. The British party system is cited as one that is separated essentially over ideology, but also one that compels the Conservatives to dampen down their extremists, and the Laborites to hold the doctrinaire Socialists in check.

The case for party realignment rests largely on the view that changing world and domestic conditions are confronting democratic government with new challenges that can be met only if the government is united, purposeful, and efficient. Presidents must work closely with Congressmen and national officials must work closely with state and local officials.

Only a united party system, it is argued, reaching into every area of the nation, can mobilize the political power that will guide and sustain and unite our leaders. And minority and individual rights can thrive best in a nation which is kept productive and strong by unified government.

It is not easy to choose between these two cases. However, many students of politics—including this one—favor party realignment because they are concerned about a central goal of democratic government: responsibility. They fear that the present party hodgepodge confuses the people and cuts them off from direct control of their leaders. An essential element of democratic government, they feel, is the presenting of relatively clear and simple alternatives to the great mass of people who are bewildered by the complexities and confusions of democratic government.

The idea of party responsibility has an even more vital implication. In a time when we desperately need a steady course abroad, we cannot afford politicians who rock the boat with shrill cries for retaliation and adventures. In a day when the crying need at home is for tolerance and harmony in the face of suspicion and witch hunts, we cannot afford politicians who earn their political living by arousing group against group.

If a party wishes to gain power on a platform of Knowlandism abroad and McCarthyism at home, it has the right to try to do so. But a party hardly has the right to beg votes for a moderate and responsible platform at the same time that it exploits the appeal of its political adventurers to malcontents.

One way to curb such political adventurers is to hold their parties responsible for them. But we can hardly hold parties responsible if we allow them to serve as simple holding companies for every group across the political spectrum. To clean house a party must stand for something—otherwise it can hardly know what to keep and what to sweep out of the house.

Ultimately, in short, party realignment means party responsibility, and this in turn means the personal responsiblity of political leaders to the people. As long as we must stake our hopes on the moderation and good sense of the American people, no better test of our party system can be devised.

VII

THE WELFARE STATE

The Case for Social Equality*

BY C. A. R. CROSLAND

[1956]

It still has to be shown that more equality would be a good thing. This cannot now be demonstrated by certain economic arguments which were often used before the war.

At any time up to 1939, the case for greater equality, at least of incomes, seemed self-evident. By making the rich less rich, the poor could be made less poor; and to all those with a social conscience this seemed a sufficient and conclusive argument. It appealed to every humanitarian sentiment, and to ordinary feelings of justice and compassion; while on the intellectual plane it was reinforced by the powerful influence of utilitarian thought. Poverty in the midst of plenty seemed obviously repugnant, and great wealth a disgrace because it appeared the cause of great poverty. To take some caviar from the rich, and distribute it in bread to the poor, was a clear moral imperative.

But we have now reached the point where further redistribution would make little difference to the standard of living of the masses; to make the rich less rich would not make the poor significantly less poor. If we distributed all surtax incomes amongst the working class, the latter would gain by at most a few shillings a week per head; and nobody supposes that even this is possible in practice. The main prop of traditional egalitarianism has been knocked away by its own success. . . .

Nevertheless the case can still rest firmly, as I believe, on certain value or ethical judgments of a non-economic character: on a belief that more equality, even though carrying few implications for the sum of economic satisfaction, would yet conduce to a 'better' society. This I believe to be so for three reasons, relating respectively to the diminution of social antagonism, to social

---

* From *The Future of Socialism*. Published by Jonathan Cape, 1956. Reprinted by permission. C. A. R. Crosland (b. 1918) is an economist and was a fellow of Trinity College, Oxford, from 1947-50. He is a former member of Parliament and a leading younger theoretician of the British Labour Party.

justice, and to the avoidance of social waste. These will be considered in turn.

Extreme inequalities can obviously give rise to antagonism by evoking purely *individual* feelings of frustration, envy, and resentment. In the past, such feelings have usually been associated with glaring inequalities of wealth. Even economists have long realised that high consumption by the rich cannot be treated in isolation as merely giving a certain quantum of satisfaction to the rich consumers; it also has consequences for other people's states of mind. 'The affluence of the rich,' wrote Adam Smith, 'excites the envy of the poor, who are often both driven by want and prompted by envy to invade his possession.' . . .

But it is sometimes said that one is doing something disgraceful, and merely pandering to the selfish clamour of the mob, by taking account of social envy and resentment. This is not so. These feelings exist, amongst people not morally inferior to those who administer such high-minded rebukes; and they are quite natural. It is no more disgraceful to take them into account than many other facts that the politician must attend to—such as the greed of the richer classes, who claim they must have higher monetary rewards and reduced taxation as an incentive to greater effort, patriotism being evidently not enough. If all envy (or all rapacity) could disappear by a wave of the wand, or by the peripatetic performance of Buchmanite plays, then well and good. But as it will not, it is a social fact of cardinal importance; and since it makes society less peaceful and contented, it is wrong not to try and adjust affairs in such a way as to minimise the provocation to it.

In fact, of course, the envy often takes a form which by no extreme of bigoted intolerance could be condemned as cupidity or selfishness, as when it is inspired by inequalities of educational opportunity. The upper and middle classes think it not reprehensible, but the mark of a good parent, to show anxious concern over a child's education and future prospects. They would be unwise, then, to censure the envy of working-class parents for the better education, the wider vistas, and the superior prospects of richer children than their own. . . .

Thus the ethical basis for the first argument for greater equality is that it will increase social contentment and diminish social resentment. Such a statement could be purely descriptive; that is, there could exist such differences in the objective conditions under which individuals lived in two societies that they manifestly constituted a difference in the contentment of those societies. In practice, this degree of objectivity is lacking; and such a statement therefore becomes, partly at least, a value-judgment with a strong recommendatory force. It is justified, first, by the ethical premiss that a contented society is better than a discontented one, and secondly by the judgment that the contentment of the community is an increasing function of the contentment of individuals. It then rests on the hypothesis, which I have argued in this chapter, that some at least of our collective discontents can be traced to social inequality, and would be diminished if that inequality were less: and on the further hypothesis that the consequent gain in contentment would outweigh the diminution in contentment of the present privileged classes. . . .

The second argument rests on a view of what constitutes a 'just' distribution of privileges and rewards. Being in essence a simple moral judgment, it is not susceptible of proof or disproof; it must be accepted or rejected according to the moral predilections of the reader. But there appear to me to be four respects in which existing inequalities offend against social justice; and this is wholly irrespective of whether or not they create resentment.

First, I suppose that most liberal people would now allow that every child had a natural 'right,' as citizen, not merely to 'life, liberty, and the pursuit of happiness,' but to that position in the social scale to which his native talents entitle him: should have, in other words, an equal opportunity for wealth, advancement, and renown. Complete achievement of this is, of course, an unattainable ideal; for the children of talented parents start with a pronounced environmental advantage. But subject only to this, all children can, if the society so decides, at least be given an equal chance of access to the best education. . . .

Secondly, a similar argument applies to the distribution of wealth. An equitable distribution (ignoring deliberately eleemosynary payments) requires first that wealth should be a reward for the performance of a definite service or function, and secondly that all should have an equal chance of performing the function, and so of earning the reward. The highest rewards would then accrue to those who, because they possess skills or services in short supply, can contribute most to national prosperity or enjoyment; provided only that the possession of these skills and services should not be artificially (that is, to a greater extent than can be explained by innate differences in talent) restricted to a privileged few.

This last condition, for the educational reason just mentioned, is not completely fulfilled even in respect of incomes from work. But it is scarcely fulfilled at all in respect of incomes from property. . . .

This would be the case only if all private property were 'earned,' in the sense of representing the individual's own accumulated savings, the fruits of his personal effort and abstinence. But in fact the greater part of it has been inherited; and its distribution is related not to the owner's present or past performance, but to the accident of birth. There is thus no equal opportunity for acquiring it. And it is in addition, as we shall see, most unequally distributed, so that a small upper class of rich citizens all but monopolises the stream of unearned income.

This aspect of inequality is, surely, unjust. It confers on a particular group of fortunate heirs, and denies to the rest of the population, the massive advantage not merely of an additional source of income, and the possibility of capital gains and spending out of capital, but also of security and freedom to take risks; and this they enjoy through no merit of their own, and with no corresponding obligation. And the injustice feeds on itself inasmuch as private capital also makes possible, by the better education which it permits and the subtle social advantage which it confers, a higher occupation and work-income than might have been gained on merit alone.

Thirdly, the greater the inequality, the heavier the concentration of power. Liberals as well as socialists have always disliked the possibility that one individual, or a small group, should wield a dominant and irresponsible power over the lives and fortunes of other individuals. No one has any obvious moral right to such untrammelled power. The temptation to abuse it is great; and it is in any case distasteful and humiliating to adult people to be completely subject to the whims and moods of a single superior. Yet such undemocratic disparities of power may easily follow from large social inequalities (though they may follow from other causes also). They may derive simply from great concentrations of wealth, as with the large private landlord, owning numerous tied cottages, or even whole villages, and perhaps the sole source of local employment and parochial patronage. But authoritarian power to-day stems less commonly from monetary wealth or private ownership than from position in a bureaucratic hierarchy. The top executives in public and private industry wield, in particular, a degree of 'remote' power, and their managerial subordinates a degree of 'face-to-face' power, which, although diminished as compared with before the war, still appears excessive. I believe that social justice would be improved if it were to be still further diminished, and the power of the worker at the point of production correspondingly increased.

Fourthly, rewards from work. No socialist (except for Shaw, and he not in later life) has disputed the need for a degree of inequality here, both because superior talent deserves some rent of ability, and because otherwise certain kinds of work, or risk, or burdensome responsibility will not be shouldered. Thus one should pay differentially high rewards to the artist, the coalminer, the innovating entrepreneur, and the top executive....

How much should be allowed as rent of ability? This is a pleasantly ambiguous concept (though Shaw characteristically defined it as 'the excess of its produce over that of ordinary stupidity'). It could be taken as a noneconomic, normative concept, expressing not the money rent which the community *needs* to pay in order to elicit the ability, but the individual's 'worth' (in some sense) to the community. But if this were made the sole criterion (which of course it never has been, since only those abilities for which a popular demand exists have ever in practice commanded high rents—those of motor tycoons and film stars, not poets or philosophers), it is quite certain that we should get not more equality, but a degree of inequality which would be furiously (and rightly) resented by everyone. This is because the scatter of human ability and inventiveness is far wider than any known scatter, in modern societies, of monetary rewards; the 'worth' to society of a Stevenson, a Faraday, a Ford, a Rutherford, or a Fleming, measured in terms of their contribution to future living standards or the abolition of disease, is not merely twenty times greater than the 'worth' of the rest of us, but some hundreds or thousands of times.

But if we reject 'worth' in this vague sense as the proper criterion, on the grounds that (assuming it to be biologically transmitted) it seems unjust and unwise to reward or penalise people to quite such a prodigious extent for inherited characteristics, we are left with rent of ability as an economic concept: that is, the additional reward which exceptional ability can in practice command from the community.

How large this should be is of course, impossible to lay down in general terms. If we believe in equality, we can only say that we shall balance the possible loss to equality agains the possible gain from exploiting the ability. The balance of loss and gain will depend on the supply price of different grades of ability; this raises the whole question of incentives, about which we still know very little. Some danger point must evidently exist at which equality begins to react really seriously on the supply of ability (and also of effort, risk-taking, and so on), and hence on economic growth. Where exactly this point lies, no one knows. . . .

The third objection to extreme social inequality is that it is wasteful and inefficient. If the determinants of class make deep incisions, and the space of free social movement is restricted, as is the case in Britain (mainly on account of the distinct layers traced by a segregated educational system), two undesirable consequences follow.

First, social intercourse between the classes is markedly inhibited, both by external differences in 'manners' and behaviour, and by subjective consciousness of class. One of the strong attractions of Swedish or American society is the extraordinary social freedom, the relaxed, informal atmosphere, the easier contacts, the natural assumption of equality, the total absence of deference, and the relative absence of snobbery and of that faint, intangible but none the less insistent sense of class that permeates social attitudes in Britain. . . . If social mobility is low, as it must be in a stratified society, and people cannot easily move up from the lower or middle reaches to the top, then the ruling élite becomes hereditary and self-perpetuating; and whatever one may concede to inherited or family advantages, this must involve a waste of talent. . . .

How far towards equality do we wish to go? I do not regard this as either a sensible or a pertinent question, to which one could possibly give, or should attempt to give, a precise reply. We need, I believe, more equality than we now have, for the reasons set out in this chapter. We can therefore describe the direction of advance, and even discern the immediate landscape ahead; but the ultimate objective lies wrapped in complete uncertainty.

This must be the case unless one subscribes to the vulgar fallacy that some ideal society can be said to exist, of which blueprints can be drawn, and which will be ushered in as soon as certain specific reforms have been achieved. The apocalyptic view that we might one day wake up to find that something called 'socialism' had arrived was born of revolutionary theories of capitalist collapse. But in Western societies change is gradual and evolutionary, and not always either foreseeable or even under political control. It is therefore futile and dangerous to think in terms of an ideal society, the shape of which can already be descried, and which will be reached at some definite date in the future. . . .

The Ethics of Redistribution*

BY BERTRAND DE JOUVENEL

[1951]

I PROPOSE to discuss a predominant preoccupation of our day: the redistribution of incomes.

The process of redistribution. In the course of a lifetime, current ideas as to what may be done in a society by political decision have altered radically. It is now generally regarded as within the proper province of the State, and indeed as one of its major functions, to shift wealth from its richer to its poorer members....

Our subject: the ethical aspect. A spirited controversy is now raging on what is termed 'the disincentive effect of excessive redistribution.' It is known from experience that in most cases, though by no means in all, men are spurred by material rewards proportional or even more than proportional to their effort, as for example in 'time and a half.' Making each increase of effort less rewarding than those which preceded it, whilst at the same time lowering, by the provision of benefits, the basic effort necessary to sustain existence, can be held to affect the pace of production and economic progress. Thus the policy of redistribution is subject to heavy fire. The attack, however, is made on grounds of expediency. Current criticism of redistribution is not based on its being undesirable but on its being, beyond a certain point, imprudent. Nor do champions of redistribution deny that there are limits to what can be achieved, if it is proposed, as they wish, to maintain economic progress. This whole conflict of which so much is made today is a borderline quarrel, involving no fundamentals.

I propose to skirt this field of combat and shall assume here that redistribution, however far it may be carried, exerts no disincentive influence, and leaves the volume and growth of production

---

* Selections from Ch. I of *The Ethics of Redistribution,* published by Cambridge University Press, 1951. Reprinted by permission. The contents of this book were first delivered as lectures at Corpus Christi College in Cambridge in 1949. Bertrand de Jouvenel (b. 1903), is a noted French writer on economic and political theory. He has written books on *Power* and on *Sovereignty.*

entirely unaffected. This assumption is made in order to centre attention upon other aspects of redistribution. To some the assumption may seem to do away with the need for discussion. If it were not going to affect production, they will say, redistribution would have to proceed to its extreme of total equality of incomes. This would be good and desirable. But would it? Why would it? And how far would it? This is my starting-point.

Dealing with redistribution purely on ethical grounds, our first concern must be to distinguish sharply between the social ideal of income equalization and others with which it is sentimentally, but not logically, associated. It is a common but ill-founded belief that ideals of social reform are somehow lineal descendants of one another. It is not so: redistributionism is not descended from socialism; nor can any but a purely verbal link be discovered between it and agrarian egalitarianism. It will greatly clarify the problem if we stress the contrasts between these ideals.

Land redistribution in perspective. What was demanded in the name of social justice over thousands of years was land redistribution. This may be said to belong to a past phase of history when agriculture was by far the major economic activity. Yet the agrarian demand comes right down to our own times: did not the First World War bring in its train an ample redistribution of land over all of Eastern Europe? ...

Land redistribution not equivalent to redistribution of income. There is a clear contrast between redistribution of land and redistribution of incomes. Agrarianism does not advocate the equalization of the produce, but of natural resources out of which the several units will autonomously provide themselves with the produce. This is justice, in the sense that inequality of rewards between units equally provided with natural resources will reflect inequality of toil. In other words, the role played by inequality of 'capital' in bringing about unequal rewards is nullified. What is equalized is the supply of 'capital.' ...

Socialism as the City of Brotherly Love. Agrarianism can be summed up under the heading of *fair rewards*. Socialism aims even higher than the establishment of 'mere' justice. It seeks to establish a new order of brotherly love. The basic socialist feeling is not that things are out of proportion and thus unjust, that reward is not proportional to effort, but an emotional revolt against the antagonisms within society, against the ugliness of men's behaviour to each other.

It is of course logically possible to minimize antagonism by minimizing the occasions on which men's paths cross. Thus, the agrarian solution lies in the economic sovereignty of each several owner on his well delimited field, which is equal in size to that of his neighbour. But this is not possible in modern societies, where interests are intertwined as in a Gordian knot. To cut the knot means reversion to a ruder state. But there is another solution: it is a new spirit of joyful acceptance of this interdependence; it is that men, called to serve one another ever increasingly by economic progress and division of labour, should do so 'in newness of spirit,' not as the 'old' man did who grudgingly measured his service against his reward, but as a 'new' man who finds his delight in the welfare of his brethren. . . .

Socialism has singled out private property as the basic 'situation' creating antagonisms: it creates firstly the essential antagonism between those with property and those without, and secondly the struggle among the propertied.

How to do away with antagonism: socialist goal, and socialist means. The socialist solution then is the destruction of private property as such. This is to erase the contrast between men's positions and thereby do away with tension. . . .

It is clear for all to see that the destruction of private property has not done away with antagonisms or given rise to a spirit of solidarity permitting men to dispense with police powers; and it is further apparent that what spirit of solidarity there is seems to have as its necessary ingredient the distrust and hatred of another society, or of another section of society. The warlike intentions of foreign powers seem to be a basic postulate of the collectivist State, and may even be attributed by one collectivist State to another, or, if the process of socialization has not been completed, to the aggressive disposition of the capitalist classes, backed by foreign capitalists. Thus the solidarity obtained is not, as intended, a solidarity of love, but, at least in part, a solidarity in strife. Clearly, this is not consonant with the basic intention of socialism: 'the fruit of righteousness is sown in peace of them that make peace.'

Yet the socialist ideal is not to be summarily dismissed. We do aspire to something more than a society of good neighbours who do not displace landmarks, who return stray sheep to their owner, and refrain from coveting their neighbour's ass. And indeed a community based not upon economic independence but upon a fraternal partaking of the common produce, and inspired by the deep-seated feeling that its members are of one family, should not be called utopian.

The inner contradiction of socialism. Such a community works. It has worked for centuries and we can see it at work under our very eyes, in every monastic community. But it is to be noticed that these are cities of brotherly love *because* they were originally cities built up by love of the Father. It is further to be noticed that material goods are shared without question *because* they are spurned. The members of the community are not anxious to increase their individual well-being at the expense of one another, but then they are not very anxious to increase it *at all*. Their appetites are not addressed to scarce material commodities, and thus competitive; they are addressed to God, who is infinite. In short, they are members of one another not because they form a social body but because they are part of a mystical body.

Socialism seeks to restore this unity without the faith which causes it. It seeks to restore sharing as amongst brothers without contempt for worldly goods, without recognition of their worthlessness. It does not accept the view that consumption is a trivial thing, to be kept down to the minimum. On the contrary it adheres to the fundamental belief of modern society that there must be ever more worldly goods to be enjoyed, the spoils of a conquest of nature which is held to be man's noblest venture. The socialist ideal is grafted on to the progressive society and adheres to this society's veneration of commodities, its encouragement of fleshly appetites and pride in technical-imperialism.

The moral seduction of socialism lies in the fact that it repudiates the methodical exploitation of the personal interest motive, of the fleshly appetites, of egoism, which held pride of place in the economic society it has undertaken to supersede; yet that, in so far as it has endorsed this society's pursuit of ever-increasing consumption, it has become a heterogeneous system, torn by an inner contradiction.

If 'more goods' are the goal to which society's efforts are to be addressed, why should 'more goods' be a disreputable objective for the individual? Socialism suffers from ambiguity in its judgment of values: if the good of society lies in greater riches, why not the good of the individual? If society should press towards that good, why not the individual? If this appetite for riches is wrong in the individual, why not in society? . . .

Further, so long as the general purpose of society is the conquest of nature and the enjoyment of its spoils, is it not logical that this purpose should determine the characteristics of that society? Is not society shaped by its predominant desire, by the end towards which it tends? Is it not possible that many unpleasant traits of society are functionally related to its basic purpose? And is not their unpleasantness inherent in the purpose, so that any different society one seeks to build up with the same purpose must display the same characteristics, possibly under a different guise?

The productivist society may be likened to the military society. That which is meant for war must in its structure show characteristics appropriate to war. An army, or a military society, embodies many traits which are indefensible, by the standards of a 'good society.' But the military hierarchy and discipline cannot be done away with as long as victory remains the purpose—though of course they can be amended. In the same manner, there may be a relation between the structure of productivist society and its purpose. And there is much to be said for the view that socialism's higher aspirations were doomed when it accepted the general purpose of modern society. . . .

Redistribution and the scandal of poverty. What has now come to the fore, as against the ideal of fair rewards, and brotherly love, is the ideal of more equal consumption. It may be regarded as compounded of two convictions: *one,* that it is good and necessary to remove want and that the surplus of some should be sacrificed to the urgent needs of others, and *two,* that inequality of means between the several members of a society is bad in itself and should be more or less radically removed.

The two ideas are not logically related. The first rests squarely upon the Christian idea of brotherhood. Man is his brother's keeper, must act as the Good Samaritan, has a moral obligation to help the unfortunate, an obligation which rests most heavily, though not exclusively, upon the most unfortunate. There is, on the other hand, no *prima facie* evidence for the current contention that justice demands near equality of material conditions. Justice means proportion. The individualist is entitled to hold that justice demands individual rewards proportionate to individual endeavours; and the socialist is entitled to hold that it demands individual rewards proportionate to the services received by the community. . . .

It is however a loose modern habit to call 'just' whatever is thought emotionally desirable. Attention was legitimately called in the nineteenth century to the sorry condition of the labouring classes. It was felt to be wrong that their human needs were so ill-satisfied. The idea of proportion then came to be applied to the relation between needs and resources. Just as it seemed improper that some should have less than what was adjudged necessary, so it also seemed improper that others should have so much more.

The first feeling was almost the only one at work in the early stage of redistributionism. The second has gained almost the upper hand in the latter stage. . . .

Socialists, at the inception of the move towards redistribution, took rather a disdainful attitude; the initial measures were in their eyes mere bribes offered to the working classes in an attempt to divert them from the higher aims of socialism. . . .

While it is difficult for men to imagine the suppression of private property, that is, of something that all desire, it is natural to them to compare their condition with that of others; the poorer can easily imagine the uses to which they would put some of the riches of others, and the richer, if once awakened to the condition of the poorer, are bound to feel some remorse on account of their luxuries.

At all times the revelation of poverty has come as a shock to the chosen few: it has impelled them to regard their personal extravagance with a sense of guilt, has driven them to distribute their riches and to mingle with the poor. In every case one knows of in the past, this has been associated with a religious experience: the mind may have been turned to God by the discovery of the poor, or to the poor by the discovery of God: in any case the two were linked, and a revulsion away from riches as evil was always implied.

However, in our century the feeling that has assailed not merely a few spirits but practically all the members of the leading classes has been of a different kind. Upon a society inordinately proud of its ever increasing riches it dawned that 'in the midst of plenty,' as the saying went, misery was still rife; and this called for action to raise the standard of the poor. While the discovery of poverty, coupled with an assumption of the impossibility of removing it, had formerly brought about a revulsion against riches, this time a deep-rooted appreciation of worldly goods, coupled with a sense of power, caused an onslaught on poverty itself. Riches had been a scandal in the face of poverty; now poverty was a scandal in the face of riches. . . . To the pace-making middle classes, profoundly committed to the religion of progress, the existence of poverty was not only emotionally but intellectually disturbing: in the same manner as is the existence of evil to the simpler sort of deist. The increasing goodness of civilization, the increasing power of man, were to be finally demonstrated by the eradication of poverty.

Thus charity and pride went hand in hand. . . .

The notions of relief and of lifting working-class standards merged. We must, however, note that redistribution appears as a novelty only in contrast to the practices immediately preceding it and in the choice of its agent, the State. It is inherent in the very notion of society that those in direct want must be taken care of. The principle is applied in every family and in every small community, and in fact went out of practice only a few generations ago as a result of the disruption of smaller communities by the Industrial Revolution. This caused the isolation of the individual, and the new 'master' he acquired did not regard himself as bound to him by the same ties as the former lord. It is characteristic that the feasts of consumption of the landed class were feasts *for all,* whereas the consumption of the rich in the new era is purely selfish. It is moreover almost needless to point out that the Church, when it enjoyed enormous gifts from the powerful and the rich, was a great redistributive agency. Between the old customs and the age of the welfare state stretch the 'hard times,' when the individual was left helpless in his need.

This cannot be ascribed to lack of feeling in generations which were fired with sympathy for slaves, for oppressed nationalities, and with indignation at the news of the 'Bulgarian atrocities.' . . .

But perhaps there was some confusion between two different notions: *one*, that the situation of the 'median' worker is best improved by the play of productive forces; and *two*, that there is no call to take care of an unfortunate 'rearguard.' . . . As long as emphasis was laid on the raising of the median by the processes of the market, there was reluctance to intervene on behalf of the unfortunate (compare the attitude of the American Federation of Labor in the first years of the Great Depression), while as soon as attention was focussed upon this rearguard, it came to be held that the median condition was also to be raised by political measures.

While relief is an unquestionable social obligation which the destruction of neighbourliness, of responsible aristocracies and of Church wealth has laid on the State for want of any other agency, it is open to discussion whether policies of redistribution are the best means of dealing with the problem of raising *median* working incomes, whether they can be effective, and whether they do not come into conflict with other legitimate social objectives.

The distinction drawn here is admittedly a difficult one. The two things are confused in practice, and it is not always clear to which end the enormous social machinery set up in our generation is actually working; this creation of ours presents a structure not easily amenable to our intellectual categories. When, through the working of the social services, a man in actual want is provided with the means of subsistence, whether it be a minimum income in days of unemployment, or basic medical care for which he could not have paid, this is a primary manifestation of solidarity. And it does not come under redistribution as we understand it here.

What does come under redistribution is everything which relieves the individual of an expenditure that he could and presumably would have undertaken out of his own purse, and which, freeing a proportion of his income, is therefore equivalent to a raising of this income. A family which would have bought the same amount of food at non-subsidized prices and gets it so much cheaper, an individual who would have sought the same medical services and gets them free, see their incomes raised. And this is what we want to discuss. . . .

The floor and the ceiling. Intellectual harmony and financial harmony. We now need a terminology which we shall keep within modest bounds. We call *floor* the minimum income regarded as necessary and *ceiling* the maximum income regarded as desirable. We call floor and ceiling 'intellectually harmonious' in so far as they are the floor and ceiling acceptable to the same mind or minds. Further, we shall call a floor and ceiling 'financially harmonious' in so far as there is sufficient surplus to be taken from 'above the ceiling' incomes to make up the deficiency in 'beneath the floor' incomes....

Redistributionism is a spontaneous feeling. And in its more naïve forms it carries with it an implied conviction that the floor and ceiling which are intellectually harmonious will also prove to be financially harmonious. This, like so many spontaneous assumptions of the human mind, is an error....

A discussion of satisfactions. Redistribution started with a feeling that some have too little and some too much. When attempts are made to express this feeling more precisely, two formulae are spontaneously offered. The first we may call objective, the second subjective. The objective formula is based upon an idea of a decent way of life beneath which no one should fall and above which other ways of life are desirable and acceptable within a certain range. The subjective formula is not based upon a notion of what is objectively good for men but can be roughly stated as follows: 'The richer would feel their loss less than the poorer would appreciate their gain'; or even more roughly: 'A certain loss of income would mean less to the richer than the consequent gain would mean to the poorer.'

Here a comparison of satisfactions is made. Can such a comparison be rendered effective? Can we with any precision come to weigh losses of satisfaction to some and gains of satisfaction to others?...

As we are still ruled by [the] principle that satisfactions and dissatisfactions of different persons are not commensurable, one falls back upon the mode of measurement which effectively prevails. It is not to be proven that the sum of individual satisfactions of people benefited is greater than the sum of dissatisfactions of people despoiled. In fact there is every reason to believe that if what is taken from a number of people were distributed among an equal number of people, the latter would gain less total satisfaction than the former were losing. But the fact is that the takings are distributed among a far greater number of people. And there will be more people pleased than displeased, more positive signs than negative: and as the intensity of the values is not to be measured, all one can do is state that there are more positive signs than negative and take the result as a gain; which is what in fact is currently done. . . .

Let us however notice a certain consequence of equalization, valid in whatever future we care to place the completion of reform. Let us grant that any differences in tastes due to social habits have been erased. Men will not however be uniform in character; some differences in tastes must exist among individuals. Economic demand will not any more be weighted by differences in individual incomes that will have been abolished: it will be weighted solely by numbers. It is clear that those goods and services in demand by greater collections of individuals will be provided to those individuals more cheaply than other goods and services wanted by smaller collections of individuals will be provided to these latter. The satisfaction of minority wants will be more expensive than the satisfaction of majority wants. Members of a minority will be discriminated against.

There is nothing novel in this phenomenon. It is a regular feature of any economic society. People of uncommon tastes are at a disadvantage for the satisfaction of their wants. But they can and do endeavour to raise their incomes in order to pay for their distinctive wants. And this by the way is a most potent incentive; its efficiency is illustrated by the more than average effort, the higher incomes and the leading positions achieved by racial and religious minorities; what is true of these well-defined minorities is just as true of individuals presenting original traits. Sociologists will readily grant that, in society where free competition obtains, the more active and the more successful are also those with the more uncommon personalities.

If, however, it is not open to those whose tastes differ from the common run to remedy their economic disadvantage by an increase in their incomes, then, in the name of equality, they will be enduring discrimination.

Four consequences deserve notice. Firstly, personal hardships for individuals of original tastes; secondly, the loss to society of the special effort these people would make in order to satisfy their special needs; thirdly, the loss to society of the variety in ways of life resulting from successful efforts to satisfy special wants; fourthly, the loss to society of those activities which are supported by minority demands....

Can we reconcile ourselves to the loss suffered by civilization if creative intellectual and artistic activities fail to find a market? We must if we follow the logic of the felicific calculus. If the two thousand guineas heretofore spent by two thousand buyers of an original piece of historical or philosophical research, are henceforth spent by forty-two thousand buyers of shilling books, aggregate satisfaction is very probably enhanced. There is therefore a gain to society, according to this mode of thought which represents society as a collection of independent consumers. Felicific calculus, counting in units of satisfactions afforded to individuals, cannot enter into its accounts the loss involved in the suppression of the piece of research. A fact which, by the way, brings to light the radically individualistic assumptions of a viewpoint usually labelled socialistic.

In fact, and although this entails an intellectual inconsistency, the most eager champions of income redistribution are highly sensitive to the cultural losses involved. And they press upon us a strong restorative. It is true that individuals will not be able to build up private libraries; but there will be bigger and better and ever more numerous public libraries. It is true that the producer of the book will not be sustained by individual buyers; but the author will be given a public grant, and so forth. All advocates of extreme redistribution couple it with most generous measures of state support for the whole superstructure of cultural activities....

The fact that redistributionists are eager to repair by State expenditure the degradation of higher activities which would result from redistribution left to itself is very significant. They want to prevent a loss of values. Does this make sense?...

Surely, when we achieve the distribution of incomes which, it is claimed, maximizes the sum of satisfactions, we must let this distribution of incomes exert its influence upon the allocation of resources and productive activities, for it is only through this adjustment that the distribution of incomes is made meaningful. And when resources are so allocated, we must not interfere with their disposition, since by doing so we shall, as a matter of course, decrease the sum of satisfactions. It is then an inconsistency, and a very blatant one, to intervene with state support for such cultural activities as do not find a market. Those who spontaneously correct their schemes of redistribution by schemes for such support are in fact denying that the ideal allocation of resources and activities is that which maximizes the sum of satisfactions. . . .

Economists as such are interested in the play of consumer's preferences through the market, and in showing how this play guides the allocation of productive resources so that it comes to correspond with the consumer's preferences. The perfection of this correspondence is general equilibrium. It is perfection of a kind: and it is quite legitimate to speak of such allocation of resources as the best, it being understood that it is the best from the angle of subjective wants, weighted by the actual distribution of incomes. . . . Calling it the best without qualification implies a value judgment which equates the good with the desired, on Hobbesian lines. Now it is quite legitimate for the economist to deal only with the desired and not with the good. But it is not legitimate to treat the optimum in relation to desires as an optimum in any other sense. And that the allocation of resources in relation to desires should fail to be optimal by other standards should not come as a surprise to us.

That a society which we may assume to have maximized the sum of subjective satisfactions should, when we survey it as a whole, strike us as falling far short of a 'good society,' could have been foreseen by anyone with a Christian background or a classical education.

To the many, however, who were apt to think so much in terms of satisfactions that the 'badness' of society seemed to them due to the uneven distribution of satisfactions, it must come as a most useful lesson that the outcome of this viewpoint leads them into an unacceptable state of affairs. The error must then lie in the original assumption that incomes are to be regarded solely as means to consumer-enjoyment. In so far as they are so regarded, the form of society which maximizes the sum of consumer-enjoyments should be best: and yet it is unacceptable. It follows that incomes are not to be so regarded.

Redistributionism the end result of utilitarian individualism. There is no doubt that incomes are currently regarded as means to consumer-enjoyment, and society as an association for the promotion of consumption. This is made clear by the character of the controversy now proceeding on the theme of redistribution. The arguments set against one another are cut from the same cloth. It is fair, some say, to equalize consumer-satisfactions. It is prudent, the others retort, to allow greater rewards to spur production and thereby provide greater means of consumption.

There is an Armenian proverb: 'The world is a pot and man a spoon in it.' In this image our two sides might choose slogans: an expanding pot with unequal spoons, or a static and possibly declining pot with equal spoons. But perhaps the world is not a pot and surely man is not a spoon. Here we have completely slipped away from any conception of the 'good life' and the 'good society.' It is quite inadmissible to consider the 'good life' as a buyer's spree or the 'good society' as a suitable queueing up of buyers. . . .

Redistributionism takes its cue wholly from the society it seeks to reform. An increased consuming power is the promise held out, and fulfilled, by capitalist mercantile society: so is it the promise of the modern reformer. And in fact the choice of right or left is to be finally regarded as not an ethical choice at all, but a bet. Taking, say, the period 1956-65, do we bet that redistributionism with its probable negative effect on economic progress will provide a majority with a higher standard of living than capitalism with its inequality? Or do we put our money—it seems the proper term—on the other horse?

There is no question of ethics here. The end-product of society is anyhow taken to be personal consumption: this is, under socialistic colours, the extremity of individualism. Finally my probable consumption under one or the other system is to be my criterion. Nothing quite so trivial has ever been made into a social ideal. But it is wrong to accuse our reformers of having invented it: they found it.

What is to be held against them is not that they are utopian, it is that they completely fail to be so; it is not their excessive imagination, but their complete lack of it; not that they wish to transform society beyond the realm of possibility, but that they have renounced any essential transformation; not that their means are unrealistic, but that their ends are flatfooted. In fact the mode of thought which tends to predominate in advanced circles is nothing but the tail-end of nineteenth-century utilitarianism.

Security and Freedom*
BY HERBERT H. LEHMAN

[1950]

... IT HAS BECOME fashionable in circles of political reaction to attack the concept of the welfare state as being prejudicial to individual liberty and freedom. These reactionaries view with fright and alarm the current and proposed activities of government in the fields of housing, health, and social security.

"These are steps on the road to communism," the alarmists cry. But . . . I could cite laws and programs by the score enacted over the violent opposition of the reactionaries—laws and programs which were assailed as communistic at the time—but which are now accepted even in the most conservative circles.

This cry of state tyranny has been raised during the last half-century whenever the community has attempted to interfere with the right of a few to destroy forests, exploit little children, operate unsanitary and unsafe shops, indulge in race or religious discrimination, and pursue other policies endangering the health, safety and welfare of the community. These few have completely ignored the fact that, when their license to exploit the community was restricted, the freedom of the many from ignorance, insecurity, and want—the freedom of the many to live the good life—was measurably enhanced.

I do not believe that our Federal Government should seek to assume functions which properly belong to the individual or to the family, to the local community, or to free organizations of individuals. But I do believe that our Federal Government should and must perform those functions which, in this complex and interdependent society, the individual, the family, or the community cannot practicably perform for themselves.

Today we in America and in the entire freedom-loving world are confronted with a world-wide threat to that principle which

---

* From an Address to the League for Industrial Democracy, New York, April 15, 1950. Reprinted by permission. Herbert H. Lehman (b. 1878), was Governor of the State of New York, 1932–1942, and United States Senator from New York, 1950–1956.

we hold most dear, the principle of individual dignity and of individual freedom. For the preservation of that principle we are willing to dedicate our lives, if it should prove necessary. But while this is a threat which we face on the world front, we face another danger here at home. That is the threat to our freedom from those within our own country who would identify individual freedom with special privilege. Any move to diminish privilege, to stamp out discrimination and to bring security to our citizens is branded by these people as un-American.

Not so long ago an American political leader said that "the governments of the past could fairly be characterized as devices for maintaining in perpetuity the place and position of certain privileged classes. The government of the United States, on the other hand, is a device for maintaining in perpetuity the rights of the people, with the ultimate extinction of all privileged classes." Was it some Communist, some irresponsible radical or reformer who made that statement? No, it was not. It was the late President Calvin Coolidge in a speech at Philadelphia in 1924.

It is my firm belief that the extinction of special privilege is an essential and basic program of the welfare state. Today the forces of special privilege provide the chief opposition and raise the wildest cries of alarm against economic security for all.

In addition to the forces of special privilege who are opposed, on principle, to all social legislation, there are some who, while paying lip service to liberalism, claim to be troubled by the expanding scope of government in its direct concern with the welfare of the individual citizen. These people, while conceding merit to the specific programs of the welfare state, and while approving the welfare state programs of the past, join with the forces of privilege in contending that if the government provides any further services, it is moving in the direction of totalitarianism.

In my opinion these men of little vision have lost sight of the most important, and to me the most obvious, truth of our times—that a government which has secured the greatest degree of welfare for its people is the government which stands most firmly against totalitarianism. The critics of the welfare state do not understand this simple fact. They spend their time looking for Communists in and out of government and at the same time attack those measures which would deprive Communists and would-be Communists of their ammunition—and of their audience. The measures which would provide for the welfare of the people are the surest weapons against totalitarianism.

The Communist international, its leaders, and their philosophy, have been responsible for many designs which we in the democratic world consider the quintessence of evil. Certainly the suppression of basic rights—the police state and the slave labor camp—constitute the most repulsive and obnoxious way of life we can imagine.

But, as a liberal, I have a *special* resentment against the Communists. I feel that one of their greatest disservices to the cause of human progress has been their identification of economic security with the suppression of freedom. It is their claim that in order to achieve the solution of the economic needs of the many, it is necessary to curb the freedoms of all. They say, in effect, that you cannot have a full stomach and a free mind at the same time.

I reject this concept! I reject it as being the ultimate in reaction. This is but another demonstration of the basic affinity between Communists and reactionaries in their thinking about man and his problems. *Both* groups believe that a nation of free men cannot possibly conquer the scourges of hunger, disease, lack of shelter, intolerance and ignorance. And they *both* have much to gain if they convince enough people that freedom and security are incompatible.

It is a strange paradox that the same conservatives and reactionaries who pose as champions of national security express the greatest antagonism toward individual security. Most of us readily acknowledge that the nations of the world cannot be free if they are not secure. It seems equally logical to me that *individuals* cannot be free if they are beset by fear and insecurity. To my mind the welfare state is simply a state in which people are free to develop their individual capacities, to receive just rewards for their talents, and to engage in the pursuit of happiness, unburdened by fear of actual hunger, actual homelessness or oppression by reason of race, creed, or color.

The fear of old age, the fear of sickness, the fear of unemployment, and the fear of homelessness are not—as some would have us believe—essential drives in a productive society. These fears are not necessary to make free competitive enterprise work. The fear of insecurity is rather a cancer upon free competitive enterprise. It is the greatest threat which confronts our economic system. I hasten to add that I believe in free competitive enterprise. I believe it is the best system yet devised by man. But it is not a goal in itself. It must always serve the public interest.

We have had [nearly] twenty years of the New Deal and the Fair Deal. Who would say that the American worker, the American farmer and the ordinary American businessman is less free than he was twenty years ago? Actually, freedom in the true sense flourishes more generally and more widely today than ever before in our history. The worker, the farmer and the businessman have vastly more freedom than they ever had before. They are freer to enjoy the fruits and benefits of a productive economy and a full life. But they are not yet free enough.

We are still far from the goal we seek. Insecurity still haunts millions. Inadequate housing poisons the wells of family life in vast numbers of cases. Inadequate schooling handicaps a great segment of our people. And the fear of sickness and old age still clutches at the hearts of many if not most of our fellow citizens. Until we solve all these problems and quiet all these fears, our people will not be truly free.

Three Ways to Live Under the Welfare State*

BY SIR ALEXANDER GRAY

[1949]

WE ARE moving, as we all know, to a world very different from nineteenth century Victorianism. . . . I need not tell you what are the promised characteristics of this new age, and of this new state. It is what, for convenience, is quite usefully described as the welfare state, a state responsive to the material needs of all its subjects; providing complete and adequate security against all the sinister contingencies of life; abolishing want and unemployment; giving education up to university standard (and beyond) to all who desire to avail themselves of it. In short it will be what I recently saw described somewhat derisively . . . as the Santa Claus state. I have also heard it depicted, with even more derision and cattiness, as a world in which the entire population will have breakfast in bed. It will also be a world in which the state, directly or at one remove, assumes responsibility for the conduct of the major basic industries; a world also in which economically the individual will have largely disappeared. Society indeed has already undergone a process of coagulation, so that, if we count for anything at all, we count solely as a member of our appropriate group. Moreover, let us not forget that these groups may have conflicting interests.

Now in such a changed world where many things, if not all things, will have been made new, the politico-economic problem may be very different from that to which we older people have been accustomed. By the politico-economic problem I mean the

---

* From "Economics: Yesterday and Tomorrow," Presidential address delivered before the Economics Section of the British Association for the Advancement of Science, Newcastle upon Tyne, September 1949. *Advancement of Science*, vol. 6, pp. 233-43. October 1949. Reprinted by permission. Sir Alexander Gray (b. 1882) is Professor of Political Economy and Mercantile Law, University of Edinburg. He is the author of *The Socialist Tradition. Moses to Lenin* (1946).

problem of how to live together (always a difficult matter) and how to keep things going. It may seem a perverse thing to say; but just as life in heaven (or in some heavens) may not be altogether easy, so possibly life in the agreeable world of the future may present peculiar difficulties of its own.

There are, I think, three ways in which we may manage to live together in the complete welfare state; or perhaps it would be better to say that there are three preliminary conditions, any one of which, if satisfied, would enable us to do the trick.

1. *Compulsion:* The first is that of relying on a degree of compulsion vastly greater than we have yet had the courage or the honesty to admit may be necessary. I am not now pointing out the horrors of the road to serfdom, or of what awaits you when you come to the end of the road. In its higher altitudes this is already a well-discussed topic. I confine myself to the superficial and indeed the platitudinous. A state cannot undertake to provide from under the counter whatever anyone may need unless simultaneously it sees that someone is putting under the counter what is required for the purpose. The state cannot promise every school child a glass of milk at eleven o'clock, unless it has directly or indirectly the corresponding number of cows standing at command. May I refer you to John Ruskin, a writer whom ordinarily I would not commend for his economic insight? "Finally I hold it for indisputable that the first duty of a state is to see that every child born therein shall be well housed, clothed, fed and educated, till it attains the years of discretion." (So far, an excellent definition of the welfare state, even if its concern with welfare is limited to the years of indiscretion.) "But," he goes on, "in order to the effecting this the government must have an authority over the people of which we now do not so much as dream." It ought to be fairly obvious that the state cannot guarantee everyone against want, unless it reserves the right, if need arises, to take anyone forcibly and pack him off to Caithness or Cornwall to do whatever requires being done there. And sooner or later the time may, probably will, come when it will have to realize that it must not be too mealy mouthed or timorous about the exercise of these powers of compulsion with which, on Ruskin's view, the welfare state must arm itself. The state cannot give what is not there. Indeed, in a sense the state, of itself, cannot guarantee anything or any standard of

life. It is an old criticism of the anarchists that the state is forever sterile. Properly understood, it is an entirely true statement. It can act only through its subjects; and in this matter of distribution, it can only redistribute what its nationals produce or what can be got from other nations in exchange for their products. And if plague comes, fortified by the Colorado beetle, potato disease, foot and mouth disease, blockade by the enemy and all the other horrors in the Malthusian repertory, a government guarantee of a standard of life will get you nowhere. The power of the government to give is forever limited by what the people themselves produce.

Indeed in this matter I am inclined to carry my pessimism still further. It is the tritest and most hackneyed of platitudes that rights must forever be accompanied by duties; but though we invariably pay lip service to the well-worn dictum, in fact our eyes in these days are morbidly fixed on our rights, whereas our duties, after a vague and perfunctory wave of the hand in their direction, are allowed to fade into the background. The Universal Declaration of Human Rights is in this respect an illuminating document. Now in the economic field a right is something that you get from someone else, whereas a duty is what we do for another. And a society in which each member concentrates on getting rather than on giving has lost the roots of its stability.

If then, in the words of the Universal Declaration of Human Rights, the welfare state is going to provide every one "a standard of living adequate for the health and well-being of himself and of his family, including food, clothing, housing" and much more, it must, in the last resort, have power to compel its subjects to see to it that the national bins are kept full, and that there are ample reserves under the counters of the national stores. . . .

II. *Incentive:* So much for the first method of meeting the future, the method of compulsion and direction—that degree of authority over the people of which John Ruskin alone was capable of dreaming. You do not like it? No more do I, though I am of a more submissive disposition than most. But economically, it might be an efficient system, if the rest of the population were as submissive as I am, which probably they are not. From all we know of the Incas and the Jesuit settlements, it might be for certain placid peoples a highly efficient system indeed, producing on the material level a remarkably high degree of comfort and well-being. If compulsion is to be condemned, it must be on moral rather than on economic grounds; it is to be abhorred, above all, because it involves a denial of personality and an abrogation of responsibility.

But if, disliking the idea of a world resting on compulsion, you ask for an alternative which will preserve our free society, I suggest, as the second of my possible devices for the future, that you might consider what can be done towards a solution of the age-long question of incentive. And in some way this is the most urgent of all our industrial problems, for never . . . has the clarion-call to work sounded so insistently as today. The idea of "Incentives," however, seems to involve the acceptance of the view that the natural man does not love work, or that he does not love it or endure it cheerfully, except in moderate doses. Yet apparently this innocuous proposition seems to be one which it is rather dangerous to advance in certain circles. For moralists tell us that we ought to love our work, and tend to think that if we have not yet reached that stage, the fault lies not in man, but in the organization of labor. . . . As you may have discovered in your miscellaneous reading, any suggestion that mankind as a whole have in their makeup a something which leads them to regard work as no more than a second-best way of passing the time, is at times roundly described as a foul slander on our fellow-men. . . .

Work, for the present, then, I assume to be something which soon comes to be avoided, except for a small happy section of the community whose work and whose play merge into each other—the small body of insincere cranks who in *Who's Who* describe their recreation as "More Work." We work because we must; we consent to do more work, because of additional inducements and incentives. But in the new world to which we are moving the question of incentive, of overcoming man's disinclination to work, will of necessity become vastly more difficult, just because the older incentives in their harsher form must become enfeebled as we seek to guarantee security and a reasonable standard of life in all circumstances.

Doubtless the problem of incentives should be considered along with the allied question of the removal of disincentives, if we may lapse into the barbarous jargon of these times. I am not sure that the authorities realize how powerfully in certain circumstances the present income tax arrangements operate to restrict effort. The fact that payments in respect of overtime may bring into the group that pays income tax a worker who would otherwise be exempt leads him to assign the whole of the tax exclusively to that portion of his earnings that comes from the extra hours worked. He complains that whereas he has from time immemorial been paid time-and-a-half for work outside normal hours, he is now being fobbed off with three-quarters time: he is in fact being paid at a lower rate for his overtime than for his ordinary day's work. There is of course a catch in it; but if you discuss the matter with him, unless you are a very good dialectician, he has a fair chance of persuading you that he is right.

Waiving the question of the removal of such "disincentives" which I mention merely to complete the picture, it remains true and regrettably true, that the only effective incentives are of a material character, with an appeal to the individualistic and competitive instincts of mankind which we are supposed to be eradicating. You may ring the changes on higher rates of pay and ingenious bonus payments; you may bring in shorter hours; but you are still moving on the material plane and appealing in one way or another to the desire of gain and comfort. Nor do you escape (entirely, or at all) this material character by giving better seats at the opera, or tours to the northern capitals, on the *Kraft durch Freude* [Strength through Joy] principle. And as I said, precisely because we are moving into a better world, this problem of incentive must become more difficult; it may indeed turn out to be the acid test of the stability of our future economy. We have the firmest assurances that the government will in future accept the task of maintaining full employment, though doubtless "full employment" may be variously defined. The testing time which will reveal whether the state has at all times and in all circumstances the power to fulfill this undertaking still lies ahead. But if full employment is defined, as it sometimes has been defined, as a world in which there are more jobs than men, then I tremble as in the presence of a nightmare. . . .

In a world where there are *more* jobs than men, it is fairly obvious that two things will happen. Firstly, and by definition, a number of jobs will remain unfilled, and it is not difficult to say which jobs these will be. [Men] . . . will avoid those tasks which, for any reason, are regarded as unattractive, unpleasant or unduly arduous, though it may quite possibly be that without these a city may not stand. And simultaneously, higher up the scale, employers in desperation will spend their days enticing away the employees of others by the offer of higher wages. . . .

It is sometimes instructive to see on a small scale what may later happen on a larger. Today you have a fairly good illustration of a world of more jobs than men, if you take the market for domestic service. Domestic servants, so far as they survive, having read their Bernard Shaw, make a bee-line for widowers' houses—elderly widowers, with no children about the place. I do not blame them; I should do the same myself. An elderly widower who spends all his evenings at the club provides a haven of peace compared to anything that can be offered by a harassed young woman with four children. Also the elderly widower is more squeezable in money matters. At present, accordingly, the survivors of the race of domestic servants flock to the houses of those who in many cases might be better dead, and avoid those homes where the need is the sorest. This is almost a parable of what would take place in a world of more jobs than men; and the only possible remedy is clear. If there is a danger that certain essential tasks may remain undone because no one is willing to undertake them, the state, even if professedly shunning compulsion, would have to devise means of coercion. And so by another route, you are brought back to compulsion which we have already considered. . . .

Am I perverse in sometimes feeling that today we are tending to be far too impatient of any suggestion of inequality in any sphere of life? I am still enough of a Saint-Simonian (and for that matter, a Fourierist) to consider that it is only by admitting a certain measure of inequality, something of the nature of a hierarchy, that we shall be able to get anywhere at all, or be able to make the machine march. The only place where there is absolute equality (just as it is the only place where there is absolute security) is in the grave. There is, I think, no harm in a certain degree of inequality, on two conditions: firstly, that the resultant disparity should not be offensive; and secondly, that there should be no barrier in the way of the somewhat-less-favored of today becoming the somewhat-more-favored of tomorrow by their own efforts, and (admittedly) by such an admixture of luck as you will never eliminate from life. Life always has been, and always will be, something of the nature of a race; and the young at least would have it so. But there is not much fun in taking part in, or in watching, a race where in advance the umpires impose handicaps which will effectively ensure that all the competitors will arrive simultaneously at the winning-post.

In fact incentives won't work, unless you are prepared to allow some inequality of one kind or another. And there is this further point. We turned to the possibility of devising effective incentives, in order to avoid compulsion and to preserve our free society. But indeed though Liberty and Equality have been for a century and a half yoked with Fraternity in a curious and uneasy trinity, nevertheless Liberty and Equality are natural enemies.... For if we are to be free, and if we exercise our freedom, we must be free to be, among other things, unequal. On the other hand, if we are to be equal, it will only be because we are forcibly compelled to be equal, and denied the liberty of surpassing our fellows.

I have offered you compulsion; but you say that if possible you would prefer to retain your freedom. I have suggested that you endeavor to solve the question of incentive in a free society; but you tell me that there may be something illogical in our new society in relying on incentives which, at their worst, . . . are unlovely, and which are of necessity material in their nature, with an individualistic appeal, inevitably differentiating if not dividing men.

III. *A Higher Type of Morality:* The third course . . . is to enroll yourself frankly among the followers of Lenin (for this purpose) and wait in faith for the emergence of a better man, for the universal prevalence of a higher order of morality than that now to be found among us; when in consequence the worker will no longer (the words are Lenin's) "calculate with the shrewdness of a Shylock whether he has not worked half an hour more than another, whether he is not getting less pay than another"; when "the necessity of observing the simple fundamental rules of human intercourse will become a habit." Admittedly this is a long-term policy and a long-term hope, and postulates, in the words of Lenin, "a person unlike the present man-in-the-street."

How far we can make ourselves fit to live in a better world by evolving a higher type of morality is presumably not an economic question. Nevertheless I may not omit it today, because in the past ... it has so often been hoped that our economic difficulties would be solved in some such way. Nearly all our ardent reformers have confidently trusted that in a better world where all things are held and operated on behalf of the community, a higher order of morality would prevail. When workers could feel that they were working for their fellow-workers and not merely for the gain of another, they would adopt a new code of behavior. Admittedly it will be a slow process, even on Lenin's own showing; and therefore perhaps we should not be impatient. ... Also we have made doubtful progress towards industrial peace; for apparently employment by the state or by a National Board does not necessarily mean bliss and content for all concerned. There is a devastating sentence in the second annual report of the Coal Board. It tells with restrained gratification that there were no official strikes during the year; "but," it adds, "there were 1,635 unofficial strikes where men stopped work in defiance of their union." One thousand six hundred and thirty five! One is reminded of the love affairs of Don Juan: "And in Spain, a thousand-and-three." Here surely, in the cold statistics of the Coal Board is an illuminating figure significant of the economic and psychological friction prevailing in our society. It will assuredly take some time before we are all prepared to crucify self to such an extent as will qualify us for admission to Lenin's heaven.

And so I leave it—somewhat inconclusively, like certain modern composers who end without even asking a question, but merely because they have exhausted their time or their paper. Compulsion, shall we say, is detestable and ultimately immoral. Incentive is essentially a species of bribery, relying on the competitive instincts and leading us back to what some would have us regard as the jungle of individualism. I am all for the evolution of a better man, and for that matter of a better woman; but it is a slow process for frail creatures whose years are three-score-and-ten. Perhaps in the end there *is* a conclusion, though it is not an economic one. It is that before we can be trusted to live in the New Jerusalem, we must first of all be fit to walk the streets of the New Jerusalem. And as applied to our transitional times, I would suggest that despite all superficial appearances, the New World into which we are moving is not going to be a world which will make everything easy for everybody by giving everybody everything. If it is to work, it will be a world which will make vastly greater demands on every one. It will demand that most difficult of all things to attain, that plant of very slow growth, a higher standard of public and private morality in all things, and in particular the suppression of self. For socialism is parading under a false name, unless it means an order of things in which we forget ourselves in our zeal for the good of society and of our fellows, and in which speculation as to our place in the queue is the last thought that occurs to us. And it is not I, but Lenin, who says so.

To Promote the General Welfare*

BY FRANKLIN D. ROOSEVELT

[1932]

I WANT to speak not of politics but of Government. I want to speak not of parties, but of universal principles. They are not political, except in that larger sense in which a great American once expressed a definition of politics, that nothing in all of human life is foreign to the science of politics. . . .

The issue of Government has always been whether individual men and women will have to serve some system of Government or economics, or whether a system of Government and economics exists to serve individual men and women. This question has persistently dominated the discussion of Government for many generations. On questions relating to these things men have differed, and for time immemorial it is probable that honest men will continue to differ.

The final word belongs to no man; yet we can still believe in change and in progress. Democracy, as a dear old friend of mine in Indiana, Meredith Nicholson, has called it, is a quest, a never-ending seeking for better things, and in the seeking for these things and the striving for them, there are many roads to follow. But, if we map the course of these roads, we find that there are only two general directions.

When we look about us, we are likely to forget how hard people have worked to win the privilege of Government. The growth of the national Governments of Europe was a struggle for the development of a centralized force in the Nation, strong enough to impose peace upon ruling barons. In many instances the victory of the central Government, the creation of a strong central Government, was a haven of refuge to the individual. The people pre-

---

* From a speech given at the Commonwealth Club at San Francisco, September 23, 1932. Franklin Delano Roosevelt (1882–1945) was the 32nd President of the United States.

ferred the master far away to the exploitation and cruelty of the smaller master near at hand.

But the creators of national Government were perforce ruthless men. They were often cruel in their methods, but they did strive steadily toward something that society needed and very much wanted, a strong central State able to keep the peace, to stamp out civil war, to put the unruly nobleman in his place, and to permit the bulk of individuals to live safely. The man of ruthless force had his place in developing a pioneer country, just as he did in fixing the power of the central Government in the development of Nations. Society paid him well for his services and its development. When the development among the Nations of Europe, however, had been completed, ambition and ruthlessness, having served their term, tended to overstep their mark.

There came a growing feeling that Government was conducted for the benefit of a few who thrived unduly at the expense of all. The people sought a balancing—a limiting force. There came gradually, through town councils, trade guilds, national parliaments, by constitution and by popular participation and control, limitations on arbitrary power.

Another factor that tended to limit the power of those who ruled was the rise of the ethical conception that a ruler bore a responsibility for the welfare of his subjects.

The American colonies were born in this struggle. The American Revolution was a turning point in it. After the Revolution the struggle continued and shaped itself in the public life of the country. There were those who because they had seen the confusion which attended the years of war for American independence surrendered to the belief that popular Government was essentially dangerous and essentially unworkable. They were honest people, my friends, and we cannot deny that their experience had warranted some measure of fear. The most brilliant, honest and able exponent of this point of view was Hamilton. He was too impatient of slow-moving methods. Fundamentally he believed that the safety of the republic lay in the autocratic strength of its Government, that the destiny of individuals was to serve that Government, and that fundamentally a great and strong group of central institutions, guided by a small group of able and public spirited citizens, could best direct all Government.

But Mr. Jefferson, in the summer of 1776, after drafting the Declaration of Independence turned his mind to the same problem and took a different view. He did not deceive himself with outward forms. Government to him was a means to an end, not an end in itself; it might be either a refuge and a help or a threat and a danger, depending on the circumstances. We find him carefully analyzing the society for which he was to organize a Government. "We have no paupers. The great mass of our population is of laborers, our rich who cannot live without labor, either manual or professional, being few and of moderate wealth. Most of the laboring class possess property, cultivate their own lands, have families and from the demand for their labor are enabled to exact from the rich and the competent such prices as enable them to feed abundantly, clothe above mere decency, to labor moderately and raise their families."

These people, he considered, had two sets of rights, those of "personal competency" and those involved in acquiring and possessing property. By "personal competency" he meant the right of free thinking, freedom of forming and expressing opinions, and freedom of personal living, each man according to his own lights. To insure the first set of rights, a Government must so order its functions as not to interfere with the individual. But even Jefferson realized that the exercise of the property rights might so interfere with the rights of the individual that the Government, without whose assistance the property rights could not exist, must intervene, not to destroy individualism, but to protect it.

You are familiar with the great political duel which followed; and how Hamilton, and his friends, building toward a dominant centralized power were at length defeated in the great election of 1800, by Mr. Jefferson's party. Out of that duel came the two parties, Republican and Democratic, as we know them today.

So began, in American political life, the new day, the day of the individual against the system, the day in which individualism was made the great watchword of American life. The happiest of economic conditions made that day long and splendid. On the Western frontier, land was substantially free. No one, who did not shirk the task of earning a living, was entirely without opportunity to do so. Depressions could, and did, come and go; but they could not alter the fundamental fact that most of the people lived partly by selling their labor and partly by extracting their livelihood from the soil, so that starvation and dislocation were practically impossible. At the very worst there was always the possibility of climbing into a covered wagon and moving west where the untilled prairies afforded a haven for men to whom the East did not provide a place. So great were our natural resources that we could offer this relief not only to our own people, but to the distressed of all the world; we could invite immigration from Europe, and welcome it with open arms. Traditionally, when a depression came a new section of land was opened in the West; and even our temporary misfortune served our manifest destiny.

It was in the middle of the nineteenth century that a new force was released and a new dream created. The force was what is called the industrial revolution, the advance of steam and machinery and the rise of the forerunners of the modern industrial plant. The dream was the dream of an economic machine, able to raise the standard of living for everyone; to bring luxury within the reach of the humblest; to annihilate distance by steam power and later by electricity, and to release everyone from the drudgery of the heaviest manual toil. It was to be expected that this would necessarily affect Government. Heretofore, Government had merely been called upon to produce conditions within which people could live happily, labor peacefully, and rest secure. Now it was called upon to aid in the consummation of this new dream. There was, however, a shadow over the dream. To be made real, it required use of the talents of men of tremendous will and tremendous ambition, since by no other force could the problems of financing and engineering and new developments be brought to a consummation.

So manifest were the advantages of the machine age, however, that the United States fearlessly, cheerfully, and, I think, rightly, accepted the bitter with the sweet. It was thought that no price was too high to pay for the advantages which we could draw from a finished industrial system. The history of the last half century is accordingly in large measure a history of a group of financial Titans, whose methods were not scrutinized with too much care, and who were honored in proportion as they produced the results, irrespective of the means they used. The financiers who pushed the railroads to the Pacific were always ruthless, often wasteful, and frequently corrupt; but they did build railroads, and we have them today. It has been estimated that the American investor paid for the American railway system more than three times over in the process; but despite this fact the net advantage was to the United States. As long as we had free land; as long as population was growing by leaps and bounds; as long as our industrial plants were insufficient to supply our own needs, society chose to give the ambitious man free play and unlimited reward provided only that he produced the economic plant so much desired.

During this period of expansion, there was equal opportunity for all and the business of Government was not to interfere but to assist in the development of industry. This was done at the request of business men themselves. The tariff was originally imposed for the purpose of "fostering our infant industry," a phrase I think the older among you will remember as a political issue not so long ago. The railroads were subsidized, sometimes by grants of money, oftener by grants of land; some of the most valuable oil lands in the United States were granted to assist the financing of the railroad which pushed through the Southwest. A nascent merchant marine was assisted by grants of money, or by mail subsidies, so that our steam shipping might ply the seven seas. Some of my friends tell me that they do not want the Government in business. With this I agree; but I wonder whether they realize the implications of the past. For while it has been American doctrine that the Government must not go into business in competition with private enterprises, still it has been traditional, particularly in Republican administrations, for business urgently to ask the Government to put at private disposal all kinds of Government assistance. The same man who tells you that he does not want to see the Government interfere in business—and he means it, and has plenty of good reasons for saying so—is the first to go to Washington and ask the Government for a prohibitory tariff on his product. When things get just bad enough, as they did two years ago, he will go with equal speed to the United States Government and ask for a loan; and the Reconstruction Finance Corporation is the outcome of it. Each group has sought protection from the Government for its own special interests, without realizing that the function of Government must be to favor no small group at the expense of its duty to protect the rights of personal freedom and of private property of all its citizens.

In retrospect we can now see that the turn of the tide came with the turn of the century. We were reaching our last frontier; there was no more free land and our industrial combinations had become great uncontrolled and irresponsible units of power within the State. Clear-sighted men saw with fear the danger that opportunity would no longer be equal; that the growing corporation, like the feudal baron of old, might threaten the economic freedom of individuals to earn a living. In that hour, our antitrust laws were born. The cry was raised against the great corporations. Theodore Roosevelt, the first great Republican Progressive, fought a Presidential campaign on the issue of "trust busting" and talked freely about malefactors of great wealth. If the Government had a policy it was rather to turn the clock back, to destroy the large combinations and to return to the time when every man owned his individual small business.

This was impossible; Theodore Roosevelt, abandoning the idea of "trust busting," was forced to work out a difference between "good" trusts and "bad" trusts. The Supreme Court set forth the famous "rule of reason" by which it seems to have meant that a concentration of industrial power was permissible if the method by which it got its power, and the use it made of that power, were reasonable.

Woodrow Wilson, elected in 1912, saw the situation more clearly. Where Jefferson had feared the encroachment of political power on the lives of individuals, Wilson knew that the new power was financial. He saw, in the highly centralized economic system, the despot of the twentieth century, on whom great masses of individuals relied for their safety and their livelihood, and whose irresponsibility and greed (if they were not controlled) would reduce them to starvation and penury. The concentration of financial power had not proceeded so far in 1912 as it has today; but it had grown far enough for Mr. Wilson to realize fully its implications. It is interesting, now, to read his speeches. What is called "radical" today (and I have reason to know whereof I speak) is mild compared to the campaign of Mr. Wilson. "No man can deny," he said, "that the lines of endeavor have more and more narrowed and stiffened; no man who knows anything about the development of industry in this country can have failed to observe that the larger kinds of credit are more and more difficult to obtain unless you obtain them upon terms of uniting your efforts with those who already control the industry of the country, and nobody can fail to observe that every man who tries to set himself up in competition with any process of manufacture which has taken place under the control of large combinations of capital will presently find himself either squeezed out or obliged to sell and allow himself to be absorbed." Had there been no World War—had Mr. Wilson been able to devote eight years to domestic instead of to international affairs—we might have had a wholly different situation at the present time. However, the then distant roar of European cannon, growing ever louder, forced him to abandon the study of this issue. The problem he saw so clearly is left with us as a legacy; and no one of us on either side of the political controversy can deny that it is a matter of grave concern to the Government.

A glance at the situation today only too clearly indicates that equality of opportunity as we have known it no longer exists. Our industrial plant is built; the problem just now is whether under existing conditions it is not overbuilt. Our last frontier has long since been reached, and there is practically no more free land. More than half of our people do not live on the farms or on lands and cannot derive a living by cultivating their own property. There is no safety valve in the form of a Western prairie to which those thrown out of work by the Eastern economic machines can go for a new start. We are not able to invite the immigration from Europe to share our endless plenty. We are now providing a drab living for our own people.

Our system of constantly rising tariffs has at last reacted against us to the point of closing our Canadian frontier on the north, our European markets on the east, many of our Latin-American markets to the south, and a goodly proportion of our Pacific markets on the west, through the retaliatory tariffs of those countries. It has forced many of ur great industrial institutions, which exported their surplus production to such countries, to establish plants in such countries, within the tariff walls. This has resulted in the reduction of the operation of their American plants, and of opportunity for employment.

Just as freedom to farm has ceased, so also the opportunity in business has narrowed. It still is true that men can start small enterprises, trusting to native shrewdness and ability to keep abreast of competitors; but area after area has been preempted altogether by the great corporations, and even in the fields which still have no great concerns, the small man starts under a handicap. The unfeeling statistics of the past three decades show that the independent business man is running a losing race. Perhaps he is forced to the wall; perhaps he cannot command credit; perhaps he is "squeezed out," in Mr. Wilson's words, by highly organized corporate competitors, as your corner grocery man can tell you. Recently a careful study was made of the concentration of business in the United States. It showed that our economic life was dominated by some six hundred-odd corporations who controlled two-thirds of American industry. Ten million small business men divided the other third. More striking still, it appeared that if the process of concentration goes on at the same rate, at the end of another century we shall have all American industry controlled by a dozen corporations, and run by perhaps a hundred men. Put plainly, we are steering a steady course toward economic oligarchy, if we are not there already.

Clearly, all this calls for a re-appraisal of values. A mere builder of more industrial plants, a creator of more railroad systems, an organizer of more corporations, is as likely to be a danger as a help. The day of the great promoter or the financial Titan, to whom we granted anything if only he would build, or develop, is over. Our task now is not discovery or exploitation of natural resources, or necessarily producing more goods. It is the soberer, less dramatic business of administering resources and plants already in hand, of seeking to re-establish foreign markets for our surplus production, of meeting the problem of underconsumption, of adjusting production to consumption, of distributing wealth and products more equitably, of adapting existing economic organizations to the service of the people. The day of enlightened administration has come.

Just as in older times the central Government was first a haven of refuge, and then a threat, so now in a closer economic system the central and ambitious financial unit is no longer a servant of national desire, but a danger. I would draw the parallel one step farther. We did not think because national Government had become a threat in the 18th century that therefore we should abandon the principle of national Government. Nor today should we abandon the principle of strong economic units called corporations, merely because their power is susceptible of easy abuse. In other times we dealt with the problem of an unduly ambitious central Government by modifying it gradually into a constitutional democratic Government. So today we are modifying and controlling our economic units.

As I see it, the task of Government in its relation to business is to assist the development of an economic declaration of rights, an economic constitutional order. This is the common task of statesman and business man. It is the minimum requirement of a more permanently safe order of things.

Happily, the times indicate that to create such an order not only is the proper policy of Government, but it is the only line of safety for our economic structures as well. We know, now, that these economic units cannot exist unless prosperity is uniform, that is, unless purchasing power is well distributed throughout every group in the Nation. That is why even the most selfish of corporations for its own interest would be glad to see wages restored and unemployment ended and to bring the Western farmer back to his accustomed level of prosperity and to assure a permanent safety to both groups. That is why some enlightened industries themselves endeavor to limit the freedom of action of each man and business group within the industry in the common interest of all; why business men everywhere are asking a form of organization which will bring the scheme of things into balance, even though it may in some measure qualify the freedom of action of individual units within the business.

The exposition need not further be elaborated. It is brief and incomplete, but you will be able to expand it in terms of your own business or occupation without difficulty. I think everyone who has actually entered the economic struggle—which means everyone who was not born to safe wealth—knows in his own experience and his own life that we have now to apply the earlier concepts of American Government to the conditions of today.

The Declaration of Independence discusses the problem of Government in terms of a contract. Government is a relation of give and take, a contract, perforce, if we would follow the thinking out of which it grew. Under such a contract rulers were accorded power, and the people consented to that power on consideration that they be accorded certain rights. The task of statesmanship has always been the redefinition of these rights in terms of a changing and growing social order. New conditions impose new requirements upon Government and those who conduct Government.

I held, for example, in proceedings before me as Governor, the purpose of which was the removal of the Sheriff of New York, that under modern conditions it was not enough for a public official merely to evade the legal terms of official wrongdoing. He owed a positive duty as well. I said in substance that if he had acquired large sums of money, he was when accused required to explain the sources of such wealth. To that extent this wealth was colored with a public interest. I said that in financial matters, public servants should, even beyond private citizens, be held to a stern and uncompromising rectitude.

I feel that we are coming to a view through the drift of our legislation and our public thinking in the past quarter century that private economic power is, to enlarge an old phrase, a public trust as well. I hold that continued enjoyment of that power by any individual or group must depend upon the fulfillment of that trust. The men who have reached the summit of American business life know this best; happily, many of these urge the binding quality of this greater social contract.

The terms of that contract are as old as the Republic, and as new as the new economic order.

Every man has right to life; and this means that he has also a right to make a comfortable living. He may by sloth or crime decline to exercise that right; but it may not be denied him. We have no actual famine or dearth; our industrial and agricultural mechanism can produce enough and to spare. Our Government, formal and informal, political and economic, owes to everyone an avenue to possess himself of a portion of that plenty sufficient for his needs, through his own work.

Every man has a right to his own property; which means a right to be assured, to the fullest extent attainable, in the safety of his savings. By no other means can men carry the burdens of those parts of life which, in the nature of things, afford no chance of labor; childhood, sickness, old age. In all thought of property, this right is paramount; all other property rights must yield to it. If, in accord with this principle, we must restrict the operations of the speculator, the manipulator, even the financier, I believe we must accept the restriction as needful, not to hamper individualism but to protect it.

These two requirements must be satisfied, in the main, by the individuals who claim and hold control of the great industrial and financial combinations which dominate so large a part of our industrial life. They have undertaken to be, not business men, but princes of property. I am not prepared to say that the system which produces them is wrong. I am very clear that they must fearlessly and competently assume the responsibility which goes with the power. So many enlightened business men know this that the statement would be little more than a platitude, were it not for an added implication.

This implication is, briefly, that the responsible heads of finance and industry, instead of acting each for himself, must work together to achieve the common end. They must, where necessary, sacrifice this or that private advantage; and in reciprocal self-denial must seek a general advantage. It is here that formal Government—political Government, if you choose—comes in. Whenever in the pursuit of this objective the lone wolf, the unethical competitor, the reckless promoter, the Ishmael or Insull whose hand is against every man's, declines to join in achieving an end recognized as being for the public welfare, and threatens to drag the industry back to a state of anarchy, the Government may properly be asked to apply restraint. Likewise, should the group ever use its collective power contrary to the public welfare, the Government must be swift to enter and protect the public interest.

The Government should assume the function of economic regulation only as a last resort, to be tried only when private initiative, inspired by high responsibility, with such assistance and balance as Government can give, has finally failed. As yet there has been no final failure, because there has been no attempt; and I decline to assume that this Nation is unable to meet the situation.

The final term of the high contract was for liberty and the pursuit of happiness. We have learned a great deal of both in the past century. We know that individual liberty and individual happiness mean nothing unless both are ordered in the sense that one man's meat is not another man's poison. We know that the old "rights of personal competency," the right to read, to think, to speak, to choose and live a mode of life, must be respected at all hazards. We know that liberty to do anything which deprives others of those elemental rights is outside the protection of any compact; and that Government in this regard is the maintenance of a balance, within which every individual may have a place if he will take it; in which every individual may attain such power as his ability permits, consistent with his assuming the accompanying responsibilty. . . .

The American System of Liberty*

BY HERBERT HOOVER

[1936]

WE HAVE three alternatives.

First: Unregulated business.
Second: Government-regulated business, which I believe is the American System.
Third: Government-dictated business, whether by dictation to business or by government in business. This is the New Deal choice. These ideas are dipped from cauldrons of European Fascism or Socialism.

Unregulated Business

While some gentlemen may not agree, we may dismiss any system of unregulated business. We know from experience that the vast tools of technology and mechanical power can be seized for purposes of oppression. They have been used to limit production and to strangle competition and opportunity. We can no more have economic power without checks and balances than we can have political power without checks and balances. Either one leads to tyranny.

And there must be regulation of the traffic even when it is honest. We have too many people and too many devices to allow them to riot all over the streets of commerce. But a traffic policeman must only enforce the rules. He will block the traffic if he stands on the corner demanding to know their business and telling them how to run it.

---

* From a speech given at Colorado Springs, Colorado, March 7. From *Addresses Upon the American Road, 1933–1938*, by Herbert Hoover, copyright 1938 by Edgar Richard. Reprinted by permission of Charles Scribner's Sons. Herbert Hoover (b. 1874) was the 31st President of the United States.

The American System of Regulation

I am one who believes that the only system which will preserve liberty and hold open the doors of opportunity is government-regulated business. And this is as far from government-dictated business as the two poles. Democracy can regulate its citizens through law and judicial bodies. No democracy can dictate and survive as a democracy. The only way to preserve individual initiative and enterprise is for the government to make the same rules for everybody and act as umpire.

But if we are to preserve freedom we must face the fact that ours is a regulatory system.

And let us be definite once and for all as to what we mean by a system of regulation. It looms up more clearly against the past three years.

1. A great area of business will regulate its own prices and profits through competition. Competition is also the restless pillow of progress. But we must compel honest competition through prevention of monopolies and unfair practices. That is indirect regulation.

2. The semi- yet natural monopolies, such as railways and utilities, must be directly regulated as to rates to prevent the misuse of their privilege.

3. Banking, finance, public markets, and other functions of trust must be regulated to prevent abuse and misuse of trust.

The failure of the States, particularly New York, to do their part during the boom years has necessitated an extension of Federal action. The New Deal regulations of stock and security promotion in various aspects have the right objectives. They were hastily and poorly formed without proper consideration by Congress. But they point right.

4. Certain groups must be appropriately regulated to prevent waste of natural resources.

5. Labor must have the right to free collective bargaining. But it must have responsibilities as well as rights.

6. At one time we relied upon the theory of "shirt sleeves to shirt sleeves in three generations" to regulate over-accumulations of wealth. This is now guaranteed by our income and inheritance taxes. Some people feel these taxes take the shirt also.

But there are certain principles that must run through these methods.

1. The first principle of regulation is the least regulation that will preserve equality of opportunity and liberty itself. We cannot afford to stifle a thousand honest men in order to smother one evil person.

2. To preserve Liberty the major burden of regulation must fall upon the States and local government. But where the States hopelessly fail or when the problem grows beyond their powers we should call upon the Federal government. Or we should invoke the machinery of interstate compacts.

3. Regulation should be by specific law, that all who run may read. That alone holds open the doors of the courts to the citizen. This must be "a government of laws and not of men."

4. And the American System of Liberty will not function solely through traffic policemen. The fundamental regulation of the nation is the Ten Commandments and the Sermon on the Mount.

Incidentally, the government might regulate its own business by some of the standards it imposes on others.

There are certain humanities which run through all business. As we become more experienced, more humane, as conditions change, we recognize things as abuses which we once passed over. There are the abuses of slums, child-labor, sweated hours, and sweated wages. They have been diminishing for decades before the New Deal. They have not been solved yet. They must be solved. We must not be afraid to use the powers of government to eliminate them.

There will be periodic unemployment in any system. It is even so in the self-declared economic heavens of Socialism and Fascism. With common sense we could provide insurance programs against it. We could go further and prevent many causes of depressions.

Out of medical and public-health discoveries we have in eighty years increased the number of people over sixty years of age from four per cent to eight per cent. That imposes another problem upon us.

This American System has sprung from the spirit of our people. It has been developing progressively over many generations. However grave its faults may be they are but marginal to a great area of human well-being. The test of a system is its comparative results with others and whether it has the impulses within to cure its faults. This system based on ordered liberty alone answers these tests.

The doors of opportunity cannot be held open by inaction. That is an ideal that must be incessantly fought for.

These doors are partly closed by every gentleman who hatches some special privilege. They are closed to somebody by every betrayal of trust. But because brickbats can be used for murder we do not need stop building houses. These doors are partly shut by every needless bureaucrat. And there is the tax collector. He stands today right in the door.

Every new invention, every new idea, every new war shifts and changes our economic life. That greatest instrument of American joy, the automobile, has in twenty years shifted regulation in a hundred directions.

Many obstructions and abuses have been added by the New Deal. Many of them are older but no worse. While the inspiration to reform comes from the human heart, it is achieved only by the intellect. Enthusiastic hearts have flooded us with illusions. Ideals without illusions are good. Ideals with illusions are no good. You may remember that youth with a banner of strange device. Was it "Excelsior" or was it "Planned Economy"? He froze to death.

Government-Dictated Economic Life

Young men and women have grave need to look into this New Deal alternative to our American System.

If any one does not believe there is a bite in that innocent term "Planned Economy," he might reread this paragraph from one of the leading New Deal spokesmen:

"It is . . . a logical impossibility to have a *planned economy* and to *have business operating its industries,* just as it is also impossible to have one within our present *constitutional* and *statutory structure*. Modifications in both, so serious as to mean *destruction* and *rebeginning*, are required."

That is involved language but if it means anything it means that both private business and the Constitution must be modified so seriously as to mean destruction and rebeginning.

The President, far from repudiating these ideas, has continuously supported "Planned Economy." On one occasion he said, "... All of the proposals and all of the legislation since the fourth of March have not been just a collection of haphazard schemes but rather the orderly component parts of a connected and logical whole."

The Supreme Court has removed some ten of the component parts. And rather than have the score raised to thirteen before an election we have seen three more quietly removed. However, if the New Deal is re-elected they will be found to have a lot of spare parts.

Do not mistake. The choice is still yours. But the New Deal has no choice. The New Deal is committed to drive ahead for government dictation of our economic life. It is committed by a thousand statements, by a thousand actions. It is committed by the supporters upon whom it is dependent.

The President assures them "we will not retreat." They did mention a breathing spell. A spell is a very limited period. I have spoken at length upon these subjects elsewhere, but I may remind you of a few examples of the choice that the New Deal offers to youth. Under that "connected and logical whole" a man could be fined and sent to jail for starting a new business of his own; for refusing to sell his own products as directed; for not reducing his production; for increasing his production if his energies found a market; for selling at prices below his competitors; or for having 101 gold dollars.

Also you might note that when you ask the man with a profit and loss motive for a job, he asks just one thing, "Can you do the job?" When you ask the government for a job, your ability is second to your politics, your delivery of votes, and your affiliations generally. That is not equality of opportunity.

And what of this managed currency and this managed credit, which threaten Liberty and opportunity with the poison of inflation? What of this governmentally raised cost of living? What of all this continued waste and folly wrought in the name of relief? What of the folly of these purchases of foreign silver? What of the debauchery of the Civil Service and the politics in relief?

What of the taxes that will ooze from this spending and debt all your lives?

Do not mistake. The new taxes of today are but part of them.

More of them are as inevitable as the first of the month. The only alternatives are repudiation or inflation. No matter what nonsense you are told about corporations and the rich paying the bill, there will be two-thirds of it for the common man to pay after the corporations and the rich are sucked dry.

Taxation enslaves as well as dictatorship. Every increased dollar in taxes is a limitation upon your opportunities. It means you have to work that many days more for the government instead of for your own advancement. Your fireside talks in the future will be with the tax collector. . . .

VIII

CIVIL RIGHTS AND FEDERALISM

The Civil Rights Cases

(109 U.S. 3)

[1883]

MR. JUSTICE BRADLEY delivered the opinion of the Court.

These cases were all founded on the first and second sections of the Act of Congress, known as the Civil Rights Act, passed March 1st, 1875, entitled "An Act to protect all citizens in their civil and legal rights." 18 Stat. 335. Two of the cases, those against Stanley and Nichols, were indictments for denying to persons of color the accommodations and privileges of an inn or hotel; two of them, those against Ryan and Singleton, were . . . for denying to individuals the privileges and accommodations of a theatre, the information against Ryan being for refusing a colored person a seat in the dress circle of Maguire's theatre in San Francisco; and the indictment against Singleton was for denying to another person, whose color was not stated, the full enjoyment of the accommodations of the theatre known as the Grand Opera House in New York. . . . The case of Robinson and wife against the Memphis & Charleston R. R. Company was an action brought . . . [because of] the refusal by the conductor of the railroad company to allow the wife to ride in the ladies' car, for the reason . . . that she was a person of African descent. The jury rendered a verdict for the defendants. . . .

It is obvious that the primary and important question in all the cases is the constitutionality of the law: for if the law is unconstitutional none of the prosecutions can stand. . . .

The essence of the law is, not to declare broadly that all persons shall be entitled to the full and equal enjoyment of the accommodations, advantages, facilities, and privileges of inns, public conveyances, and theatres; but that such enjoyment shall not be subject to any conditions applicable only to citizens of a particular race or color, or who had been in a previous condition of servitude. In other words, it is the purpose of the law to declare that, in the enjoyment of the accommodations and privileges of inns, public conveyances, theatres, and other places of public amusement, no distinction shall be made between citizens of different race or color, or between those who have, and those who have not, been slaves. . . .

Has Congress constitutional power to make such a law? Of course, no one will contend that the power to pass it was contained in the Constitution before the adoption of the last three amendments. The power is sought, first, in the Fourteenth Amendment, and the views and arguments of distinguished Senators, advanced whilst the law was under consideration, claiming authority to pass it by virtue of that amendment, are the principal arguments adduced in favor of the power. . . .

It is State action of a particular character that is prohibited. Individual invasion of individual rights is not the subject-matter of the amendment. It has a deeper and broader scope. It nullifies and makes void all State legislation, and State action of every kind, which impairs the privileges and immunities of citizens of the United States, or which injures them in life, liberty or property without due process of law, or which denies to any of them the equal protection of the laws. It not only does this, but, in order that the national will, thus declared, may not be a mere *brutum fulmen,* the last section of the amendment invests Congress with power to enforce it by appropriate legislation. To enforce what? To enforce the prohibition. To adopt appropriate legislation for correcting the effects of such prohibited State laws and State acts, and thus to render them effectually null, void, and innocuous. This is the legislative power conferred upon Congress, and this is the whole of it. . . .

And so in the present case, until some State law has been passed, or some State action through its officers or agents has been taken, adverse to the rights of citizens sought to be protected by the Fourteenth Amendment, no legislation of the United States under said amendment, nor any proceeding under such legislation, can be called into activity: for the prohibitions of the amendment are against State laws and acts done under State authority. Of course, legislation may, and should be, provided in advance to meet the exigency when it arises; but it should be adapted to the mischief and wrong which the amendment was intended to provide against; and that is, State laws, or

State action of some kind, adverse to the rights of the citizen secured by the amendment. Such legislation cannot properly cover the whole domain of rights appertaining to life, liberty and property, defining them and providing for their vindication. That would be to establish a code of municipal law regulative of all private rights between man and man in society. It would be to make Congress take the place of the State legislatures and to supersede them. It is absurd to affirm that, because the rights of life, liberty and property (which include all civil rights that men have), are by the amendment sought to be protected against invasion on the part of the State without due process of law, Congress may therefore provide due process of law for their vindication in every case; and that, because the denial by a State to any persons, of the equal protection of the laws, is prohibited by the amendment, therefore Congress may establish laws for their equal protection. . . .

An inspection of the law shows that it makes no reference whatever to any supposed or apprehended violation of the Fourteenth Amendment on the part of the States. It is not predicated on any such view. It proceeds *ex directo* to declare that certain acts committed by individuals shall be deemed offences, and shall be prosecuted and punished by proceedings in the courts of the United States. It does not profess to be corrective of any constitutional wrong committed by the States; it does not make its operation to depend upon any such wrong committed. It applies equally to cases arising in States which have the justest laws respecting the personal rights of citizens, and whose authorities are ever ready to enforce such laws, as to those which arise in States that may have violated the prohibition of the amendment. In other words, it steps into the domain of local jurisprudence, and lays down rules for the conduct of individuals in society towards each other, and imposes sanctions for the enforcement of those rules, without referring in any matter to any supposed action of the State or its authorities.

If this legislation is appropriate for enforcing the prohibitions of the amendment, it is difficult to see where it is to stop. Why may not Congress with equal show of authority enact a code of laws for the enforcement and vindication of all rights of life, liberty, and property. . . . The truth is, that the implication of a power to legislate in this manner is based upon the assumption that if the States are forbidden to legislate or act in a particular way on a particular subject, and power is conferred upon Congress to enforce the prohibition, this gives Congress power to legislate generally upon that subject, and not merely power to provide modes of redress against such State legislation or action. The assumption is certainly unsound. It is repugnant to the Tenth Amendment of the Constitution, which declares that powers not delegated to the United States by the Constitution, nor prohibited by it to the States, are reserved to the States respectively or to the people. . . .

In this connection it is proper to state that civil rights, such as are guaranteed by the Constitution against State aggression, cannot be impaired by the wrongful acts of individuals, unsupported by State authority in the shape of laws, customs, or judicial or executive proceedings. The wrongful act of an individual, unsupported by any such authority, is simply a private wrong, or a crime of that individual; an invasion of the rights of the injured party, it is true, whether they affect his person, his property, or his reputation; but if not sanctioned in some way by the State, or not done under State authority, his rights remain in full force, and may presumably be vindicated by resort to the laws of the State for redress. An individual cannot deprive a man of his right to vote, to hold property, to buy and sell, to sue in the courts, or to be a witness or a juror; he may, by force or fraud, interfere with the enjoyment of the right in a particular case; he may commit an assault against the person, or commit murder, or use ruffian violence at the polls, or slander the good name of a fellow citizen; but, unless protected in these wrongful acts by some shield of State law or State authority, he cannot destroy or injure the right; he will only render himself amenable to satisfaction or punishment; and amenable therefor to the laws of the State where the wrongful acts are committed. Hence, in all those cases where the Constitution seeks to protect the rights of the citizen against discriminative and unjust laws of the State by prohibiting such laws, it is not individual offences, but abrogation and denial of rights, which it denounces, and for which it clothes the Congress with power to provide a remedy. . . .

Of course, these remarks do not apply to those cases in which

Congress is clothed with direct and plenary powers of legislation over the whole subject, accompanied with an express or implied denial of such power to the States, as in the regulation of commerce with foreign nations, among the several States, and with the Indian tribes, the coining of money, the establishment of post offices and post roads, the declaring of war, etc. In these cases Congress has power to pass laws for regulating the subjects specified in every detail, and the conduct and transactions of individuals in respect thereof. . . .

But the power of Congress to adopt direct and primary, as distinguished from corrective legislation, on the subject in hand, is sought, in the second place, from the Thirteenth Amendment, which abolishes slavery. . . .

This amendment, as well as the Fourteenth, is undoubtedly self-executing without any ancillary legislation, so far as its terms are applicable to any existing state of circumstances. By its own unaided force and effect it abolished slavery, and established universal freedom. Still, legislation may be necessary and proper to meet all the various cases and circumstances to be affected by it, and to prescribe proper modes of redress for its violation in letter or spirit. And such legislation may be primary and direct in its character; for the amendment is not a mere prohibition of State laws establishing or upholding slavery, but an absolute declaration that slavery or involuntary servitude shall not exist in any part of the United States.

It is true, that slavery cannot exist without law, any more than property in lands and goods can exist without law: and, therefore, the Thirteenth Amendment may be regarded as nullifying all State laws which establish or uphold slavery. But it has a reflex character also, establishing and decreeing universal civil and political freedom throughout the United States; and it is assumed, that the power vested in Congress to enforce the article by appropriate legislation, clothes Congress with power to pass all laws necessary and proper for abolishing all badges and incidents of slavery in the United States: and upon this assumption it is claimed, that this is sufficient authority for declaring by law that all persons shall have equal accommodations and privileges in all inns, public conveyances, and places of amusement; the argument being, that the denial of such equal accommodations and privileges is, in itself, a subjection to a species of servitude within the meaning of the amendment. . . .

The only question under the present head, therefore, is, whether the refusal to any persons of the accommodations of an inn, or a public conveyance, or a place of public amusement, by an individual, and without any sanction or support from any State law or regulation, does inflict upon such persons any manner of servitude, or form of slavery, as those terms are understood in this country? . . .

. . . It would be running the slavery argument into the ground to make it apply to every act of discrimination which a person may see fit to make as to the guests he will entertain, or as to the people he will take into his coach or cab or car, or admit to his concert or theatre, or deal with in other matters of intercourse of business. Innkeepers and public carriers, by the laws of all the States, so far as we are aware, are bound, to the extent of their facilities, to furnish proper accommodation to all unobjectionable persons who in good faith apply for them. If the laws themselves make any unjust discrimination, amenable to the prohibitions of the Fourteenth Amendment, Congress has full power to afford a remedy under that amendment and in accordance with it.

When a man has emerged from slavery, and by the aid of beneficent legislation has shaken off the inseparable concomitants of that state, there must be some stage in the progress of his elevation when he takes the rank of a mere citizen, and ceases to be the special favorite of the laws, and when his rights as a citizen, or a man, are to be protected in the ordinary modes by which other men's rights are protected. There were thousands of free colored people in this country before the abolition of slavery, enjoying all the essential rights of life, liberty and property the same as white citizens; yet no one, at that time, thought that it was any invasion of his personal status as a freeman because he was not admitted to all the privileges enjoyed by white citizens, or because he was subjected to discriminations in the enjoyment of accommodations in inns, public conveyances and places of amusement. Mere discriminations on account of race or color were not regarded as badges of slavery. . . .

On the whole we are of opinion, that no countenance of authority for the passage of the law in question can be found in either the Thirteenth or Fourteenth Amendment of the Constitution;

and no other ground of authority for its passage being suggested, it must necessarily be declared void, at least so far as its operation in the several States is concerned. . . .

Mr. Justice Harlan dissenting.

The opinion in these cases proceeds, it seems to me, upon grounds entirely too narrow and artificial. I cannot resist the conclusion that the substance and spirit of the recent amendments of the Constitution have been sacrificed by a subtle and ingenious verbal criticism. "It is not the words of the law but the internal sense of it that makes the law: the letter of the law is the body; the sense and reason of the law is the soul." Constitutional provisions, adopted in the interest of liberty, and for the purpose of securing, through national legislation, if need be, rights inhering in a state of freedom, and belonging to American citizenship, have been so construed as to defeat the ends the people desired to accomplish, which they attempted to accomplish, and which they supposed they had accomplished by changes in their fundamental law. By this I do not mean that the determination of these cases should have been materially controlled by considerations of mere expediency of policy. I mean only, in this form, to express an earnest conviction that the court has departed from the familiar rule requiring, in the interpretation of constitutional provisions, that full effect be given to the intent with which they were adopted.

The purpose of the first section of the act of Congress of March 1, 1875, was to prevent *race* discrimination in respect of the accommodations and facilities of inns, public conveyances, and places of public amusement. It does not assume to define the general conditions and limitations under which inns, public conveyances, and places of public amusement may be conducted, but only declares that such conditions and limitations, whatever they may be, shall not be applied so as to work a discrimination solely because of race, color, or previous condition of servitude. The second section provides a penalty against any one denying, or aiding or inciting the denial, to any citizen, of that equality of right given by the first section, except for reasons by law applicable to citizens of every race or color and regardless of any previous condition of servitude.

There seems to be no substantial difference between my brethren and myself as to the purpose of Congress. . . .

The court adjudges, I think erroneously, that Congress is without power, under either the Thirteenth or Fourteenth Amendment, to establish such regulations, and that the first and second sections of the statute are, in all their parts, unconstitutional and void.

. . . [In] *Sinking Fund Cases*, 99 U.S., 718, we said: "It is our duty when required in the regular course of judicial proceedings, to declare an act of Congress void if not within the legislative power of the United States, but this declaration should never be made except in a clear case. Every possible presumption is in favor of the validity of a statute, and this continues until the contrary is shown beyond a rational doubt. One branch of the government cannot encroach on the domain of another without danger. The safety of our institutions depends in no small degree on a strict observance of this salutary rule."

Before considering the language and scope of these amendments it will be proper to recall the relations subsisting, prior to their adoption, between the national government and the institution of slavery, as indicated by the provisions of the Constitution, the legislation of Congress, and the decisions of this court. In this mode we may obtain keys with which to open the mind of the people, and discover the thought intended to be expressed.

In section 2 of article IV of the Constitution it was provided that "no person held to service or labor in one State, under the laws thereof, escaping into another, shall, in consequence of any law or regulation therein, be discharged from such service or labor, but shall be delivered up on claim of the party to whom such service or labor may be due." Under the authority of this clause Congress passed the Fugitive Slave Law of 1793, establishing a mode for the recovery of fugitive slaves, and prescribing a penalty against any person who should knowingly and willingly obstruct or hinder the master, his agent, or attorney, in seizing, arresting, and recovering the fugitive, or who should rescue the fugitive from him, or who should harbor or conceal the slave after notice that he was a fugitive.

In *Prigg* v. *Commonwealth of Pennsylvania*, 16 Pet. 539, this court . . . said: "The fundamental principle, applicable to all cases of this sort, would seem to be that when the end is required the means are given, and when the duty is enjoined the ability to perform it is contemplated to exist on the part of the functionary to whom it is entrusted." Again: "It would be a strange anomaly and forced construction to suppose that the national government meant to rely for the due fulfilment of its own proper duties, and the rights which it intended to secure, upon State legislation, and not upon that of the Union. A *fortiori,* it would be more objectionable to suppose that a power which was to be the same throughout the Union, should be confided to State sovereignty which could not rightfully act beyond its own territorial limits."

The act of 1793 was, upon these grounds, adjudged to be a constitutional exercise of the powers of Congress. . . .

We next come to the Fugitive Slave Act of 1850, the constitutionality of which rested, as did that of 1793, solely upon the implied power of Congress to enforce the master's rights. The provisions of that act were far in advance of previous legislation. They placed at the disposal of the master seeking to recover his fugitive slave, substantially the whole power of the nation. It invested commissioners, appointed under the act, with power to summon the *posse comitatus* for the enforcement of its provisions, and commanded all good citizens to assist in its prompt and efficient execution whenever their services were required as part of the *posse comitatus.* Without going into the details of that act, it is sufficient to say that Congress omitted from it nothing which the utmost ingenuity could suggest as essential to the successful enforcement of the master's claim to recover his fugitive slave. And this court, in *Ableman* v. *Booth,* 21 How. 506, adjudged it to be "in all of its provisions fully authorized by the Constitution of the United States." . . .

We have seen that the power of Congress, by legislation, to enforce the master's right to have his slave delivered up on claim was *implied* from the recognition of that right in the national Constitution. But the power conferred by the Thirteenth Amendment does not rest upon implication or inference. Those who framed it were not ignorant of the discussion, covering many years of our country's history, as to the constitutional power of Congress to enact the Fugitive Slave Laws of 1793 and 1850. When, therefore, it was determined, by a change in the fundamental law, to uproot the institution of slavery wherever it existed in the land, and to establish universal freedom, there was a fixed purpose to place the authority of Congress in the premises beyond the possibility of a doubt. Therefore, *ex industria*, power to enforce the Thirteenth Amendment, by appropriate legislation, was expressly granted. Legislation for that purpose, my brethren concede, may be direct and primary. But to what specific ends may it be directed? This court has uniformly held that the national government has the power, whether expressly given or not, to secure and protect rights conferred or guaranteed by the Constitution. . . . That doctrine ought not now to be abandoned when the inquiry is not as to an implied power to protect the master's rights, but what may Congress, under powers expressly granted, do for the protection of freedom and the rights necessarily inhering in a state of freedom.

The Thirteenth Amendment, it is conceded, did something more than to prohibit slavery as an *institution*, resting upon distinctions of race, and upheld by positive law. My brethren admit that it established and decreed universal *civil freedom* throughout the United States. But did the freedom thus established involve nothing more than exemption from actual slavery? Was nothing more intended than to forbid one man from owning another as property? Was it the purpose of the nation simply to destroy the institution, and then remit the race, theretofore held in bondage, to the several States for such protection, in their civil rights, necessarily growing out of freedom, as those States, in their discretion, might choose to provide? Were the States against whose protest the institution was destroyed, to be left free, so far as national interference was concerned, to make or allow discriminations against that race, as such, in the enjoyment of those fundamental rights which by universal concession, inhere in a state of freedom? . . .

That there are burdens and disabilities which constitute badges of slavery and servitude, and that the power to enforce by appropriate legislation the Thirteenth Amendment may be exerted by legislation of a direct and primary character, for the eradication, not simply of the institution, but of its badges and incidents, are propositions which ought to be deemed indisputable. They lie at the foundation of the Civil Rights Act of 1866. Whether that act was authorized by the Thirteenth Amendment alone, without the support which it subsequently received from the Fourteenth Amendment, after the adoption of which it was re-enacted with some additions, my brethren do not consider it necessary to inquire. But I submit, with all respect to them, that its constitutionality is shown by their opinion. They admit, as I have said, that the Thirteenth Amendment established freedom; that there are burdens and disabilities, the necessary incidents of slavery, which constitute its substance and visible form; that Congress, by the act of 1866, passed in view of the Thirteenth Amendment, before the Fourteenth was adopted, undertook to remove certain burdens and disabilities, the necessary incidents of slavery, and to secure to all citizens of every race and color, and without regard to previous servitude, those fundamental rights which are the essence of civil freedom, namely, the same right to make and enforce contracts, to sue, be parties, give evidence, and to inherit, purchase, lease, sell, and convey property as is enjoyed by white citizens; that under the Thirteenth Amendment, Congress has to do with slavery and its incidents; and that legislation, so far as necessary or proper to eradicate all forms of slavery and involuntary servitude, may be direct and primary, operating upon the acts of individuals, whether sanctioned by State legislation or not. These propositions being conceded, it is impossible, as it seems to me, to question the constitutional validity of the Civil Rights Act of 1866. I do not contend that the Thirteenth Amendment invests Congress with authority, by legislation, to define and regulate the entire body of the civil rights which citizens enjoy, or may enjoy, in the several States. But I hold that slavery, as the court has repeatedly declared . . . was the moving or principal cause of the adoption of that amendment, and since that institution rested wholly upon the inferiority, as a race, of those held in bondage, their freedom necessarily involved immunity from, and protection against, all discrimination against them, because of their race, in respect of such civil rights as belong to freemen of other races. Congress, therefore, under its express power to enforce that amendment, by appropriate legislation, may enact laws to protect that people against the deprivation, *because of their race*, of any civil rights granted to other freemen in the same State; and such legislation may be of a direct and primary character, operating upon States, their officers and agents, and also, upon, at least, such individuals and corporations as exercise public functions and wield power and authority under the State.

To test the correctness of this position, let us suppose that, prior to the adoption of the Fourteenth Amendment, a State had passed a statute denying to freemen of African descent, resident within its limits, the same right which was accorded to white persons, of making and enforcing contracts, and of inheriting, purchasing, leasing, selling and conveying property; or a statute subjecting colored people to severer punishment for particular offences than was prescribed for white persons, or excluding that race from the benefit of the laws exempting homesteads from execution. Recall the legislation of 1865-6 in some of the States, of which this court, in *Slaughter-House Cases,* said, that it imposed upon the colored race onerous disabilities and burdens; curtailed their rights in the pursuit of life, liberty and property to such an extent that their freedom was of little value; forbade them to appear in the towns in any other character than menial servants; required them to reside on and cultivate the soil, without the right to purchase or own it; excluded them from many occupations of gain; and denied them the privilege of giving testimony in the courts where a white man was a party. 16 Wall. 57. Can there be any doubt that all such enactments might have been reached by direct legislation upon the part of Congress under its express power to enforce the Thirteenth Amendment? Would any court have hesitated to declare that such legislation imposed badges of servitude in conflict with the civil freedom ordained by that amendment? That it would have been also in conflict with the Fourteenth Amendment, because inconsistent with the fundamental rights of American citizenship, does not prove that it would have been consistent with the Thirteenth Amendment.

What has been said is sufficient to show that the power of Congress under the Thirteenth Amendment is not necessarily restricted to legislation against slavery as an institution upheld by positive law, but may be exerted to the extent, at least, of protecting the liberated race against discrimination, in respect of legal rights belonging to freemen, where such discrimination is based upon race.

It remains now to inquire what are the legal rights of colored persons in respect of the accommodations, privileges and facilities of public conveyances, inns and places of public amusement?

. . . The sum of the adjudged cases is that a railroad corporation is a governmental agency, created primarily for public purposes, and subject to be controlled for the public benefit. . . .

Such being the relations these corporations hold to the public, it would seem that the right of a colored person to use an improved public highway, upon the terms accorded to freemen of other races, is as fundamental, in the state of freedom established in this country, as are any of the rights which my brethren concede to be so far fundamental as to be deemed the essence of civil freedom. "Personal liberty consists," says Blackstone, "in the power of locomotion, of changing situation, or removing one's person to whatever places one's own inclination may direct, without restraint, unless by due course of law." But of what value is this right of locomotion, if it may be clogged by such burdens as Congress intended by the act of 1875 to remove? They are burdens which lay at the very foundation of the institution of slavery as it once existed. They are not to be sustained, except upon the assumption that there is in this land of universal liberty, a class which may still be discriminated against, even in respect of rights of a character so necessary and supreme, that, deprived of their enjoyment in common with others, a freeman is not only branded as one inferior and infected, but, in the competitions of life, is robbed of some of the most essential means of existence; and all this solely because they belong to a particular race which the nation has liberated. The Thirteenth Amendment alone obliterated the race line, so far as all rights fundamental in a state of freedom are concerned. . . .

[A] keeper of an inn is in the exercise of a quasi public employment. The law gives him special privileges and he is charged with certain duties and responsibilities to the public. The public nature of his employment forbids him from discriminating against any person asking admission as a guest on account of the race or color of that person.

. . . As to places of public amusement. It may be argued that the managers of such places have no duties to perform with which the public are, in any legal sense, concerned, or with which the public have any right to interfere; and, that the exclusion of a black man from a place of public amusement, on account of his race, or the denial to him, on that ground, of equal accommodations at such places, violates no legal right for the vindication of which he may invoke the aid of the courts. My answer is, that places of public amusement, within the meaning of the act of 1875, are such as are established and maintained under direct license of the law. The authority to establish and maintain them comes from the public. The colored race is a part of that public. The local government granting the license represents them as well as all other races within its jurisdiction. A license from the public to establish a place of public amusement, imports, in law, equality of right, at such places, among all the members of that public. This must be so, unless it be—which I deny—that the common municipal government of all the people may, in the exertion of its powers, conferred for the benefit of all, discriminate or authorize discrimination against a particular race, solely because of its former condition of servitude. . . .

Congress has not, in these matters, entered the domain of State control and supervision. It does not, as I have said, assume to prescribe the general conditions and limitations under which inns, public conveyances, and places of public amusement, shall be conducted or managed. It simply declares, in effect, that since the nation has established universal freedom in this country, for all time, there shall be no discrimination, based merely upon race or color, in respect of the accommodations and advantages of public conveyances, inns, and places of public amusement.

I am of the opinion that such discrimination practised by corporations and individuals in the exercise of their public or quasi-public functions is a badge of servitude the imposition of which Congress may prevent under its power, by appropriate legislation, to enforce the Thirteenth Amendment; and, consequently, without reference to its enlarged power under the Fourteenth Amendment, the act of March 1, 1875, is not, in my judgment, repugnant to the Constitution.

It remains now to consider these cases with reference to the power Congress has possessed since the adoption of the Fourteenth Amendment. Much that has been said as to the power of Congress under the Thirteenth Amendment is applicable to this branch of the discussion, and will not be repeated.

Before the adoption of the recent amendments, it had become . . . the established doctrine of this court that negroes, whose ancestors had been imported and sold as slaves, could not become citizens of a State, or even of the United States, with the rights and privileges guaranteed to citizens by the national constitution; further, that one might have all the rights and privileges of a citizen of a State without being a citizen in the sense in which that word was used in the national Constitution, and without being entitled to the privileges and immunities of citizens of the several States. Still, further, between the adoption of the Thirteenth Amendment and the proposal by Congress of the Fourteenth Amendment, on June 16, 1866, the statute books of several of the States, as we have seen, had become loaded down with enactments which, under the guise of Apprentice, Vagrant, and Contract regulations, sought to keep the colored race in a condition, practically, of servitude. It was openly announced that whatever might be the rights which persons of that race had, as freemen, under the guarantees of the national Constitution, they could not become citizens of a State, with the privileges belonging to citizens, except by the consent of such State; consequently, that their civil rights, as citizens of the State, depended entirely upon State legislation. To meet this new peril to the black race, that the purposes of the nation might not be doubted or defeated, and by way of further enlargement of the power of Congress, the Fourteenth Amendment was proposed for adoption.

Remembering that this court, in the *Slaughter-House Cases*, declared that the one pervading purpose found in all the recent amendments, lying at the foundation of each, and without which none of them would have been suggested—was "the freedom of the slave race, the security and firm establishment of that freedom, and the protection of the newly-made freeman and citizen from the oppression of those who had formerly exercised unlimited dominion over him"—that each amendment was addressed primarily to the grievances of that race—let us proceed to consider . . . the Fourteenth Amendment. . . .

. . . [My] brethren concede, that positive rights and privileges were intended to be secured, and are in fact secured, by the Fourteenth Amendment.

But when, under what circumstances and to what extent, may

Congress, by means of legislation, exert its power to enforce the provisions of this amendment? The theory of the opinion of the majority of the court—the foundation upon which their reasoning seems to rest—is, that the general government cannot, in advance of hostile State laws or hostile State proceedings, actively interfere for the protection of any of the rights, privileges, and immunities secured by the Fourteenth Amendment. It is said that such rights, privileges and immunities are secured by way of *prohibition* against State laws and State proceedings affecting such rights and privileges, and by power given to Congress to legislate for the purpose of carrying *such prohibition* into effect; also, that congressional legislation must necessarily be predicated upon such supposed State laws or State proceedings, and be directed to the correction of their operation and effect. . . .

The assumption that this amendment consists wholly of prohibitions upon State laws and State proceedings in hostility to its provisions is unauthorized by its language. The first clause of the first section—"All persons born or naturalized in the United States, and subject to the jurisdiction thereof, are citizens of the United States, and of the State wherein they reside"—is of a distinctly affirmative character. In its application to the colored race, previously liberated, it created and granted, as well citizenship of the United States, as citizenship of the State in which they respectively resided. It introduced all of that race, whose ancestors had been imported and sold as slaves, at once, into the political community known as the "People of the United States." They became, instantly, citizens of the United States, *and* of their respective States. . . .

The citizenship thus acquired, by that race, in virtue of an affirmative grant from the nation, may be protected, not alone by the judicial branch of the government, but by congressional legislation of a primary direct character; this, because the power of Congress is not restricted to the enforcement of prohibitions upon State laws or State action. It is, in terms distinct and positive, to enforce "the *provisions* of *this article*" of amendment; not simply those of a prohibitive character, but the provisions—*all* of the provisions—affirmative and prohibitive, of the amendment. It is, therefore, a grave misconception to suppose that the fifth section of the amendment has reference exclusively to express prohibitions upon State laws or State action. If any right was created by that amendment, the grant of power, through appropriate legislation, to enforce its provisions, authorizes Congress, by means of legislation, operating throughout the entire Union, to guard, secure, and protect that right.

It is, therefore, an essential inquiry what, if any, right, privilege or immunity was given, by the nation, to colored persons, when they were made citizens of the State in which they reside? Did the constitutional grant of State citizenship to that race, of its own force, invest them with any rights, privileges and immunities whatever? That they became entitled, upon the adoption of the Fourteenth Amendment, "to all privileges and immunities of citizens in the several States," within the meaning of section 2 of article IV of the Constitution, no one, I suppose, will for a moment question. What are the privileges and immunities to which, by that clause of the Constitution, they became entitled? To this it may be answered, generally, upon the authority of the adjudged cases, that they are those which are fundamental in citizenship in a free republican government, such as are "common to the citizens in the latter States under their constitutions and laws by virtue of their being citizens." Of that provision it has been said, with the approval of this court, that no other one in the Constitution has tended so strongly to constitute the citizens of the United States one people. . . .

. . . [No] State can sustain her denial to colored citizens of other States, while within her limits, of privileges or immunities, fundamental in republican citizenship upon the ground that she accords such privileges and immunities only to her white citizens and withholds them from her colored citizens. The colored citizens of other States, within the jurisdiction of that State, could claim, in virtue of section 2 of article IV of the Constitution, every privilege and immunity which that State secures to her white citizens. Otherwise, it would be in the power of any State, by discriminating class legislation against its own citizens of a particular race or color, to withhold from citizens of other States, belonging to that proscribed race, when within her limits, privileges and immunities of the character regarded by all courts as fundamental in citizenship. . . .

But what was secured to colored citizens of the United States—

as between them and their respective States—by the national grant to them of State citizenship? With what rights, privileges, or immunities did this grant invest them? There is one, if there be no other—exemption from race discrimination in respect of any civil right belonging to citizens of the white race in the same State. That, surely, is their constitutional privilege when within the jurisdiction of other States. And such must be their constitutional right, in their own State, unless the recent amendments be splendid baubles, thrown out to delude those who deserved fair and generous treatment at the hands of the nation. Citizenship in this country necessarily imports at least equality of civil rights among citizens of every race in the same State. It is fundamental in American citizenship that, in respect of such rights, there shall be no discrimination by the State, or its officers, or by individuals or corporations exercising public functions or authority, against any citizen because of his race or previous condition of servitude. . . .

The language of this court with reference to the Fifteenth Amendment, adds to the force of this view. In *United States* v. *Cruikshank,* it was said: ". . . [We have] held that the Fifteenth Amendment has invested the citizens of the United States with a new constitutional right, which is exemption from discrimination in the exercise of the elective franchise, on account of race, color, or previous condition of servitude. From this it appears that the right of suffrage is not a necessary attribute of national citizenship, but that exemption from discrimination in the exercise of that right on account of race, &c., is. The right to vote in the States comes from the States; but the right of exemption from the prohibited discrimination comes from the United States. The first has not been granted or secured by the Constitution of the United States, but the last has been."

Here, in language at once clear and forcible, is stated the principle for which I contend. It can scarcely be claimed that exemption from race discrimination, in respect of civil rights, against those to whom State citizenship was granted by the nation, is any less, for the colored race, a new constitutional right, derived from and secured by the national Constitution, than is exemption from such discrimination in the exercise of the elective franchise. It cannot be that the latter is an attribute of national citizenship, while the other is not essential in national citizenship, or fundamental in State citizenship.

If, then, exemption from discrimination, in respect of civil rights, is a new constitutional right, secured by the grant of State citizenship to colored citizens of the United States—and I do not see how this can now be questioned—why may not the nation, by means of its own legislation of a primary direct character, guard, protect and enforce that right? It is a right and privilege which the nation conferred. It did not come from the States in which those colored citizens reside. It has been the established doctrine of this court during all its history, accepted as essential to the national supremacy, that Congress, in the absence of a positive delegation of power to the State legislatures, may, by its own legislation, enforce and protect any right derived from or created by the national Constitution. . . .

This court has always given a broad and liberal construction to the Constitution, so as to enable Congress, by legislation, to enforce rights secured by that instrument. The legislation which Congress may enact, in execution of its power to enforce the provisions of this amendment, is such as may be appropriate to protect the right granted. The word appropriate was undoubtedly used with reference to its meaning, as established by repeated decisions of this court. Under given circumstances, that which the court characterizes as corrective legislation might be deemed by Congress appropriate and entirely sufficient. Under the other circumstances primary direct legislation may be required. But it is for Congress, not the judiciary, to say what legislation is appropriate—that is—best adapted to the end to be attained. The judiciary may not, with safety to our institutions, enter the domain of legislative discretion, and dictate the means which Congress shall employ in the exercise of its granted powers. That would be sheer usurpation of the functions of a co-ordinate department, which, if often repeated, and permanently acquiesced in, would work a radical change in our system of government. . . . "The sound construction of the Constitution," said Chief Justice Marshall, "must allow to the national legislature that discretion, with respect to the means by which the powers it confers are to be carried into execution, which will enable that body to perform the high duties assigned to it in the manner most beneficial to the people. Let the end be legitimate, let it be within the scope of the Constitution, and all means which are appropriate, which are plainly adapted to that end, which are not prohibited, but consist with the letter and spirit of the Constitution, are constitutional." *McCulloch* v. *Maryland*, 4 Wh. 421.

Must these rules of construction be now abandoned? Are the powers of the national legislature to be restrained in proportion as the rights and privileges, derived from the nation, are valuable? Are constitutional provisions, enacted to secure the dearest rights of freemen and citizens, to be subjected to that rule of construction, applicable to private instruments, which requires that the words to be interpreted must be taken most strongly against those who employ them? Or, shall it be remembered that "a constitution of government, founded by the people for themselves and their posterity, and for objects of the most momentous nature—for perpetual union, for the establishment of justice, for the general welfare, and for a perpetuation of the blessings of liberty—necessarily requires that every interpretation of its powers should have a constant reference to these objects? No interpretation of the words in which those powers are granted can be a sound one, which narrows down their ordinary import so as to defeat those objects." I Story Const. § 422.

. . . If the grant to colored citizens of the United States of citizenship in their respective States, imports exemption from race discrimination, in their States, in respect of such civil rights as belong to citizenship, then, to hold that the amendment remits that right to the States for their protection, primarily, and stays the hands of the nation, until it is assailed by State laws or State proceedings, is to adjudge that the amendment, so far from enlarging the powers of Congress—as we have heretofore said it did —not only curtails them, but reverses the policy which the general government has pursued from its very organization. Such an interpretation of the amendment is a denial to Congress of the power, by appropriate legislation, to enforce one of its provisions. In view of the circumstances under which the recent amendments were incorporated into the Constitution, and especially in view of the peculiar character of the new rights they created and secured, it ought not to be presumed that the general government has abdicated its authority, by national legislation, direct and primary in its character, to guard and protect privileges and immunities secured by that instrument. Such an interpretation of the Constitution ought not to be accepted if it be possible to avoid it. Its acceptance would lead to this anomalous result: that whereas, prior to the amendments, Congress, with the sanction of this court, passed the most stringent laws—operating directly and primarily upon States and their officers and agents, as well as upon individuals—in vindication of slavery and the right of the master, it may not now, by legislation of a like primary and direct character, guard, protect, and secure the freedom established, and the most essential right of the citizenship granted, by the constitutional amendments. With all respect for the opinion of others, I insist that the national legislature may, without transcending the limits of the Constitution, do for human liberty and the fundamental rights of American citizenship, what it did, with the sanction of this court, for the protection of slavery and the rights of the masters of fugitive slaves. If fugitive slave laws, providing modes and prescribing penalties, whereby the master could seize and recover his fugitive slave, were legitimate exertions of an implied power to protect and enforce a right recognized by the Constitution, why shall the hands of Congress be tied, so that—under an express power, by appropriate legislation, to enforce a constitutional provision granting citizenship—it may not, by means of direct legislation, bring the whole power of this nation to bear upon States and their officers, and upon such individuals and corporations exercising public functions as assume to abridge, impair, or deny rights confessedly secured by the supreme law of the land?

It does not seem to me that the fact that, by the second clause of the first section of the Fourteenth Amendment, the States are expressly prohibited from making or enforcing laws abridging the privileges and immunities of citizens of the United States, furnishes any sufficient reason for holding or maintaining that the amendment was intended to deny Congress the power, by general, primary, and direct legislation, of protecting citizens of the several States, being also citizens of the United States, against all discrimination, in respect of their rights as citizens, which is founded on race, color, or previous condition of servitude.

Such an interpretation of the amendment is plainly repugnant to its fifth section, conferring upon Congress power, by appropriate legislation, to enforce not merely the provisions containing prohibitions upon the States, but all of the provisions of the amendment, including the provisions, express and implied, in the first clause of the first section of the article granting citizenship. This alone is sufficient for holding that Congress is not restricted to the enactment of laws adapted to counteract and redress the operation of State legislation, or the action of State officers, of the character prohibited by the amendment. It was perfectly well known that the great danger to the equal enjoyment by citizens of their rights, as citizens, was to be apprehended not altogether from unfriendly State legislation, but from the hostile action of corporations and individuals in the States. And it is to be presumed that it was intended, by that section, to clothe Congress with power and authority to meet that danger. If the rights intended to be secured by the act of 1875 are such as belong to the citizen, in common or equally with other citizens in the same State, then it is not to be denied that such legislation is peculiarly appropriate to the end which Congress is authorized to accomplish, viz., to protect the citizen, in respect of such rights, against discrimination on account of his race. . . .

It is said that any interpretation of the Fourteenth Amendment different from that adopted by the majority of the court, would imply that Congress had authority to enact a municipal code for all the States, covering every matter affecting the life, liberty, and property of the citizens of the several States. Not so. Prior to the adoption of that amendment the constitutions of the several States, without perhaps an exception, secured all *persons* against deprivation of life, liberty, or property, otherwise than by due process of law, and, in some form, recognized the right of all *persons* to the equal protection of the laws. Those rights, therefore, existed before that amendment was proposed or adopted, and were not created by it. If, by reason of that fact, it be assumed that protection in these rights of persons still rests primarily with the States, and that Congress may not interfere except to enforce, by means of corrective legislation, the prohibitions upon State laws or State proceedings inconsistent with those rights, it does not at all follow, that privileges which has been *granted by the nation*, may not be protected by primary legislation upon the part of Congress. The personal rights and immunities recognized in the prohibitive clauses of the amendment were, prior to its adoption, under the protection, primarily, of the States, while rights, created by or derived from the United States, have always been, and, in the nature of things, should always be, primarily, under the protection of the general government. Exemption from race discrimination in respect of the civil rights which are fundamental in *citizenship* in a republican government, is, as we have seen, a new right, created by the nation, with express power in Congress, by legislation, to enforce the constitutional provision from which it is derived. If, in some sense, such race discrimination is, within the letter of the last clause of the first section, a denial of that equal protection of the laws which is secured against State denial to all persons, whether citizens or not, it cannot be possible that a mere prohibition upon such State denial, or a prohibition upon State laws abridging the privileges and immunities of citizens of the United States, takes from the nation the power which it has uniformly exercised of protecting, by direct primary legislation, those privileges and immunities which existed under the Constitution before the adoption of the Fourteenth Amendment, or have been created by that amendment in behalf of those thereby made *citizens* of their respective States.

This construction does not in any degree intrench upon the just rights of the States in the control of their domestic affairs. It simply recognizes the enlarged powers conferred by the recent amendments upon the general government. In the view which I take of those amendments, the States possess the same authority which they have always had to define and regulate the civil rights which their own people, in virtue of State citizenship, may enjoy within their respective limits; except that its exercise is now subject to the expressly granted power of Congress, by legislation, to enforce the provisions of such amendments—a power which necessarily carries with it authority, by national legislation, to protect and secure ·the privileges and immunities which are created by or are derived from those amendments. That exemption of citizens from discrimination based on race or color, in respect of civil rights, is one of those privileges or immunities, can no longer be deemed an open question in this court. . . .

My brethren say, that when a man has emerged from slavery,

and by the aid of beneficent legislation has shaken off the inseparable concomitants of that state, there must be some stage in the progress of his elevation when he takes the rank of a mere citizen, and ceases to be the special favorite of the laws, and when his rights as a citizen, or a man, are to be protected in the ordinary modes by which other men's rights are protected. It is, I submit, scarcely just to say that the colored race has been the special favorite of the laws. The statute of 1875, now adjudged to be unconstitutional, is for the benefit of citizens of every race and color. What the nation, through Congress, has sought to accomplish in reference to that race, is—what had already been done in every State of the Union for the white race—to secure and protect rights belonging to them as freemen and citizens; nothing more. It was not deemed enough "to help the feeble up, but to support him after." The one underlying purpose of congressional legislation has been to enable the black race to take the rank of mere citizens. The difficulty has been to compel a recognition of the legal right of the black race to take the rank of citizens, and to secure the enjoyment of privileges belonging, under the law, to them as a component part of the people for whose welfare and happiness government is ordained. At every step, in this direction, the nation has been confronted with class tyranny, which a contemporary English historian says is, of all tyrannies, the most intolerable, "for it is ubiquitous in its operation, and weighs, perhaps, most heavily on those whose obscurity or distance would withdraw them from the notice of a single despot." To-day, it is the colored race which is denied, by corporations and individuals wielding public authority, rights fundamental in their freedom and citizenship. At some future time, it may be that some other race will fall under the ban of race discrimination. If the constitutional amendments be enforced, according to the intent with which, as I conceive, they were adopted, there cannot be, in this republic, any class of human beings in practical subjection to another class, with power in the latter to dole out to the former just such privileges as they may choose to grant. The supreme law of the land has decreed that no authority shall be exercised in this country upon the basis of discrimination, in respect of civil rights, against freemen and citizens because of their race, color, or previous condition of servitude. To that decree—for the due enforcement of which, by appropriate legislation, Congress has been invested with express power—every one must bow, whatever may have been, or whatever now are, his individual views as to the wisdom or policy, either of the recent changes in the fundamental law, or of the legislation which has been enacted to give them effect.

For the reasons stated I feel constrained to withhold my assent to the opinion of the Court.

Plessy v. Ferguson

(163 U.S. 537)

[1896]

MR. JUSTICE BROWN delivered the opinion of the Court.

... The object of the [Fourteenth] amendment was undoubtedly to enforce the absolute equality of the two races before the law, but in the nature of things it could not have been intended to abolish distinctions based upon color, or to enforce social, as distinguished from political equality, or a commingling of the two races upon terms unsatisfactory to either. Laws permitting, and even requiring their separation in places where they are liable to be brought into contact do not necessarily imply the inferiority of either race to the other, and have been generally, if not universally, recognized as within the competency of the state legislatures in the exercise of their police power. The most common instance of this is connected with the establishment of separate schools for white and colored children, which has been held to be a valid exercise of the legislative power even by courts of States where the political rights of the colored race have been longest and most earnestly enforced.

One of the earliest of these cases is that of *Roberts* v. *City of Boston,* 5 Cush. 198, in which the Supreme Judicial Court of Massachusetts held that the general school committee of Boston had power to make provision for the instruction of colored children in separate schools established exclusively for them, and to prohibit their attendance upon the other schools. ... Similar laws have been enacted by Congress under its general power of legislation over the District of Columbia, ... as well as by the legislatures of many of the States, and have been generally, if not uniformly, sustained by the courts. ...

Laws forbidding the intermarriage of the two races may be said in a technical sense to interfere with the freedom of contract, and yet have been universally recognized as within the police power of the State. *State* v. *Gibson,* 36 Indiana, 389.

The distinction between laws interfering with the political equality of the negro and those requiring the separation of the two races in schools, theatres and railway carriages has been frequently drawn by this court. . . .

In this connection, it is also suggested by the learned counsel for the plaintiff in error that the same argument that will justify the state legislature in requiring railways to provide separate accommodations for the two races will also authorize them to require separate cars to be provided for people whose hair is of a certain color, or who are aliens, or who belong to certain nationalities, or to enact laws requiring colored people to walk upon one side of the street, and white people upon the other, or requiring white men's houses to be painted white, and colored men's black, or their vehicles or business signs to be of different colors, upon the theory that one side of the street is as good as the other, or that a house or vehicle of one color is as good as one of another color. The reply to all this is that every exercise of the police power must be reasonable, and extend only to such laws as are enacted in good faith for the promotion for the public good, and not for the annoyance or oppression of a particular class. . . .

So far, then, as a conflict with the Fourteenth Amendment is concerned, the case reduces itself to the question whether the statute of Louisiana is a reasonable regulation, and with respect to this there must necessarily be a large discretion on the part of the legislature. In determining the question of reasonableness it is at liberty to act with reference to the established usages, customs and traditions of the people, and with a view to the promotion of their comfort, and the preservation of the public peace and good order. Gauged by this standard, we cannot say that a law which authorizes or even requires the separation of the two races in public conveyances is unreasonable, or more obnoxious to the Fourteenth Amendment than the acts of Congress requiring separate schools for colored children in the District of Columbia, the constitutionality of which does not seem to have been questioned, or the corresponding acts of state legislatures.

We consider the underlying fallacy of the plaintiff's argument to consist in the assumption that the enforced separation of the two races stamps the colored race with a badge of inferiority. If this be so, it is not by reason of anything found in the act, but solely because the colored race chooses to put that construction upon it. The argument necessarily assumes that if, as has been more than once the case, and is not unlikely to be so again, the colored race should become the dominant power in the state legislature, and should enact a law in precisely similar terms, it would thereby relegate the white race to an inferior position. We imagine that the white race, at least, would not acquiesce in this assumption. The argument also assumes that social prejudices may be overcome by legislation, and that equal rights cannot be secured to the negro except by an enforced commingling of the two races. We cannot accept this proposition. If the two races are to meet upon terms of social equality, it must be the result of natural affinities, a mutual appreciation of each other's merits and a voluntary consent of individuals. . . . Legislation is powerless to eradicate racial instincts or to abolish distinctions based upon physical differences, and the attempt to do so can only result in accentuating the difficulties of the present situation. If the civil and political rights of both races be equal one cannot be inferior to the other civilly or politically. If one race be inferior to the other socially, the Constitution of the United States cannot put them upon the same plane. . . .

The judgment of the court below is, therefore, *Affirmed*.

Mr. Justice Harlan dissenting.

By the Louisiana statute, the validity of which is here involved, all railway companies (other than street railroad companies) carrying passengers in that State are required to have separate but equal accommodations for white and colored persons, "by providing two or more passenger coaches for each passenger train, *or* by dividing the passenger coaches by a *partition* so as to secure separate accommodations." Under this statute, no colored person is permitted to occupy a seat in a coach assigned to white persons; nor any white person, to occupy a seat in a coach assigned to colored persons. The managers of the railroad are not allowed to exercise any discretion in the premises, but are required to assign each passenger to some coach or compartment set apart for the exclusive use of his race. If a passenger insists upon going into a coach or compartment not set apart for persons of his race, he is subject to be fined, or to be imprisoned in the parish jail. Penalties are prescribed for the refusal or neglect of the officers, directors, conductors and employés of railroad companies to comply with the provisions of the act. . . .

While there may be in Louisiana persons of different races who are not citizens of the United States, the words in the act, "white and colored races," necessarily include all citizens of the United States of both races residing in that State. So that we have before us a state enactment that compels, under penalties, the separation of the two races in railroad passenger coaches, and makes it a crime for a citizen of either race to enter a coach that has been assigned to citizens of the other race.

Thus the State regulates the use of a public highway by citizens of the United States solely upon the basis of race.

However apparent the injustice of such legislation may be, we have only to consider whether it is consistent with the Constitution of the United States.

That a railroad is a public highway, and that the corporation which owns or operates it is in the exercise of public functions, is not, at this day, to be disputed. . . .

In respect of civil rights, common to all citizens, the Constitution of the United States does not, I think, permit any public authority to know the race of those entitled to be protected in the enjoyment of such rights. Every true man has pride of race, and under appropriate circumstances when the rights of others, his equals before the law, are not to be affected, it is his privilege to express such pride and to take such action based upon it as to him seems proper. But I deny that any legislative body or judicial tribunal may have regard to the race of citizens when the civil rights of those citizens are involved. Indeed, such legislation, as that here in question, is inconsistent not only with that equality of rights which pertains to citizenship, National and State, but with the personal liberty enjoyed by every one within the United States. . . .

The white race deems itself to be the dominant race in this country. And so it is, in prestige, in achievements, in education, in wealth and in power. So, I doubt not, it will continue to be for all time, if it remains true to its great heritage and holds fast to the principles of constitutional liberty. But in view of the

Constitution, in the eye of the law, there is in this country no superior, dominant, ruling class of citizens. There is no caste here. Our Constitution is color-blind, and neither knows nor tolerates classes among citizens. In respect of civil rights, all citizens are equal before the law. The humblest is the peer of the most powerful. The law regards man as man, and takes no account of his surroundings or of his color when his civil rights as guaranteed by the supreme law of the land are involved. It is, therefore, to be regretted that this high tribunal, the final expositor of the fundamental law of the land, has reached the conclusion that it is competent for a State to regulate the enjoyment by citizens of their civil rights solely upon the basis of race.

In my opinion, the judgment this day rendered will, in time, prove to be quite as pernicious as the decision made by this tribunal in the *Dred Scott case.* It was adjudged in that case that the descendants of Africans who were imported into this country and sold as slaves were not included nor intended to be included under the word "citizens" in the Constitution, and could not claim any of the rights and privileges which that instrument provided for and secured to citizens of the United States; that at the time of the adoption of the Constitution they were "considered as a subordinate and inferior class of beings, who had been subjugated by the dominant race, and, whether emancipated or not, yet remained subject to their authority, and had no rights or privileges but such as those who held the power and the government might choose to grant them." 19 How. 393, 404. The recent amendments of the Constitution, it was supposed, had eradicated these principles from our institutions. But it seems that we have yet, in some of the States, a dominant race—a superior class of citizens, which assumes to regulate the enjoyment of civil rights, common to all citizens, upon the basis of race. The present decision, it may well be apprehended, will not only stimulate aggressions, more or less brutal and irritating, upon the admitted rights of colored citizens, but will encourage the belief that it is possible, by means of state enactments, to defeat the beneficent purposes which the people of the United States had in view when they adopted the recent amendments of the Constitution, by one of which the blacks of this country were made citizens of the United

States and of the States in which they respectively reside, and whose privileges and immunities, as citizens, the States are forbidden to abridge. Sixty millions of whites are in no danger from the presence here of eight millions of blacks. The destinies of the two races, in this country, are indissolubly linked together, and the interests of both require that the common government of all shall not permit the seeds of race hate to be planted under the sanction of law. What can more certainly arouse race hate, what more certainly create and perpetuate a feeling of distrust between these races, than state enactments, which, in fact, proceed on the ground that colored citizens are so inferior and degraded that they cannot be allowed to sit in public coaches occupied by white citizens? That, as all will admit, is the real meaning of such legislation as was enacted in Louisiana.

The sure guarantee of the peace and security of each race is the clear, distinct, unconditional recognition by our governments, National and State, of every right that inheres in civil freedom, and of the equality before the law of all citizens of the United States without regard to race. State enactments, regulating the enjoyment of civil rights, upon the basis of race, and cunningly devised to defeat legitimate results of the war, under the pretence of recognizing equality of rights, can have no other result than to render permanent peace impossible, and to keep alive a conflict of races, the continuance of which must do harm to all concerned. This question is not met by the suggestion that social equality cannot exist between the white and black races in this country. That argument, if it can be properly regarded as one, is scarcely worthy of consideration; for social equality no more exists between two races when travelling in a passenger coach or a public highway than when members of the same races sit by each other in a street car or in the jury box, or stand or sit with each other in a political assembly, or when they use in common the streets of a city or town, or when they are in the same room for the purpose of having their names placed on the registry of voters, or when they approach the ballot-box in order to exercise the high privilege of voting. . . .

The arbitrary separation of citizens, on the basis of race, while they are on a public highway, is a badge of servitude wholly inconsistent with the civil freedom and the equality before the law established by the Constitution. It cannot be justified upon any legal grounds.

If evils will result from the commingling of the two races upon public highways established for the benefit of all, they will be infinitely less than those that will surely come from state legislation regulating the enjoyment of civil rights upon the basis of race. We boast of the freedom enjoyed by our people above all other peoples. But it is difficult to reconcile that boast with a state of the law which, practically, puts the brand of servitude and degradation upon a large class of our fellow-citizens, our equals before the law. The thin disguise of "equal" accommodations for passengers in railroad coaches will not mislead any one, nor atone for the wrong this day done. . . .

I am of opinion that the statute of Louisiana is inconsistent with the personal liberty of citizens, white and black, in that State, and hostile to both the spirit and letter of the Constitution of the United States. If laws of like character should be enacted in the several States of the Union, the effect would be in the highest degree mischievous. Slavery, as an institution tolerated by law would, it is true, have disappeared from our country, but there would remain a power in the States, by sinister legislation, to interfere with the full enjoyment of the blessings of freedom; to regulate civil rights, common to all citizens, upon the basis of race; and to place in a condition of legal inferiority a large body of American citizens, now constituting a part of the political community called the People of the United States, for whom, and by whom through representatives, our government is administered. Such a system is inconsistent with the guarantee given by the Constitution to each State of a republican form of government, and may be stricken down by Congressional action, or by the courts in the discharge of their solemn duty to maintain the supreme law of the land, anything in the constitution or laws of any State to the contrary notwithstanding.

For the reasons stated, I am constrained to withhold my assent from the opinion and judgment of the majority.

Federalism and the Administration of Justice*

BY BURKE MARSHALL

[1964]

A NECESSARY COROLLARY of Negro disenfranchisement, in the limited areas where that corruption of representative government is practiced, is the double standard in favor of whites, because of their race. It is not only that qualified Negroes are rejected. Whites who are unqualified under any interpretation of state law are registered, in large numbers, solely because they are white.

What view of the impartiality of justice, of the administration of law by public officials is held by the society that countenances such practices? There is involved, for one thing, a great gap between the demands of federal law and the practices of the states. That gap is tolerated publicly, at least in home territory, on historical and constitutional grounds. But also involved is the acceptance of a double standard in the daily administration of law—in many cases clearly beyond the very large limits of permissibility set by federal constitutional standards. This double standard is presently almost outside the reach of federal action unless state criminal convictions come up for review by the United States Supreme Court. Registration and voting is basic but sporadic. A double standard of law enforcement is routine and immediate, and affects not only the citizens involved, but the concept of government held by all the public officials concerned.

An incident brought to the attention of the Justice Department in early 1961 raised the question of federal responsibility for the administration of justice by state officials. A Negro Air Force captain, accompanied by another Negro who had formerly been

---

* From *Federalism and Civil Rights* (New York & London: Columbia University Press, 1964). Copyright © 1964 Columbia University Press. Reprinted by permission. Burke Marshall is the Head of the Civil Rights Division of the Justice Department.

an officer in the Air Force, visited the home, in a large southern city, of a white major in the Air Force with whom they had done duty. A neighbor complained, and all three were arrested and charged with disturbing the peace. On the same day, in the same city, a Negro civilian employee at a nearby Air Force base was arrested and charged with disorderly conduct because he went to the home of a white co-worker to discuss official business.

There was no justification for any of the arrests, under either state or federal law. Federal employees were involved. The entire incident smacked of police-state tactics imposing segregationist rules not only on Negroes but also on individual whites who obviously did not agree with them. The only alternatives were to complain to the local authorities or to bring criminal charges against the police officers for taking action, under orders, that violated federal constitutional rights. The Justice Department took the former step, but neither course carried any real promise of changing police conduct.

What was at issue was plain abuse of police power by state officials, directed in these cases against persons who had no point to prove but who had simply acted contrary to accepted racial patterns. Quite often the citizens who are the victims of such police action are trying to prove a point, but that only aggravates the situation because the point being proved is that Negroes are the victims of a caste system. In such a case, there may be a double imposition on federal law: individual citizens are subjected to an unconstitutional exercise of police power for trying to claim constitutional rights. There are many examples, not all of them in southern counties where the vote is also denied; in most of them no direct federal action is possible, or what is attempted proves futile. . . .

[Here is recited a series of incidents, occurring in 1963, in which authorities in various southern communities appear to have failed to protect individuals in the exercise of their constitutional rights, or to have harassed them, in different ways; or in which local authorities appear to have directed their powers to prosecute or punish against those offended instead of those offending.]

These examples cannot be disposed of as sporadic abuses of official power by minor bureaucrats. In each case, the official action was taken by leading citizens of the community. The incidents received wide attention throughout their communities and were defended by the local newspapers. Some cases attracted national interest for a time.

Since 1960, such incidents have been multiplied by hundreds of constitutionally dubious arrests in sit-in incidents and massive denials of First Amendment rights following demonstrations protesting segregation policies. In many cases, however, the constitutional limits on police action were sufficiently vague, or so much in flux at the time that no implication of deliberate double standard is provable.

This has led in the past three years to the greatest single source of frustration with and misunderstanding of the federal government, particularly among young people. They cannot understand federal inaction in the face of what they consider, often quite correctly, as official wholesale local interference with the exercise of federal constitutional rights. Apparently their schools and universities have not taught them much about the working of the federal system. In their eyes the matter is simple. Local authorities are depriving certain people of their federal rights, often in the presence of federal officials from the Justice Department. [Federal rights] . . . should be protected.

What is wrong with this analysis? Is the federal government simply failing to meet a clear responsibility for enforcing federal law?

The question embraces all the deepest complexities of the federal system. It is surrounded by some basic constitutional notions which have worked, and worked well, in other contexts, preserving the dilution of powers intended by the framers of the Constitution, and at the same time protecting individuals against deprivation of their freedoms.

The most fundamental, primary notion, of course, is that the constitutional rights involved are individual and personal, to be asserted by private citizens as they choose, in court, speaking through their chosen counsel. If the matter is one of unjustified criminal charges, the individual's rights are protected by the court system and by the right of trial by jury. If an unjust or unconstitutional conviction is obtained, it can be appealed. If the federal system of justice is not recognized and followed by the state courts, then recourse is had from review by the United

States Supreme Court or in the federal courts through *habeas corpus*. In this fashion individual rights are protected on an individual case-by-case basis, as they should be. All that is involved is a question of time. Even that is not of major importance as long as reasonable bail is allowed while the questions are in litigation.

Two other fundamental concepts flow from this structure of protecting federal rights. One is that rights must be asserted by individuals. The other is that federal courts will not interfere while the system is at work—that is, they will not enjoin a *pending* or *future* state criminal proceeding.

In general the first of these concepts means that federal government itself has no right to bring suit in federal court to protect federal rights guaranteed to individuals. For most of our history, this has not been a matter of debate. The fact of *en masse* deprival of rights has been limited, for various reasons, to Negroes. The principal one is that they are the only Americans who have been lumped together everlastingly by race into a solid caste openly treated differently from everyone else by law, and not recognized as individuals, in one significant section of the nation. Indeed, from 1896 to 1954 the system had a sort of sanction in federal law as well, under the myth that a society based on white supremacy would provide equal, if separate, schools and other facilities for the Negro. Again, for various reasons including mainly the denial of the right to vote, discrimination against Negroes has been largely ignored by most whites until very recently.

Recently this particular deference to the states has been under attack. . . . A three-judge court in Louisiana recently said that where the complaint "is based on a state law which is contrary to the superior authority of the United States Constitution, the Nation, as well as the aggrieved individuals, is injured." In technical terms, this would mean that the United States, through the Department of Justice, would have standing to sue to attack statutes presumptively used on a large-scale basis to deny to individuals their constitutional rights.

This is a matter of speculation. . . . The Congress has thus far refused to pass a statute to authorize the Justice Department, in a broad sense, to seek injunctions in federal courts to prevent any denial by state officials of federally protected rights, although parts of the pending Civil Rights Bill [*] would give the Department standing to attack school segregation, and discrimination in other municipal facilities such as parks and playgrounds. . . .

In any event, under existing law, federal courts strongly resist interfering with state court criminal proceedings. . . .

Thus more is at issue than whether the federal government has any responsibility at all, at least in court, to try to prevent unconstitutional state police action. In the entire course of our federal legal history, there has been but a handful of occasions where a federal court enjoined a state from prosecuting criminal charges for any reason. In most of those, the court has acted to protect one of two overriding, sometimes complementary, interests: enforcing its own orders, or preventing use of an unconstitutional criminal statute. . . .

The civil rights movement and public protests against segregation and racial discrimination have put a strain on these assumptions and rules of federal-state comity that is far greater than it has ever been. The massive number of arrests in the past four years is one factor. Another is the amount of bail required, making unreal the supposition that anyone convicted of a state crime can be free while the legality of his conviction under federal law is being tested. But the greatest impetus is from the immediacy and urgency of the protest movement, whose members do not understand their rights, vastly overestimate the scope of the First Amendment, and in large part do not care. These factors converge as part of the forces pulling towards a polarization between Negro leaders and their government, causing a loss in faith among young people that the federal government has the ability or will to protect constitutional rights in the South, and fostering the belief that gains can be made only through continued protest demonstrations in the streets, creating, as the Reverend Martin Luther King, Jr., has put it, a situation of total crisis which cannot be escaped.

Does this mean that the rules of comity, of federal tolerance of the administration of justice by states, should be reexamined? The concepts are not, after all, unalterable. At least two modes of

---

([*] Since passed, and commonly referred to as the Civil Rights Act of 1965.)

basic change have been proposed: one to enable the Department of Justice to seek federal court injunctions against any deprivations of federally protected rights; and the other to permit removal to the federal courts of any trials claimed to infringe constitutional guarantees. The second proposal has many difficulties: it would require a federal court to decide in advance when the state system will not work; it conceives of the federal judge as presiding over state law enforcement proceedings; it assumes a fairness in federal judges and federal juries above that available in state courts; and it ignores the administrative burden that would be thrust upon an overloaded federal judiciary unprepared for new strains. Nevertheless, that proposal in essence contemplates no more than a change in the forum of trial. The first envisages no trial at all, and cuts—as will be seen—much more deeply into the law enforcement structure of the nation. Yet both are legally conceivable, and have been vigorously advanced. In short, the insulation of state criminal processes from federal interference is not itself required by the Constitution. Should it be eliminated, or at least penetrated and changed?

It is difficult for anyone concerned with corruption of the law to say corrections are not needed. Negro disenfranchisement over decades has created a system of all-white courts, staffed entirely by white officials. The apparent inability of the bar to bring itself to provide counsel in cases involving racial implications is alone one proof that our basic assumptions about the workings of justice in state courts are wrong. The unavailability of normal sources of bail is another. Examples of abuse of authority such as I have cited are a third. They are compounded by repeated exclusion of Negroes from juries, enforced segregation and racial abuses in courtrooms, and other evidences of the weight of state authority thrusting imbalance into the processes of justice where racial customs are threatened.

All of this is, as the Civil Rights Commission has said, an affront to the conscience of the nation. Such inequity cannot be tolerated indefinitely in a free society. The question facing President Kennedy in June, 1963, and still confronting the country today, is whether it warrants or requires, as many have agreed, fundamental alterations in the relationships between the state and the federal courts, in the administration of justice.

The complete consequences of even limited adjustments of this sort can be seen only after years of experience. Some are inevitable, however, and can at least be identified and described.

For one thing, constitutional rights are indivisible. They cannot be segmented to deal with a particular problem, even that of race. The frustration and fury at state-controlled administration of justice is directed, first, at the resistance in southern states to the constitutional outlawing of official segregation by the Supreme Court, and secondly at the suppression of protests directed against the unofficial caste system in employment practices and public places which accompanied official segregation. But the personal constitutional rights affected are the rights to the equal protection of the laws, and to freedom of speech.

The wrongs are limited geographically, are hopefully transitional, and are in any event capable of definition. None of this is true of the [impact upon] constitutionally protected rights which would be affected by adjustments in the rules of comity presently controlling federal-state relations. If the federal courts are to be granted general powers to prevent state prosecutions from depriving a defendant of rights granted under the Fourteenth Amendment, those powers will not be limited to enjoining the prosecution of student civil rights workers in Georgia under that state's insurrection statute. They would include, among others in a wide range, power to stop criminal proceedings where a constitutional issue is raised as to the conduct of the trial (such as lack of counsel or racial bias in jury selection), and power to prevent prosecutions where claims are made under the First Amendment (such as censorship and interference with freedom of religion or speech).

So basic a change in the administration of the federal-state court systems would have such a clearly foreseeable impact on the speedy processing of criminal justice in the state courts that it has always met with deserved resistance. As the system now works, federal constitutional issues are tested after trial and conviction, upon the basis of the record at the trial. They are tested in the first instance in the state court. To permit them to be tested prior to trial in the federal courts, based upon a record which would be different than the record of the criminal proceedings in the state court, would permit defendants to delay their prosecution indefinitely. Furthermore these effects cannot be limited to matters arising from the distortions of the federal system linked to the civil rights crisis. It is too easy to raise at least a substantial question of a constitutional claim. The prospect of this litigation in the federal courts, prior to state trial, is by itself such a disadvantage that if there is a solution, it should be sought elsewhere.

These are purely administrative difficulties, however. There is a more fundamental objection to the suggestion that the Attorney General be given power to seek federal court injunctions against general denials of constitutional rights to individuals. There is no way of depriving him then of the power of choice. If the full range of constitutional rights were to be protected in this fashion, the Attorney General would have the power of choosing among them. He could decide either to use the resources and power of the Justice Department to protect economic rights, or not. He could at his will enter or avoid the field of censorship. He could defend or ignore plain deprivation of religious rights, including all of the rights of children in public schools. Within each of these fields, the powers of choice would be immense.

It is no answer to say that the Attorney General always has had the power to decide what cases to bring. It is one thing to make decisions in enforcing tax laws, or even to choose between conflicting economic considerations in enforcing antitrust statutes. It is another to give the government choices in advancing or protecting human rights. Decisions in protecting all constitutional rights demand choices among religions, or between religions and atheism. They permit choices among political movements, between integrationists and segregationists, between peace marchers and militarists—among, in short, all competing views as to the proper future course of the United States and of mankind.

Such problems, of course, are the result of the breadth of what has been suggested. Some of them can be eliminated, or at least narrowed in scope, to cut back on the practical possibility of any real danger to the society. There would yet remain a matter of power, the physical power of law enforcement and the responsibility for keeping order, which could not be avoided by draftsmanship or lawyers' work. It derives from interrelationship between the administration of justice in state courts and the use of police power in the streets. It is unavoidable.

A fair test of whether the problem of abuse of police power and criminal prosecution in the states can be dealt with directly at all under the federal system is to assume that a statute could be so drawn as to give the Attorney General the power to enjoin interference with constitutionally protected protests against racial injustices, and to do nothing else. That is, after all, the heart of the matter. What would be the consequences of that limited a use of the federal courts to prevent arrests and criminal prosecution that are, in any event, constitutionally prohibited?

A vivid piece of recent history suggests the answer. This was the sequence of events that took place in Alabama and Mississippi during the Freedom Rides of May, 1961.

In December, 1960, the Supreme Court decided that passengers using interstate bus lines had a federal right to be free from racial discrimination in bus terminals. The Freedom Rides followed to test whether this federal right was truly recognized in southern terminals. The rides were scheduled to end in New Orleans on May 17, the anniversary of the school cases [decision] of 1954, traversing on the way Virginia, North Carolina, South Carolina, Georgia, Alabama, and Mississippi.

The trip proceeded without attracting national attention to Atlanta, Georgia, where the riders divided into two groups, one traveling by Greyhound, and the other by Trailways, to Birmingham, Alabama, on May 14. The Greyhound bus was attacked in Anniston, Alabama, by a gang of men carrying clubs, chains, and blackjacks. They broke windows, slashed tires, and burned the bus. The Trailways bus was met at the Birmingham station by other men who severely beat several members of the Freedom Ride group. Birmingham police did not arrive until several minutes after the violence occurred, although the scheduled arrival of the bus had been well publicized and was observed by a number of newsmen.

During the next week, President Kennedy and the Attorney General attempted to persuade Governor Patterson of Alabama and local authorities to accept the police responsibility for the safety of the buses and the passengers. Until assurances of protection could be obtained, it was impossible to find a bus driver who was willing to take the bus to Montgomery, its next scheduled stop. In the meantime, the original group of Freedom Riders had been replaced by others who sat each day in buses at the

Birmingham terminal waiting to be taken to Montgomery. Governor Patterson was at first unavailable, even to the President. Finally, on Friday, May 20, the governor agreed to see a representative of the President from the Justice Department.

He gave assurances that he had the will, the force, the men, and the equipment to protect everyone in Alabama, including interstate travelers.

On Saturday, May 21, a Greyhound bus carrying a Freedom Ride group proceeded from Birmingham to Montgomery under escort of Alabama State Police. The bus was met in Montgomery by another mob of about 1,000 persons who rioted and attacked the group of riders, beating them with pipes, sticks, clubs, and fists. The President's representative who had met with Governor Patterson was himself struck down while attempting to assist one of the Freedom Riders, a girl, to escape the mob. He lay on the sidewalk for twenty-five minutes before police took him to a hospital. A federal court later found that the Montgomery police had engaged in willful and deliberate failure to provide protection before or after the bus had arrived.

As the situation developed during the week, the Department of Justice made plans for federal action in the event that state authorities continued to fail to protect American citizens traveling interstate and seeking to exercise federal rights despite local resentment. When state and local authorities failed to protect the Freedom Riders in Montgomery, despite the assurances given by Governor Patterson, these plans were put into effect. On Saturday, the Department sought and obtained a restraining order from the federal court in Montgomery against Ku Klux Klan groups involved in the violence to prevent their continued interference with interstate travel. The Department also sought and obtained an order preventing the Montgomery police from continuing to fail to provide protection for interstate travelers.

In this way the groundwork was laid for the use of direct federal police action to enforce the court order. Some 600 federal officers from the Border Patrol, the Bureau of Prisons, the Alcohol and Tobacco Tax Unit of the Treasury, and various United States marshals' offices in a number of states, all of whom had been on an alert since the previous Thursday, were instructed to proceed to Montgomery, where they were sworn in as deputy marshals for the Middle District of Alabama. The men were put under orders from the then Deputy Attorney General, Byron White, whose instructions were to assist state and local officials in seeing that the orders of the federal court were complied with and that the laws of the United States were executed. The entire plan was based upon the President's instructions to the Attorney General to take such measures as were necessary to suppress domestic violence, and protect the right of citizens of the United States to travel among the states.

On Sunday night, May 21, and the early morning of May 22, the deputized marshals were required to perform police duty to protect a Negro church in which Martin Luther King, Jr., was holding a mass meeting. A large mob gathered outside the church, burned a car in the area, and advanced upon the church throwing rocks and bottles. The mob was broken up by the federal officers with tear gas and billy clubs, and a disaster was prevented. Control was finally established when the local police arrived and joined forces with the federal agents, and after Governor Patterson put Montgomery under martial law and called out the National Guard. The federal agents went on duty at the church at about 6 o'clock Sunday evening and were withdrawn about 2 a.m. Monday. During the day a few had been placed at bus terminals to see that interstate travelers were not mobbed.

The Freedom Riders were determined, as was their right, to proceed from Montgomery to Jackson, Mississippi, by bus, and then on to New Orleans. The Attorney General asked that order be preserved by state and local authorities in Mississippi and received assurances from Governor Barnett and Mayor Thompson of Jackson. It had to be decided whether to accept the word of these officials in view of events in Montgomery following the assurances given by Governor Patterson. There was considerable evidence that there would be violence in Mississippi. Nevertheless, it was doubtful whether the President had any constitutional choice about refusing in advance to accept the word of the governor of a state, and it was clearly necessary in any event to try immediately to re-establish the responsibility of the states to use their constitutional police powers to maintain order. The federal decision was made on that basis. The bus was completely unaccompanied by any federal officials when it proceeded from Montgomery to the Alabama border under escort of Alabama officials and when it traveled across Mississippi to Jackson, it was guarded solely by the Mississippi Highway Patrol.

In Jackson, order was maintained, but federal law was not. The Freedom Riders were arrested when they attempted to use the bus terminal facilities on an integrated basis. The constitutionality of those arrests is still, almost three years later, undetermined. Conviction of one of the Riders was recently affirmed by the Supreme Court of Mississippi on the grounds that the defendant came to Jackson for the deliberate purpose of inciting violence, even though the violence incited was from local whites who resented his efforts to exercise his federal right to be free from racial discrimination in the Jackson bus terminal. During the summer that followed, a large number of Negro and white citizens who traveled to Jackson and attempted to use bus terminal facilities on an integrated basis were also arrested and jailed. Efforts to obtain federal court orders to enjoin these arrests were unsuccessful.

The constitutional and doctrinal basis for the use of federal deputy marshals in Montgomery is sufficiently complex to justify a set of lectures by itself. The incident resulted in a coalescing of the functions of state and federal governments with the responsibilities of each unclear for a period of time. The chief of police at Montgomery asked, when federal officials started arriving early Sunday morning, whether they intended to take over the traffic and fire control duties in the city. His remark shows the dilemma which the federal system imposes on the nation when a clash between federal and local law, and federal and local standards, is so deep on a substantive issue that the police machinery of the state is not available to perform its proper function.

In effect, in Anniston, in Birmingham, and in Montgomery, the white people and the authorities of Alabama tried to deal with the problem of federal rights which they found unacceptable by abdicating police responsibility. The expectation was that the matter would be resolved, as Reconstruction was finally resolved, and as the entrance of a Negro girl into the University of Alabama in 1956 had been resolved, by terror and violence which the federal government would not stop. Federal rights were protected finally only because the federal government did act, at the cost of a temporary and localized, though constitutionally permitted, alteration of the federal system as it usually operates, with federal police performing functions that were the responsibility of the state.

In Mississippi, in contrast, the state authorities met their police responsibilities, but preserved order by refusing to give any effect in their police action to the federally protected rights that were at issue. Their assumption apparently was that the time consumed between arrest and the final vindication of federal rights by the United States Supreme Court, the amount of bail required, the unavailability of counsel, and the other delays and imperfections in the constitutional mechanics of protecting federal rights in state criminal proceedings would prove so discouraging that the federal rights would atrophy. It was for this reason that the city of Jackson tried every case separately. This was done even though the facts did not vary and the legal issues could have been tested in a single case. Accordingly, every Freedom Rider who had gone to Jackson during the spring and summer had to retain a lawyer, put up bail, and return for trial in an endless procession.

The acceptance by the federal government of a responsibility directly to protect persons exercising federal rights in Alabama when the state failed to do so had wide implications. Failure to take action was unjustified then and intolerable in the long run. Other Freedom Rides would have followed at best, simply creating a new situation with each day. At worst, federal authorities would have permitted state inaction and mob violence to inhibit or destroy the willingness of Negroes to exercise federally protected rights. The action that was taken unquestionably led to a renewed acceptance of responsibility for order by both the states of Alabama and Mississippi.

On the other hand, what happened in Mississippi was not satisfactory. Persons going to Mississippi to exercise federal rights were arrested. Many were jailed; others had to meet a heavy burden of raising cash bonds and litigating their causes. The federal courts refused to enjoin the arrests. Should the system of law which permits this result be allowed to continue?

It would be possible to devise authority for the federal courts to enjoin such arrests. There is no constitutional or doctrinal difficulty involved. But the consequences would be to destroy the means by which Mississippi maintained order. An injunction could be justified as a matter of equity and under the constitution against the city authorities to prohibit arrests of persons protesting interference with the use of the bus terminal facilities on an integrated basis. What would be the practical effect?

The sequence of events in Mississippi and Alabama strongly suggests that the result would have been chaotic and more destructive of the federal system than what happened in Mississippi. The assurances given by Governor Barnett and the Mississippi authorities were based on the assumption that the police could arrest the demonstrators rather than control the mob. It is doubtful that they could have been obtained had it been clear that the arrests would be prohibited.

The issue at least can be stated clearly. On the one hand, state police thought, or were told by their lawyers, that they could arrest any Negro or mixed group attempting to use a white waiting room, and charge them with disturbing the peace, or inciting to riot, or other local offenses. The arrests might be wrong, but that is not the question. The question is whether a federal court should be empowered in advance to decide they are unlawful, and to use its process to prevent them. If that is done, what is the order that the court issues? How large a crowd should be permitted to gather before police action is permissible, under the court decision? What is constitutional in a town with almost no police, and what in a city if the police stay home? Who will advocate the federal police force that would be necessary if these decisions had to be made within minutes or, at the most, hours? The questions are easier to answer in theory than by those having responsibility for the domestic tranquility. They are less debatable where the rights are clearly defined than where demonstrations and the defensible limits of protest are at stake.

It may be said that these difficulties are a result of the existence of federal rights, and the requirements of the constitution. They are nonetheless what would have to be faced, and are not faced under the accepted roles of the federal and state governments under any existing legislation, or by the supporters of direct federal court control over police action. The deputy marshals in Montgomery were drawn from various federal agencies, assigned on a temporary basis, and could not possibly have operated for any extended period of time to maintain order in the city. There are no available federal police resources, and their use would not be tolerable over a period of time in any event. The dilemma is in this sense insoluble. There is no satisfactory way to prevent police interference with the exercise of federally protected rights under such circumstances. And it should be noted that this is not necessarily a loss to our national interest. The time spent under the federal system in protecting the rights of demonstrators under law performs a function of its own by creating a period in which to attack directly the substantive problem of discrimination (in this case segregation in the bus terminal) instead of attempting to prevent the police from interfering with demonstrations protesting the discrimination.

This is what happened finally in Jackson, and elsewhere in Mississippi.

Eventually federal court orders enjoined the police and city authorities from maintaining segregation in terminals, on the same basis as an injunction against segregation in schools or other public facilities. Official segregation signs in front of the terminals were removed. There were hours of crisis, particularly in McComb, Mississippi, when the need for direct federal intervention to protect persons availing themselves of a court order hung in the balance. But in the end the local authorities in McComb and other cities met their responsibilities. No direct federal interference with the police function as such was necessary, and the question of law enforcement that remained was the more general one of the ability of the federal courts to change the structure of Alabama and Mississippi society through injunctions against official racial discrimination.

For the present, this solution has worked. Persuasion, or federal legal pressure, or changes in local law in one way or another has resulted in accommodations, quickly enough, in particular places, to the driving need for racial fairness and some equality of treatment. For the reasons just discussed, it is fortunate that this is so. We are still at the edges of the legal problem of equality under law in most of the southern states. Yet official discrimination has become barely tolerable over the shortest period. Many of the current methods of public protest against racial injustice are not constitutionally protected, but neither are many of the repressive police tactics taken in retaliation. There are controls over the violations of local law, but only delayed, legalistic, theoretical restraints on police.

Under these circumstances, how long will the inescapable dilemma of the federal system continue to permit resistance to demands for direct federal controls over local police action? There are several factors at work.

A principal one, as in any federal-state conflict, is the future conduct of state officials, as the requirements of federal law become clearer. It has been possible for city attorneys and other lawyers for police officials in recent months to rely on lack of clarity in the federal law. After all, there are many Supreme Court decisions on freedom of expression and the right of protest, and over a period of time the opinions have not been unanimous or capable of resolution into a simple set of rules. . . .

As the protests against racial discrimination have increased, these ambiguities decrease. The Supreme Court has been required to decide more and more cases arising from official repression of assertions of racial equality. The Court's recent decisions overturning mass arrests of racial demonstrators in South Carolina are one example. Its summary reversal of a contempt citation of a Negro woman who refused to answer questions put to her when addressed by her first name is another. The Court has already defined at least some limitations on the power of police to make arrests in sit-in cases, where the state participated in racial discrimination by privately owned business. There are other cases pending now which will establish what further limitations, if any, exist.

The point is not to analyze the decisions themselves. It is enough that they are being made, inescapably, by reason of the function of the Court, and that once made, they clarify federal law, the rights of the people, and the constitutional limitations on the power of the police. To the extent these decisions are ignored by police officials, as the school decisions of 1954 have been ignored by most school boards in eleven states, cynicism, distrust of government, and disbelief in any kind of law will increase. So will the inescapable pressures for federal control of local police, by court injunction or otherwise.

Recent events in various southern towns suggest that the danger of overwhelming abuses of police authority will be overcome by local restraint and experience. The idea of racial protest was unthinkable, only months ago, in many places, and the first who protested suffered accordingly. In at least some cities, this was caused by bad legal advice, or none, and by inexperience. The area of total resistance to any protest has in any event become limited.

The important factor, however, is not the question of abuse of police authority as such. Our national objective is not [only] to protect the First Amendment right to protest racial discrimination, but to end the discrimination itself. This is the purpose of the pending bill. It assumes, as do the present voting laws, that this purpose can be accomplished, under law, through the processes of the federal courts, and without further federal interference with local police, educational, or other functions. The main issue is whether that effort can succeed.

In part, this turns on the ability of the federal government to see to it that court orders are enforced. The technical side of the question has been resolved wherever it has arisen. . . .

The dangerous aspect of the cases is the notion of perpetual relitigation of issues, on the grounds that some court may change its mind. . . .

The difficulty is—and it is a major one—that once the notion is accepted that everyone can retest basic constitutional decisions, no one obeys the law. This is what endangers the authority of local and state police in the long run. They may decide, in the pattern of massive resistance, to continue to ignore the requirements of federal law until their actions go beyond the limits of tolerance by the nation at large. . . .

Those who say that civil rights issues cut into the fabric of federalism are correct. They cut most deeply where police power is involved, for the police as well as for those in conflict with the police. There would be vast problems in any attempts at federal control of the administration of justice, even through the moderate method of federal court injunctions. Yet vast problems have been created already by police indifference to Negro rights in the South, and they will grow if the trend is not turned. The loss of faith in law—the usefulness of federal law and the fairness of local law—is gaining very rapidly among Negro and white civil rights workers. The consequences in the future cannot be foreseen.

Brown v. Board of Education of Topeka

(347 U.S. 483)

[1954]

MR. CHIEF JUSTICE WARREN delivered the opinion of the Court.

These cases come to us from the States of Kansas, South Carolina, Virginia, and Delaware. They are premised on different facts and different local conditions, but a common legal question justifies their consideration together in this consolidated opinion.

In each of the cases, minors of the Negro race, through their legal representatives, seek the aid of the courts in obtaining admission to the public schools of their community on a nonsegregated basis. In each instance, they had been denied admission to schools attended by white children under laws requiring or permitting segregation according to race. This segregation was alleged to deprive the plaintiffs of the equal protection of the laws under the Fourteenth Amendment. . . .

The plaintiffs contend that segregated public schools are not "equal" and cannot be made "equal," and that hence they are deprived of the equal protection of the laws. Because of the obvious importance of the question presented, the Court took jurisdiction. Argument was heard in the 1952 Term, and reargument was heard this Term on certain questions propounded by the Court.

Reargument was largely devoted to the circumstances surrounding the adoption of the Fourteenth Amendment in 1868. It covered exhaustively consideration of the Amendment in Congress, ratification by the states, then existing practices in racial segregation, and the views of proponents and opponents of the Amendment. This discussion and our own investigation convince us that, although these sources cast some light, it is not enough to resolve the problem with which we are faced. At best, they are inconclusive. The most avid proponents of the post-War Amendments undoubtedly intended them to remove all legal distinctions among "all persons born or naturalized in the United States." Their opponents, just as certainly, were antagonistic to both the letter and the spirit of the Amendments and wished them to have the most limited effect. What others in Congress and the state legislatures had in mind cannot be determined with any degree of certainty.

An additional reason for the inconclusive nature of the Amendment's history, with respect to segregated schools, is the status of public education at that time. In the South, the movement toward free common schools, supported by general taxation, had not yet taken hold. Education of white children was largely in the hands of private groups. Education of Negroes was almost nonexistent, and practically all of the race were illiterate. In fact, any education of Negroes was forbidden by law in some states. Today, in contrast, many Negroes have achieved outstanding success in the arts and sciences as well as in the business and professional world. It is true that public education had already advanced further in the North, but the effect of the Amendment on Northern States was generally ignored in the congressional debates. Even in the North, the conditions of public education did not approximate those existing today. The curriculum was usually rudimentary; ungraded schools were common in rural areas; the school term was but three months a year in many states; and compulsory school attendance was virtually unknown. As a consequence, it is not surprising that there should be so little in the history of the Fourteenth Amendment relating to its intended effect on public education.

In the first cases in this Court construing the Fourteenth Amendment, decided shortly after its adoption, the Court interpreted it as proscribing all state-imposed discriminations against the Negro race. The doctrine of "separate but equal" did not make its appearance in this Court until 1896 in the case of *Plessy* v. *Ferguson* involving not education but transportation. American courts have since labored with the doctrine for over half a century. In this Court, there have been six cases involving the "separate but equal" doctrine in the field of public education. In *Cumming* v. *Board of Education of Richmond County*, 175 U.S. 528, and *Gong Lum* v. *Rice*, 275 U.S. 78, the validity of the doctrine itself was not challenged. In more recent cases, all on the graduate school level, inequality was found in that specific benefits enjoyed by white students were denied to Negro students of the same educational qualifications. *State of Missouri* ex rel. *Gaines* v. *Canada,* 305 U.S. 337; *Sipuel* v. *Board of Regents of University of Oklahoma,* 332 U.S. 631; *Sweatt* v. *Painter,* 339 U.S. 629; *McLaurin* v. *Oklahoma State Regents,* 339 U.S. 637. In none of these cases was it necessary to reexamine the doctrine to grant relief to the Negro plaintiff. And in *Sweatt* v. *Painter* the Court expressly reserved decision on the question whether *Plessy* v. *Ferguson* should be held inapplicable to public education.

In the instant cases, that question is directly presented. Here, unlike *Sweatt* v. *Painter,* there are findings below that the Negro and white schools involved have been equalized, or are being equalized, with respect to buildings, curricula, qualifications and salaries of teachers, and other "tangible" factors. Our decision, therefore, cannot turn on merely a comparison of these tangible factors in the Negro and white schools involved in each of the cases. We must look instead to the effect of segregation itself on public education.

In approaching this problem, we cannot turn the clock back to 1868 when the Amendment was adopted, or even to 1896 when *Plessy* v. *Ferguson* was written. We must consider public education in the light of its full development and its present place in American life throughout the nation. Only in this way can it be determined if segregation in public schools deprives these plaintiffs of the equal protection of the laws.

Today, education is perhaps the most important function of state and local governments. Compulsory school attendance laws and the great expenditures for education both demonstrate our recognition of the importance of education to our democratic society. It is required in the performance of our most basic public responsibilities, even service in the armed forces. It is the very foundation of good citizenship. Today it is a principal instrument in awakening the child to cultural values, in preparing him for later professional training, and in helping him to adjust normally to his environment. In these days, it is doubtful that any child may reasonably be expected to succeed in life if he is denied the opportunity of an education. Such an opportunity, where the state has undertaken to provide it, is a right which must be made available to all on equal terms.

We come then to the question presented: Does segregation of children in public schools solely on the basis of race, even though the physical facilities and other "tangible" factors may be equal, deprive the children of the minority group of equal educational opportunities? We believe that it does.

In *Sweatt* v. *Painter,* in finding that a segregated law school for Negroes could not provide them equal educational opportunities, this Court relied in large part on "those qualities which are incapable of objective measurement but which make for greatness in a law school." In *McLaurin* v. *Oklahoma State Regents,* the Court, in requiring that a Negro admitted to a white graduate school be treated like all other students, again resorted to intangible considerations: ". . . his ability to study, to engage in discussions and exchange views with other students, and, in general, to learn his profession." Such considerations apply with added force to children in grade and high schools. To separate them from others of similar age and qualifications solely because of their race generates a feeling of inferiority as to their status in the community that may affect their hearts and minds in a way unlikely ever to be undone. The effect of this separation on their educational opportunities was well stated by a finding in the Kansas case by a court which nevertheless felt compelled to rule against the Negro plaintiffs:

"Segregation of white and colored children in public schools has a detrimental effect upon the colored children. The impact is greater when it has the sanction of the law; for the policy of separating the races is usually interpreted as denoting the inferiority of the Negro group. A sense of inferiority affects the motivation of a child to learn. Segregation with the sanction of law, therefore, has a tendency to retard the educational and mental development of Negro children and to deprive them of some of the benefits they would receive in a racially integrated school system." Whatever may have been the extent of psychological knowledge at the time of *Plessy* v. *Ferguson,* this finding is amply supported by modern authority. Any language in *Plessy* v. *Ferguson* contrary to this finding is rejected.

We conclude that in the field of public education the doctrine of "separate but equal" has no place. Separate educational facilities are inherently unequal. Therefore, we hold that the plaintiffs and others similarly situated for whom the actions have been brought are, by reason of the segregation complained of, deprived of the equal protection of the laws guaranteed by the Fourteenth Amendment. This disposition makes unnecessary any discussion whether such segregation also violates the Due Process Clause of the Fourteenth Amendment.

Because these are class actions, because of the wide applicability of this decision, and because of the great variety of local conditions, the formulation of decrees in these cases presents problems of considerable complexity. On reargument, the consideration of appropriate relief was necessarily subordinated to the primary question—the constitutionality of segregation in public education. We have now announced that such segregation is a denial of the equal protection of the laws. In order that we may have the full assistance of the parties in formulating decrees, the cases will be restored to the docket, and the parties are requested to present further argument [on questions as to an appropriate decree]. The Attorney General of the United States is again invited to participate. The Attorneys General of the states requiring or permitting segregation in public education will also be permitted to appear as *amici curiae* upon request to do so by September 15, 1954, and submission of briefs by October 1, 1954.

It is so ordered.

The Moral Crisis*

BY JOHN F. KENNEDY

[1963]

GOOD EVENING, my fellow citizens.

This afternoon, following a series of threats and defiant statements, the presence of the Alabama National Guardsmen was required at the University of Alabama to carry out the final and unequivocal order of the United States District Court of the Northern District of Alabama.

That order called for the admission of two clearly qualified young Alabama residents who happened to have been born Negro.

That they were admitted peacefully on the campus is due in good measure to the conduct of the students of the University of Alabama who met their responsibilities in a constructive way.

I hope that every American, regardless of where he lives, will stop and examine his conscience about this and other related incidents.

This nation was founded by men of many nations and backgrounds. It was founded on the principle that all men are created equal, and that the rights of every man are diminished when the rights of one man are threatened.

Today we are committed to a worldwide struggle to promote and protect the rights of all who wish to be free. And when Americans are sent to Vietnam or West Berlin we do not ask for whites only.

It ought to be possible, therefore, for American students of any color to attend any public institution they select without having to be backed up by troops. It ought to be possible for American consumers of any color to receive equal service in places of public accommodation, such as hotels and restaurants, and theaters and retail stores without being forced to resort to demonstrations in the street.

---

* From a national radio and television address given June 11, 1963. John F. Kennedy, (1917–1963) was the 35th President of the United States.

And it ought to be possible for American citizens of any color to register and to vote in a free election without interference or fear of reprisal.

It ought to be possible, in short, for every American to enjoy the privileges of being American without regard to his race or his color.

In short, every American ought to have the right to be treated as he would wish to be treated, as one would wish his children to be treated. But this is not the case.

The Negro baby born in America today, regardless of the section or the state in which he is born, has about one-half as much chance of completing a high school as a white baby, born in the same place, on the same day; one-third as much chance of completing college; one-third as much chance of becoming a professional man; twice as much chance of becoming unemployed; about one-seventh as much chance of earning $10,000 a year; a life expectancy which is seven years shorter and the prospects of earning only half as much.

This is not a sectional issue. Difficulties over segregation and discrimination exist in every city, in every state of the Union, producing in many cities a rising tide of discontent that threatens the public safety.

Nor is this a partisan issue. In a time of domestic crisis, men of goodwill and generosity should be able to unite regardless of party or politics.

This is not even a legal or legislative issue alone. It is better to settle these matters in the courts than on the streets, and new laws are needed at every level. But law alone cannot make men see right.

We are confronted primarily with a moral issue. It is as old as the Scriptures and is as clear as the American Constitution. The heart of the question is whether all Americans are to be afforded equal rights and equal opportunities; whether we are going to treat our fellow Americans as we want to be treated.

If an American, because his skin is dark, cannot eat lunch in a restaurant open to the public; if he cannot send his children to the best public school available; if he cannot vote for the public officials who represent him; if, in short, he cannot enjoy the full and free life which all of us want, then who among us would be content to have the color of his skin changed and stand in his place?

Who among us would then be content with the counsels of patience and delay. One hundred years of delay have passed since President Lincoln freed the slaves, yet their heirs, their grandsons, are not fully free. They are not yet freed from the bonds of injustice; they are not yet freed from social and economic oppression.

And this nation, for all its hopes and all its boasts, will not be fully free until all its citizens are free.

We preach freedom around the world, and we mean it. And we cherish our freedom here at home. But are we to say to the world—and much more importantly to each other—that this is the land of the free, except for the Negroes; that we have no second-class citizens, except Negroes; that we have no class or caste system, no ghettos, no master race, except with respect to Negroes.

Now the time has come for this nation to fulfill its promise. The events in Birmingham and elsewhere have so increased the cries for equality that no city or state or legislative body can prudently choose to ignore them.

The fires of frustration and discord are burning in every city, North and South. Where legal remedies are not at hand, redress is sought in the streets in demonstrations, parades and protests, which create tensions and threaten violence—and threaten lives.

We face, therefore, a moral crisis as a country and a people. It cannot be met by repressive police action. It cannot be left to increased demonstrations in the streets. It cannot be quieted by token moves or talk. It is a time to act in the Congress, in your state and local legislative body, and, above all, in all of our daily lives.

It is not enough to pin the blame on others, to say this is a problem of one section of the country or another, or deplore the facts that we face. A great change is at hand, and our task, our obligation is to make that revolution, that change peaceful and constructive for all.

Those who do nothing are inviting shame as well as violence. Those who act boldly are recognizing right as well as reality.

Next week I shall ask the Congress of the United States to

act, to make a commitment it has not fully made in this century to the propositions that race has no place in American life or law.

The Federal judiciary has upheld that proposition in a series of forthright cases. The Executive Branch has adopted that proposition in the conduct of its affairs, including the employment of Federal personnel, and the use of Federal facilities, and the sale of Federally financed housing.

But there are other necessary measures which only the Congress can provide, and they must be provided at this session.

The old code of equity law under which we live commands for every wrong a remedy. But in too many communities, in too many parts of the country wrongs are inflicted on Negro citizens and there is no remedy in law.

Unless the Congress acts their only remedy is the street.

I am, therefore, asking the Congress to enact legislation giving all Americans the right to be served in facilities which are open to the public—hotels, restaurants and theaters, retail stores and similar establishments. This seems to me to be an elementary right.

Its denial is an arbitrary indignity that no American . . . should have to endure, but many do.

I have recently met with scores of business leaders, urging them to take voluntary action to end this discrimination. And I've been encouraged by their response. . . . But many are unwilling to act alone. And for this reason nationwide legislation is needed, if we are to move this problem from the streets to the courts.

I'm also asking Congress to authorize the Federal Government to participate more fully in lawsuits designed to end segregation in public education. We have succeeded in persuading many districts to desegregate voluntarily. Dozens have admitted Negroes without violence.

Today a Negro is attending a state-supported institution in every one of our fifty states, but the pace is very slow.

Too many Negro children entering segregated grade schools at the time of the Supreme Court's decision nine years ago will enter segregated high schools this fall, having suffered a loss which can never be restored.

The lack of an adequate education denies the Negro a chance

to get a decent job. The orderly implementation of the Supreme Court decision, therefore, cannot be left solely to those who may not have the economic resources to carry their legal actions or who may be subject to harassment.

Other features will be requested, including greater protections for the right to vote.

But legislation, I repeat, cannot solve this problem alone. It must be solved in the homes of every American in every community across our country.

In this respect I want to pay tribute to those citizens, North and South, who have been working in their communities to make life better for all.

They are acting not out of a sense of legal duty but out of a sense of human decency. Like our soldiers and sailors in all parts of the world, they are meeting freedom's challenge on the firing line and I salute them for their honor—their courage.

My fellow Americans, this is a problem which faces us in every city of the North as well as the South.

Today there are Negroes unemployed—two or three times as many compared to whites—inadequate education; moving into the large cities, unable to find work; young people particularly out of work, without hope, denied equal rights, denied the opportunity to eat at a restaurant or a lunch counter, or to go to a movie theater; denied the right to a decent education. . . .

This is one country. It has become one country because all of us and all the people who came here had an equal chance to develop their talents.

We cannot say to ten per cent of the population that "you can't have that right. Your children can't have the chance to develop whatever talents they have, that the only way that you're going to get your rights is to go into the streets and demonstrate."

I think we owe them and ourselves a better country than that. . . .

. . . Not every child has an equal talent or an equal ability or equal motivation. But they should have the equal right to develop their talent and their ability and their motivation to make something of themselves.

We have a right to expect that every Negro will be responsible, will uphold the law. But they have a right to expect the law will be fair, that the Constitution will be color blind, as Justice Harlan said at the turn of this century.

This is what we're talking about. This is a matter which concerns this country and what it stands for, and in meeting it I ask the support of all our citizens.

IX

AMERICA IN A WORLD OF NATIONS

An Asylum for Mankind*

BY THOMAS PAINE

[1776]

I HAVE HEARD it asserted by some, that as America has flourished under her former connection with Great Britain, the same connection is necessary towards her future happiness, and will always have the same effect. Nothing can be more fallacious than this kind of argument. We may as well assert that because a child has thrived upon milk, that it is never to have meat, or that the first twenty years of our lives is to become a precedent for the next twenty. But even this is admitting more than is true; for I answer roundly that America would have flourished as much, and probably much more, had no European power taken any notice of her. The commerce by which she hath enriched herself are the necessaries of life, and will always have a market while eating is the custom of Europe.

But she has protected us, say some. That she hath engrossed us is true, and defended the continent at our expense as well as her own is admitted; and she would have defended Turkey from the same motive, viz. for the sake of trade and dominion.

Alas! we have been long led away by ancient prejudices and made large sacrifices to superstition. We have boasted the protection of Great Britain without considering that her motive was *interest*, not *attachment;* and that she did not protect us from *our enemies* on *our account,* but from her enemies on her own account, from those who had no quarrel with us on any *other account,* and who will always be our enemies on the *same account*. Let Britain waive her pretensions to the continent, or the continent throw off the dependence, and we should be at peace with France and Spain were they at war with Britain. The miseries of Hanover's last war ought to warn us against connections.

---

* Selections from *Common Sense,* written in January, 1776. Thomas Paine (1737–1809) was a pamphleteer and writer during the American and French Revolution. His *Common Sense* did much to sway public opinion in favor of independence for America.

It hath lately been asserted in parliament, that the colonies have no relation to each other but through the parent country, i.e. that Pennsylvania and the Jerseys, and so on for the rest, are sister colonies by the way of England; this is certainly a very roundabout way of proving relationship, but it is the nearest and only true way of proving enmity (or enemyship, if I may so call it). France and Spain never were, nor perhaps ever will be, our enemies as *Americans*, but as our being the *subjects of Great Britain*.

But Britain is the parent country, say some. Then the more shame upon her conduct. Even brutes do not devour their young, nor savages make war upon their families; wherefore, the assertion, if true, turns to her reproach; but it happens not to be true, or only partly so, and the phrase *parent* or *mother country* hath been jesuitically adopted by the king and his parasites, with a low papistical design of gaining an unfair bias on the credulous weakness of our minds. Europe, and not England, is the parent country of America. This new world hath been the asylum for the persecuted lovers of civil and religious liberty from *every part* of Europe. Hither have they fled, not from the tender embraces of the mother, but from the cruelty of the monster; and it is so far true of England, that the same tyranny which drove the first emigrants from home pursues their descendants still. . . .

Much hath been said of the united strength of Britain and the colonies, that in conjunction they might bid defiance to the world. But this is mere presumption, the fate of war is uncertain; neither do the expressions mean anything, for this continent would never suffer itself to be drained of inhabitants to support the British arms in either Asia, Africa, or Europe.

Besides, what have we to do with setting the world at defiance? Our plan is commerce, and that, well attended to, will secure us the peace and friendship of all Europe; because it is the interest of all Europe to have America a *free port*. Her trade will always be a protection, and her barrenness of gold and silver secure her from invaders.

I challenge the warmest advocate for reconciliation to show a single advantage that this continent can reap, by being connected with Great Britain: I repeat the challenge, not a single advantage is derived. Our corn will fetch its price in any market in Europe, and our imported goods must be paid for, buy them where we will.

But the injuries and disadvantages which we sustain by that connection are without number; and our duty to mankind at large, as well as to ourselves, instruct us to renounce the alliance: because any submission to, or dependence on, Great Britain, tends directly to involve this continent in European wars and quarrels, and set us at variance with nations who would otherwise seek our friendship, and against whom we have neither anger nor complaint. As Europe is our market for trade, we ought to form no partial connection with any part of it. 'Tis the true interest of America to steer clear of European contentions, which she never can do while by her dependence on Britain she is made the makeweight in the scale of British politics.

Europe is too thickly planted with kingdoms to be long at peace, and whenever a war breaks out between England and any foreign power, the trade of America goes to ruin, *because of her connection with Britain.* The next war may not turn out like the last, and should it not, the advocates for reconciliation now will be wishing for separation then, because neutrality in that case would be a safer convoy than a man of war. Everything that is right or reasonable pleads for separation. The blood of the slain, the weeping voice of nature cries, *'tis time to part.* Even the distance at which the Almighty hath placed England and America is a strong and natural proof that the authority of the one over the other, was never the design of heaven. The time likewise at which the continent was discovered, adds weight to the argument, and the manner in which it was peopled, increases the force of it. The Reformation was preceded by the discovery of America, as if the Almighty graciously meant to open a sanctuary to the persecuted in future years, when home should afford neither friendship nor safety....

As to government matters, it is not in the power of Britain to do this continent justice: the business of it will soon be too weighty and intricate to be managed with any tolerable degree of convenience, by a power so distant from us, and so very ignorant of us; for if they cannot conquer us they cannot govern us. To be always running three or four thousand miles with a tale or a petition, waiting four or five months for an answer, which, when obtained, requires five or six more to explain it in, will in a few years be looked upon as folly and childishness. There was a time when it was proper, and there is a proper time for it to cease.

Small islands not capable of protecting themselves are the proper objects for government to take under their care; but there is something absurd in supposing a continent to be perpetually governed by an island. In no instance hath nature made the satellite larger than its primary planet; and as England and America, with respect to each other, reverse the common order of nature, it is evident that they belong to different systems. England to Europe: America to itself.

I am not induced by motives of pride, party, or resentment to espouse the doctrine of separation and independence; I am clearly, positively, and conscientiously persuaded that 'tis the true interest of this continent to be so; that everything short of *that* is mere patchwork, that it can afford no lasting felicity—that it is leaving the sword to our children, and shrinking back at a time when a little more, a little further, would have rendered this continent the glory of the earth. . . .

Ye that tell us of harmony and reconciliation, can ye restore to us the time that is past? Can ye give to prostitution its former innocence? Neither can ye reconcile Britain and America. The last cord now is broken, the people of England are presenting addresses against us. There are injuries which nature cannot forgive; she would cease to be nature if she did. As well can the lover forgive the ravisher of his mistress, as the continent forgive the murders of Britain. The Almighty hath implanted in us these inextinguishable feelings for good and wise purposes. They are the guardians of his image in our hearts. They distinguish us from the herd of common animals. The social compact would dissolve, and justice be extirpated from the earth, or have only a casual existence, were we callous to the touches of affection. The robber and the murderer would often escape unpunished, did not the injuries which our tempers sustain, provoke us into justice.

O ye that love mankind! Ye that dare oppose not only the tyranny but the tyrant, stand forth! Every spot of the old world is overrun with oppression. Freedom hath been hunted round the globe. Asia and Africa have long expelled her. Europe regards her like a stranger, and England hath given her warning to depart. O receive the fugitive, and prepare in time an asylum for mankind.

Manifest Destiny*

BY CARL SCHURZ

[1893]

WHENEVER there is a project on foot to annex foreign territory to this republic the cry of "manifest destiny" is raised to produce the impression that all opposition to such a project is a struggle against fate. Forty years ago this cry had a peculiar significance. The slaveholders saw in the rapid growth of the free States a menace to the existence of slavery. In order to strengthen themselves in Congress they needed more slave States, and looked therefore to the acquisition of foreign territory on which slavery existed—in the first place, the island of Cuba. Thus to the pro-slavery man "manifest destiny" meant an increase of the number of slave States by annexation. There was still another force behind the demand for territorial expansion. It consisted in the youthful optimism at that time still inspiring the minds of many Americans with the idea that this republic, being charged with the mission of bearing the banner of freedom over the whole civilized world, could transform any country, inhabited by any kind of population, into something like itself simply by extending over it the magic charm of its political institutions. Such sentiments had been strengthened by the revolutionary movements of 1848 in Europe, which invited a comparison between American and European conditions, and stimulated in the American a feeling of assured superiority, as well as of generous sympathy with other less-favored nations. There was, indeed, no lack of sober-minded men in the United States who, although by no means devoid of high ambition for their country nor of warm sympathy with others, did not lose sight of the limits of human possibility. But they could not prevent a large number of their more enthusiastic and less discriminating fellow-citizens from

---

* From *Harper's New Monthly Magazine*, October 1893. Carl Schurz (1829–1906), born in Germany, came to this country after the Revolution of 1848; he became a Republican politician and rose to the rank of United States Senator from Missouri, 1869–1875, and U. S. Secretary of the Interior, 1877–1881.

cherishing the dream of a pan-American republic to be realized in a lifetime. It was, however, the Southern "manifest-destiny" movement, with a strong organized interest behind it and well-defined purposes in view, that exercised the greater influence upon the politics of the country. But as these purposes became more apparent, and the slavery question was by the Kansas-Nebraska bill thrust upon the country as the dominant political issue of the period, the merely sentimental conception of "manifest destiny" gradually vanished, and many of those who had entertained it turned squarely against the acquisition of foreign soil for the benefit of slavery.

The civil war weakened the demand for territorial expansion in two ways. With the abolition of slavery the powerful interest which had stood behind the annexation policy disappeared forever. And as to the sentimental movement, the great crisis which brought the Union so near to destruction rudely staggered the jubilant Fourth-of-July optimism of former days and reminded the American people of the inherent inadequateness of mere political institutions to the solution of all problems of human society. The troubles and perplexities left behind by the civil war sobered the minds of the most sanguine. A healthy scepticism took the place of youthful over-confidence. It stimulated earnest inquiry into existing conditions, and brought forth a strong feeling among our people that we should rather make sure of what we had, and improve it, than throw our energies into fanciful foreign ventures.

Only very few of the public men of the time still delighted in "manifest-destiny" dreams. The most prominent among them was Seward, who in 1868 predicted that "in thirty years the city of Mexico would be the capital of the United States," and whose brain was constantly busy with schemes of annexation. But public opinion received his projects with marked coldness. The purchase of Alaska found very scant favor with the people, and it would have failed but for Sumner's efforts and the popular impression that Russia had in some way done us a service in critical times, and that it would be ungracious to repel an arrangement agreeable to this friendly power. Moreover, Alaska being a part of the American continent in a high northern latitude, its acquisition appeared less objectionable than that of non-continental territory, especially in the tropics. Seward's treaty with Denmark for the purchase of

St. Thomas died of inanition in the Senate, where everything of the kind was received with instinctive apprehension. When President Grant sought to effect the annexation of Santo Domingo, neither the gorgeous pictures drawn of the advantages to be gained, nor General Grant's personal prestige, nor the determined efforts of his powerful administration, could prevail against the adverse current of public opinion, or save the treaty from defeat in the Senate.

The recent attempt made by President Harrison to precipitate the Hawaiian Islands into our Union has again stirred up the public interest in the matter of territorial expansion, and called forth the cry of "manifest destiny" once more. This attempt would no doubt already have been buried under popular disapproval had not Republican politicians and newspaper writers seen fit, for the purpose of making party capital, to defend President Harrison's action, and to discredit the cautious course of President Cleveland with deceptive appeals to American pride. To draw a matter of importance so far-reaching into the ordinary game of party politics is an act of recklessness much to be deprecated. While in all probability it will have no serious practical effect at the present time, it may result in spreading among well-meaning people misleading impressions about matters of the highest consequence to the future of the republic.

The new "manifest-destiny" precept means, in point of principle, not merely the incorporation in the United States of territory contiguous to our borders, but rather the acquisition of such territory, far and near, as may be useful in enlarging our commercial advantages, and in securing to our navy facilities desirable for the operations of a great naval power. Aside from the partisan declaimers whose interest in the matter is only that of political effect, this policy finds favor with several not numerically strong but very demonstrative classes of people—Americans who have business ventures in foreign lands, or who wish to embark in such; citizens of an ardent national ambition who think that the conservative traditions of our foreign policy are out of date, and that it is time for the United States to take an active part and to assert their power in the international politics of the world, and to this end to avail themselves of every chance for territorial aggrandizement; and lastly, what may be called the navy interest—officers of the navy and others taking especial pride in the development of our naval force, many of whom advocate a large increase of our war-fleet to support a vigorous foreign policy, and a vigorous foreign policy to give congenial occupation and to secure further increase to our war-fleet. These forces we find bent upon exciting the ambition of the American people whenever a chance for the acquisition of foreign territory heaves in sight.

As to the first of these classes, it is certainly not to be denied that among the American adventures in foreign parts there are many respectable characters, whose interests are entitled to consideration, and may be, under certain circumstances, entitled also to active protection by our government. But when they ask, under whatever pretext, that for the advancement or protection of their interests the countries in which they are engaged in private business should be incorporated in this republic, the apparent patriotism of their demand should be received with due distrust. If it were once understood that a combination of Americans engaged in business abroad could at any time start a serious annexation movement in the United States, there would be no end of wild attempts to drive the American people into the most reckless enterprises.

The patriotic ardor of those who would urge this republic into the course of indiscriminate territorial aggrandizement to make it the greatest of the great powers of the world deserves more serious consideration. To see his country powerful and respected among the nations of the earth, and to secure to it all those advantages to which its character and position entitle it, is the natural desire of every American. In this sentiment we are all agreed. There may, however, be grave differences of opinion as to how this end can be most surely, most completely, and most worthily attained. This is not a mere matter of patriotic sentiment, but a problem of statesmanship. No conscientious citizen will think a moment of incorporating a single square mile of foreign soil in this Union without most earnestly considering how it will be likely to affect our social and political condition at home as well as our relations with the world abroad.

According to the spirit of our constitutional system, foreign territory should be acquired only with a view to its admission, at no very distant day, into this Union as one or more States on an equal footing with the other States. The population inhabiting such territory, and admitted into the Union with it, would have to be endowed with certain rights and powers, and the United States would have to undertake certain obligations with regard to them. The people of the new States would not only govern themselves as to their home concerns, but also take part in the government of the whole country through the Senators and Representatives sent by them to Congress, as well as through the votes cast in the elections of our Presidents and in adopting or rejecting constitutional amendments. More than this: as the party managers would study and humor their likes and dislikes in order to obtain their votes, the new-comers would soon exercise a considerable influence upon the conduct of our political parties. The United States, on the other hand, would be bound to guarantee to them a republican form of government, to protect them against invasion, and, upon proper application, against domestic violence. In other words, this republic would admit them as equal members to its national household, to its family circle, and take upon itself all the responsibilities for them which this admission involves. To do this safely it would have to act with keen discrimination.

If the people of Canada should some day express a desire to be incorporated in this Union, there would, as to the character of the country and of the people, be no reasonable doubt of the fitness, or even the desirability, of the association. Their country has those attributes of soil and climate which are most apt to stimulate and keep steadily at work all the energies of human nature. The people are substantially of the same stock as ours, and akin to us in their traditions, their notions of law and morals, their interests and habits of life. They are accustomed to the peaceable and orderly practices of self-government. They would mingle and become one with our people without difficulty. The new States brought by them into the Union would soon be hardly distinguishable from the old in any point of importance. Their accession would make our national household larger, but it would not seriously change its character. It might take place—and, in fact, it should take place only in that way—as a result of a feeling common to both sides that the two countries and peoples naturally belong together in their sympathies as well as their interests. Nor would the union of the two countries excite among us any ambition of further aggrandizement in the same direction, for the acquisition of the Canadian Dominion would give to the United States the whole of the northern part of the continent.

Very unlike would be the situation produced by the acquisition of territory to the south of us. In the first place, it would spring from motives of a different kind—not the feeling of naturally belonging together, but the desire on our part to gain certain commercial advantages; to get possession of the resources of other countries, and by exploiting them to increase our wealth; to occupy certain strategical positions which in case of war might be of importance, and so on. It is evident that if we once are fairly started in the annexation policy for such purposes, the appetite will grow with the eating. There will always be more commercial advantages to be gained, the riches of more countries to be made our own, more strategical positions to be occupied to protect those already in our hands. Not only a taste for more, but interest, the logic of the situation, would push us on and on.

The consequences which inevitably would follow the acquisition of Cuba, which is especially alluring to the annexationist, may serve as an example. Cuba, so they tell us, possesses rich natural resources worth having. It is in the hands of a European power that may, under certain circumstances, become hostile to us. It is only a few miles from the coast of Florida. It "threatens" that coast. It "commands" also the Gulf of Mexico, with the mouths of the Mississippi and the Caribbean Sea. Its population is discontented; it wishes to cut loose from Spain and join us. If we do not take Cuba "some other power will take it." That power may be hostile. Let us take it ourselves. What then? Santo Domingo is only a few miles distant from Cuba; also a country of rich resources; other powers several times tried to get it; if in the hands of a hostile power it would "threaten" Cuba; it also "commands" the Caribbean Sea; the Dominican Republic, occupying the larger part of the island, offered to join us once, and will wish to do so again; to acquire the Haitian Republic we shall have to fight; it will cost men and money, but we can easily beat the negroes. We must have Santo Domingo. Puerto Rico will come as a matter of course with Cuba. The British possessions of Jamaica will still be there to "threaten" and "command" everything else. It will be difficult to get it and the other little islands from the clutch of the British lion.

Thus all the more necessary will it be to have possession of the mainland bordering and "commanding" the Gulf of Mexico and the Caribbean Sea on the western side. We must have all the "keys" to the seas and to the land, or at least as many as we can possibly get, one to protect another. In fact, when once well launched on this course, we shall hardly find a stopping-place north of the Gulf of Darien; and we shall have an abundance of reasons, one as good as another, for not stopping even there.

Let us admit, for argument's sake, that there is something dazzling in the conception of a great republic embracing the whole continent and the adjacent islands, and that the tropical part of it would open many tempting fields for American enterprise; let us suppose—a violent supposition, to be sure—that we could get all these countries without any trouble or cost. But will it not be well to look beyond? If we receive those countries as States of this Union, as we eventually shall have to do in case we annex them, we shall also have to admit the people inhabiting them as our fellow-citizens on a footing of equality. As our fellow-citizens they will not only govern themselves in their own States as best they can—the United States undertaking to guarantee them a republican form of government, and to protect them against invasion and domestic violence—but they will, through their Senators and Representatives in Congress, and through their votes in Presidential elections, and through their influence upon our political parties, help in governing the whole republic, in governing us. And what kind of people are those we take in as equal members of our national household, our family circle?

It is a matter of universal experience that democratic institutions have never on a large scale prospered in tropical latitudes. The so-called republic existing under the tropical sun constantly vibrate between anarchy and despotism. . . .

Democratic government cannot long endure without the maintenance of peace and order through the ready acquiescence of the minority in the verdict of public opinion as expressed in the manner provided by law—the minority, if it continues to consider that verdict wrong, reserving to itself only the right of seeking to change it by another appeal to public opinion through the means of peaceable discussion. This presupposes a state of society in which peace and order are felt by the masses of the people to be needed for their every-day occupations, their regular activities—in other words, a state of society in which everybody, or nearly everybody, being steadily at work for his own sustenance or benefit, feels himself interested in the maintenance of peace and order to insure to himself and those dependent upon him the fruit of his labor. Such a state of society is not found where, on the one hand, nature is so bountiful as to render steady work unnecessary, and where, on the other hand, the climatic conditions are such as to render steady work, especially burdensome and distasteful. This is the case in the tropics. . . .

We are frequently told that this is not a mere matter of climate, but of race, and that if those countries were under the control of Anglo-Saxons the result would be different. There are tropical countries under the control of Anglo-Saxons. But what do we see? History teaches us that the Anglo-Saxon takes and holds possession of foreign countries in two ways—as a conqueror, and as a colonizer. In his character as a conqueror he founds governments to rule the conquered. In his character as a colonizer he founds democracies to govern themselves. The governments to rule the conquered he founds in the tropics. The democracies to govern themselves he founds in the temperate zone. . . .

Imagine now fifteen or twenty, or even more, States inhabited by a people so utterly different from ours in origin, in customs and habits, in traditions, language, morals, impulses, ways of thinking—in almost everything that constitutes social and political life—and these people remaining under the climatic influences which in a great measure have made them what they are, and render an essential change of their character impossible—imagine a large number of such States to form part of this Union, and through dozens of Senators and scores of Representatives in Congress, and millions of votes in our Presidential elections, to participate in making our laws, in filling the executive places of our government, and in impressing themselves upon the spirit of our political life. The mere statement of the case is sufficient to show that the incorporation of the American tropics in our national system would essentially transform the constituency of our government, and be fraught with incalculable dangers to the vitality of our democratic institutions. Many of our fellow-citizens are greatly disturbed by the immigration into this country of a few hundred thousand Italians, Slavs, and Hungarians. But if these few hundred thousand cause apprehension as to the future of the republic, although under the inspiring influence of active American life in our bracing climate the descendants of the most ignorant of them in the second or third generation are likely to be Americanized to the point of being hardly distinguishable from other Americans in the same social sphere, what should we fear from the admission to full political fellowship of many millions of the inhabitants of the tropics whom under the influence of their climatic condition the process of true Americanization can never reach? It was a happy intuition which suggested to Mr. Seward that the policy of annexation would transfer the capital of the United States to the city of Mexico, for after the annexation of the American tropics there would certainly be an abundance of Mexican politics in that capital.

The annexation of the Hawaiian Islands would be liable to objections of a similar nature. Their population, according to the census of 1890, consists of 34,436 natives, 6186 half-castes, 7495 born in Hawaii of foreign parents, 15,301 Chinese, 12,360 Japanese, 8602 Portuguese, 1928 Americans, 1344 British, 1034 Germans, 227 Norwegians, 70 French, 588 Polynesians, and 419 other foreigners. If there ever was a population unfit to constitute a State of the American Union, it is this. But it is the characteristic population of the islands in that region—a number of semicivilized natives crowded upon by a lot of adventurers flocked together from all parts of the globe to seek their fortunes, some to stay, many to leave again after having accomplished their purpose, among them Chinese and Japanese making up nearly one-fourth of the aggregate. The climate and the products of the soil are those of the tropics, the system of labor corresponding. If attached to the United States, Hawaii would always retain a colonial character. It would be bound to this republic not by a community of interest or national sentiment, but simply by the protection against foreign aggression given to it and by certain commercial advantages. No candid American would ever think of making a State of this Union out of such a group of islands with such a population as it has and is likely to have. It would always be to this republic a mere dependency, an outlying domain, to be governed as such. ...

... The Hawaiian Islands are distant two thousand miles from our nearest seaport. Their annexation is advocated partly on commercial grounds, partly for the reason that the islands would furnish very desirable locations for naval depots, coaling-stations, and similar conveniences, and that Hawaii is the "key" to something vast and important in that region. Thus we find in favor of the scheme a combination of the interest of commercial adventure with the ambition to make this republic a great naval power which is to play an active and commanding part in the international politics of the world. Leaving aside the question whether the occupation of this "key" would not require for its protection the acquisition of further "keys," admitting for argument's sake all that is claimed for this project, might we not still ask ourselves whether the possession of such an outlying domain two thousand miles away would really be an element of strength to us as against other powers?

In our present condition we have over all the great nations of the world one advantage of incalculable value. We are the only one that is not in any of its parts threatened by powerful neighbors; the only one not under any necessity of keeping up a large armament either on land or water for the security of its possessions; the only one that can turn all the energies of its population to productive employment; the only one that has an entirely free hand. This is a blessing for which the American people can never be too thankful. It should not be lightly jeoparded.

This advantage, I say, we have *in our present condition*. We occupy a compact part of the American Continent, bounded by great oceans on the east and west, and on the north and south by neighbors neither hostile in spirit nor by themselves formidable in strength. We have a population approaching seventy millions and steadily growing, industrious, law-abiding, and patriotic; not a military, but, when occasion calls for it, a warlike, people, ever ready to furnish to the service of the country an almost unlimited supply of vigorous, brave, and remarkably intelligent soldiers. Our national wealth is great, and increases rapidly. Our material resources may, compared with those of other nations, be called inexhaustible. Our territory is large, but our means of interior communication are such as to minimize the inconveniences of distance. . . . In our compact continental stronghold we are substantially unassailable. We present no vulnerable point of importance. There is nothing that an enemy can take away from us and hope to hold. We can carry on a defensive warfare indefinitely without danger to ourselves, and meanwhile, with our enormous resources in men and means, prepare for offensive operations.

The prospect of such a war will be to any European nation, or any league of European nations, extremely discouraging, especially as not one of them has the same free hand that we have. Every one of them is within the reach of dangerous rivals, whom a favorable opportunity might tempt to proceed to hostilities, and such an opportunity would certainly be presented by a long and exhausting war with the United States. And this very circumstance would afford to this republic in such a case the possibility of alliances which would enable it to pass from its defensive warfare to a most vigorous offensive one. . . .

This advantage will be very essentially impaired if we present to a possible enemy a vulnerable point of attack which we have to defend, but cannot defend without going out of our impregnable stronghold, away from the seat of our power, to fight on ground on which the enemy may appear in superior strength, and have the conditions in *his* favor. Such a vunerable point will be presented by the Hawaiian Islands if we annex them, as well as by any outlying possession of importance. It will not be denied that in case of war with a strong naval power the defence of Hawaii would require very strong military and naval establishments there, and a fighting fleet as large and efficient as that of the enemy; and in case of a war with a combination of great naval powers, it might require a fleet much larger than that of any of them. Attempts of the enemy to gain an important advantage by a sudden stroke, which would be entirely harmless if made on our continental stronghold, might have an excellent chance of success if made on our distant insular possession, and then the whole war could be made to turn upon that point, where the enemy might concentrate his forces as easily as we, or even more easily, and be our superior on the decisive field of operations. It is evident that thus the immense advantage we now enjoy of a substantially unassailable defensive position would be lost. We would no longer possess the inestimable privilege of being stronger and more secure than any other nation without a large and costly armament. Hawaii, or whatever other outlying domain, would be our Achilles' heel. Other nations would observe it, and regard us no longer as invulnerable. If we acquire Hawaii, we acquire not an addition to our strength, but a dangerous element of weakness.

It is said that we need a large navy in any case for the protection of our commerce, and that if we have it for this purpose it may at the same time serve for the protection of outlying national domains without much extra expense. The premise is false. We need no large navy for the protection of our commerce. Since the extinction of the Barbary pirates and of the Western buccaneers, the sea is the safest public highway in the world, except, perhaps, in the Chinese waters. Our commerce is not threatened by anybody or anything, unless it be the competition of other nations and the errors of our own commercial policy; and against these influences war-ships avail nothing. Nor do we need any war-ships to obtain favorable commercial arrangements with other nations. Our position of power under existing circumstances is such that no foreign nation will, at the risk of a quarrel with us, deny our commerce any accommodation we can reasonably lay claim to. Nor would our situation as a neutral in case of a war between foreign nations be like that we occupied during the French-English wars at the beginning of this century. Then we were a feeble neutral whom every belligerent thought he could kick and cuff with impunity. Now the United States would be the most formidable neutral ever seen, whom every belligerent would be most careful not to offend. When our maritime commerce was most flourishing we had no navy worth speaking of to protect it, and nobody thought that one was needed. The pretence that we need one now for that purpose reminds one of the Texas colonel, who thinks he must arm himself with a revolver when walking on Broadway because he might be insulted by a salesman.

Nor are we under any necessity to prepare for war by building a large navy. For the reasons given, every nation will avoid war with us, and we should not seek it with any one....

In another respect a large navy might prove to the American people a most undesirable luxury. It would be a dangerous plaything. Its possession might excite an impatient desire to use it, and lead us into strong temptations to precipitate a conflict of arms in case of any difference with a foreign government, which otherwise might easily be settled by amicable adjustment. The little new navy we have has already perceptibly stimulated such a spirit among some of our navy officers and civilian navy enthusiasts, who are spoiling for an opportunity to try the new guns. . . . No great power can do so much among the nations of the world for the cause of international peace by the moral force of its example as the United States. The United States will better fulfil their mission and more exalt their position in the family of nations by indoctrinating their navy officers in the teachings of Washington's farewell address than by flaunting in the face of the world the destructive power of rams and artillery.

Nothing could be more foolish than the notion we hear frequently expressed that so big a country should have a big navy. Instead of taking pride in the possession of a big navy, the American people ought to be proud of not needing one. This is their distinguishing privilege, and it is their true glory.

The advocates of the annexation policy advance some arguments which require but a passing notice. They say that unless we take a certain country offered to us—Hawaii, for instance—some other power will take it, and that, having refused ourselves, we cannot object. This is absurd. Having shown ourselves unselfish, we shall have all the greater moral authority in objecting to an arrangement which would be obnoxious to our interests.

We are told that unless we take charge of a certain country it will be ill-governed and get into internal trouble. This is certainly no inducement. This republic cannot take charge of all countries that are badly governed. On the contrary, a country apt to get into internal trouble would be no desirable addition to our national household.

We are told that the people of a certain country wish to join us, and it would be wrong to repel them. But the question whether a stranger is to be admitted as a member of our family it is our right and our duty to decide according to our own view of the family interest.

We are told that we need coaling stations in different parts of the world for our navy, also if it be a small one, and that the rich resources of the countries within our reach should be open to American capital and enterprise. There is little doubt that we can secure by amicable negotiation sites for coaling stations which will serve us as well as if we possessed the countries in which they are situated. In the same manner we can obtain from and within them all sorts of commercial advantages. We can own plantations and business houses in the Hawaiian Islands. In the American tropics we can build and control railroads; we can purchase mines, and have them worked for our benefit; we can keep up commercial establishments in their towns—in fact, we are now doing many of these things—and all this without taking those countries into our national household on an equal footing with the States of our Union, without exposing our political institutions to the deteriorating influence of their participation in our government, without assuming any responsibilities for them which would oblige us to forego the inestimable privilege of being secure in our possessions without large and burdensome armaments. Surely the advantages we might gain by incorporating the countries themselves in the Union appear utterly valueless compared with the price this republic would have to pay for them.

The fate of the American people is in their own wisdom and will. If they devote their energies to the development of what they possess within their present limits, and look for territorial expansion only to the north, where some day a kindred people may freely elect to cast their lot with this republic, their "manifest destiny" will be the preservation of the exceptional and invaluable advantages they now enjoy, and the growth on a congenial soil of a vigorous nationality in freedom, prosperity, and power. If they yield to the allurements of the tropics and embark in a career of indiscriminate aggrandizement, their "manifest destiny" points with equal certainty to a total abandonment of their conservative traditions of policy, to a rapid deterioration in the character of the people and their political institutions, and to a future of turbulence, demoralization, and final decay.

The Star of Empire*

BY ALBERT J. BEVERIDGE

[1900]

"WESTWARD the Star of Empire takes its Way." Not the star of kingly power, for kingdoms are everywhere dissolving in the increasing rights of men; not the star of autocratic oppression, for civilization is brightening and the liberties of the people are broadening under every flag. But the star of empire, as Washington used the word, when he called this Republic an "empire"; as Jefferson understood it, when he declared our form of government ideal for extending "our empire"; as Marshall understood it, when he closed a noble period of an immortal constitutional opinion by naming the domain of the American people "our empire."

This is the "empire" of which the prophetic voice declared "Westward the Star of Empire takes its Way"—the start of the empire of liberty and law, of commerce and communication, of social order and the Gospel of our Lord—the star of the empire of the civilization of the world. Westward *that* star of empire takes its course. And to-day it illumines our path of duty across the Pacific into the islands and lands where Providence has called us.

In that path the American government is marching forward, opposed at every step by those who deny the right of the Republic to plant the institutions of the Flag where Events have planted that Flag itself. For this is our purpose, to perform which the Opposition declares that the Republic has no warrant in the Constitution, in morals or in the rights of man. And I mean to examine to-night every argument they advance for their policy of reaction and retreat.

It is not true, as the Opposition asserts, that every race without instruction and guidance is naturally self-governing. If so, the In-

---

* From *The Meaning of the Times and Other Speeches* (1908). Albert J. Beveridge (1862–1911) was Senator from Indiana, 1899–1911. He is the author of biographies of Chief Justice John Marshall and Abraham Lincoln. This speech opened the Republican Campaign for the West in the Auditorium at Chicago, September 25, 1900, in reply to William Jennings Bryan's Indianapolis speech accepting his second Democratic nomination for President. This speech was used by the Republicans as a National Campaign document.

dians were capable of self-government. America belonged to them whether they were or were not capable of self-government. If they were capable of self-government it was not only wrong, but it was a crime to set up our independent government on their land without their consent. If this is true, the Puritans, instead of being noble, are despicable characters; and the patriots of 1776, to whom the Opposition compares the Filipinos, were only a swarm of land pirates. If the Opposition is right, the Zulus who owned the Transvaal were capable of self-government; and the Boers who expelled them, according to the Opposition, deserve the abhorrence of righteous men.

But while the Boers took the lands they occupy from the natives who peopled them; while we peopled this country in spite of the Indian who owned it; and while this may be justified by the welfare of the world which those events advanced, that is not what is to be done in the Philippines. The American government, as a government, will not appropriate the Filipinos' land or permit Americans as individuals to seize it. It will protect the Filipinos in their possessions. If any American secures real estate in the Philippines, it will be because he buys it from the owner. Under American administration the Filipino who owns his little plot of ground will experience a security in the possession of his property that he has never known before....

Grant, for the purposes of argument, the Opposition's premise that the white man can not people the Philippines. Grant, also, that the Malays of those islands can not, unaided, establish civilization there; build roads, open mines, erect schools, maintain social order, repress piracy and administer safe government throughout the archipelago. And this must be granted; for they are the same race which inhabits the Malay Peninsula. What, then, is the conclusion demanded by the general welfare of the world?

Surely not that this land, rich in all that civilized man requires, and these people needing the very blessings they ignorantly repel, should be remanded to savagery and the wilderness! If you say this, you say that barbarism and undeveloped resources are better than civilization and the earth's resources developed. What is the conclusion, then, which the logic of civilization compels from these admitted premises? It is that the reign of law must be established throughout these islands, their resources developed and their people civilized by those in whose blood resides the genius of administration.

Such are all Teutonic and Celtic peoples. Such are the Dutch; behold their work in Java. Such are the English; behold their work all around the world. Such the German; behold his advance into the fields of world-regeneration and administration. Such were the French before Napoleon diverted their energies; behold their work in Canada, Louisiana and our great Northwest. And such, more than any people who ever lived, are the Americans, into whose hands God has given the antipodes to develop their resources, to regenerate their people and to establish there the civilization of law-born liberty and liberty-born law.

If the Opposition declares that we ought to set up a separate government over the Philippines because we are setting up a separate government over Cuba, I answer that such an error in Cuba does not justify the same error in the Philippines. I am speaking for myself alone, but speaking thus, I say, that for the good of Cuba more even than for the good of the United States, a separate government over Cuba, uncontrolled by the American Republic, *never should have been promised.*

Cuba is a mere extension of our Atlantic coast-line. It commands the ocean entrances to the Mississippi and the Isthmian Canal. Jefferson's dearest dream was that Cuba should belong to the United States. To possess this extension of American soil has been the wish of every far-seeing statesman from Jefferson to Blaine. Annexation to the greatest nation the world has ever seen is a prouder Cuban destiny than separate nationality. As an American possession, Cuba might possibly have been fitted for statehood in a period not much longer than that in which Louisiana was prepared for statehood.

Even now the work of regeneration—of cleansing cities, building roads, establishing posts, erecting a system of universal education and the action of all the forces that make up our civilization—is speeding forward faster than at any time or place in human history—American administration! But yesterday there were less than ten thousand Cuban children in school; to-day there are nearly one hundred and fifty thousand Cuban children in school—

American administration! But yesterday Havana was the source of our yellow-fever plagues; to-day it is nearly as healthy as New Orleans—American administration!

When we stop this work and withdraw our restraint, revolution will succeed revolution in Cuba, as in the Central and South American countries; Havana again fester with the yellow death; systematic education again degenerate into sporadic instances; and Cuba, which under our control should be a source of profit, power and glory to the Republic and herself, will be a source of irritation and of loss, of danger and disease to both. The United States needs Cuba for our protection; *but Cuba needs the United States for Cuba's salvation.*

The resolution for Cuban independence, hastily passed by all parties in Congress, at an excited hour, was an error which years of time, propinquity of location, common commerce, mutual interests and similar dangers surely will correct. The President, jealous of American honor, considers that resolution a promise. And American promise means performance. And so the unnatural experiment is to be tried. What war and nature—aye, what God hath joined together—is to be put asunder.

I speak for myself alone, but speaking thus, I say that it will be an evil day for Cuba when the Stars and Stripes come down from Morro Castle. I speak for myself alone, but I believe that in this my voice is the voice of the American millions, as it is the voice of the ultimate future, when I say that Porto Rico is ours and ours for ever; the Philippines are ours and ours for ever; and Cuba ought to have been ours, and by the free choice of her people some day will be ours, and ours for ever.

We have a foreign nation on our north; another on our southwest; and now to permit another foreign nation within cannon shot of our southeast coast, will indeed create conditions which will require that militarism which the Opposition to the Government pretends to fear. Think of Cuba in alliance with England or Germany or France! Think of Cuba a naval station and ally of one of the great foreign powers, every one of whom is a rival of America! And so my answer to Mr. Bryan's comparison is that, if we have made a mistake in Cuba, we ought not to make the same mistake in the Philippines.

I *predict that within ten years we shall again be forced to as-*

sume the government of Cuba, but only after our commerce has again been paralyzed by revolution, after internal dissension has again spilled Cuban blood, after the yellow fever has threatened our southern coast from its hot-bed in Havana harbor. Cuba independent! Impossible! I predict that at the very next session of Congress we shall pass some kind of law giving this Republic control of Cuba's destiny. If we do not we fail in our duty.

Consider, now, the Opposition's proposed method of procedure in the Philippines: It is to establish a stable government there, turn that government over to the Filipinos, and protect them and their government from molestation by any other nation.

Suppose the Opposition's plan in operation. Suppose a satisfactory government is established, turned over to the Filipinos and American troops withdrawn. The new government must experience feuds, factions and revolution. This is the history of every new government. It was so even with the American people. Witness Shays' Rebellion against the National Government, almost shaking its foundations; witness the Whiskey Rebellion in Pennsylvania, which required the first exercise of armed national power to maintain order with a state of the Union. And we were of a self-governing race—at that period we were almost wholly Anglo-Saxon.

How can we expect the Philippine Malays to escape this common fate of all new governments? Remember that as a race they have not that civil cohesion which binds a people into a nation. . . .

Again governments must have money. That is their first necessity; money for salaries, money for the army, money for public buildings, money for improvements. Before the revenues are established, the government must have money. If the revenues are inadequate, nevertheless the government must have money. Therefore, all governments are borrowers. Even the government of the American people—the richest people of history—is a borrower. Even the government of the British people, who for centuries have been accumulating wealth, must borrow; its bonds are in our own bank vaults. Much more, then, must little governments borrow money.

If, then, we "establish a stable government," as the Opposition demands, and turn that government over to the Filipinos, they also must borrow money. But suppose the Philippine government can not pay its debt when it falls due, as has been the case in many instances on our own continent within the last quarter of a century, as is the case to-day with one of the governments of Central America. If that loan is an English loan, England would seize the revenues of the Philippines for the payment of her debt, as she has done before and is doing now. So would France or Germany or whoever was the creditor nation. Should we have a right to interfere? Of course not, unless we were willing to guarantee the Philippine debt. If, then, the first purpose of the Opposition candidate is carried out, we must:

Keep "stable" the government which we first "*establish*," or the very purpose of the establishment of that government is defeated.

If the second proposition of the Opposition is performed, we must:

First: Control the finances of the Philippines perpetually; or,

Second: Guarantee the loans the Philippine government makes with other nations; or,

Third: Go to war with those nations to defeat their collection of their just debts.

Is this sound policy? Is it profitable? Is it moral? Is it just to the Filipinos, to the world, to ourselves? Is it humane to the masses of those children who need first of all, and more than all, order, law and peace? Is it prudent, wise, far-seeing statesmanship? *And does the adoption of a similar course in Cuba justify it in the Philippines?*

No. Here is the program of reason and righteousness, and Time and Events will make it the program of the Republic:

First: We have given Porto Rico such a civil government as her situation demands, under the Stars and Stripes.

Second: We will put down the rebellion and then give the Philippines such a civil government as the situation demands, under the Stars and Stripes.

Third: We are regenerating Cuba, and when our preparatory work is done, we should have given Cuba such a civil government as her situation may demand, under the Stars and Stripes.

The sovereignty of the Stars and Stripes can be nothing but a blessing to any people and to any land.

I do not advocate this course for commercial reasons, though these have their weight. All men who understand production and exchange, understand the commercial advantage resulting from our ownership of these rich possessions. But I waive this large consideration as insignificant, compared with the master argument of the progress of civilization, which under God, the American people are henceforth to lead until our day is done. For henceforward in the trooping of the colors of the nations they shall cluster around and follow the Republic's banner.

The mercantile argument is mighty with Americans in merely mercantile times, and it should be so; but the argument of destiny is the master argument in the hour of destiny, and it should be so. The American people never yet entered on a great movement for merely mercantile reasons. Sentiment and duty have started and controlled every noble current of American history. And at this historic hour, destiny is the controlling consideration in the prophetic statesmanship which conditions require of the American people.

It is destiny that the world shall be rescued from its natural wilderness and from savage men. Civilization is no less an evolution than the changing forms of animal and vegetable life. Surely and steadily the reign of law which is the very spirit of liberty, takes the place of arbitrary caprice. Surely and steadily the methods of social order are bringing the whole earth under their subjection. And to deny that this is right, is to deny that civilization should increase. In this great work the American people must have their part. They are fitted for the work as no people have ever been fitted; and their work lies before them.

If the Opposition say that they grant this, but that the higher considerations of abstract human rights demand that the Philippines shall have such a government as they wish, regardless of the remainder of the world, I answer that the desire of the Filipinos is not the only factor in determining their government, just as the desire of no individual man is the only factor determining his conduct. It is written in the moral law of individuals that "No man liveth to himself alone"; and it is no less written in the moral law of peoples that "No people liveth to itself alone."

The world is interested in the Philippines, and it has a right to be. The world is interested in India, and it has a right to be. Civilization is interested in China and its government, and that is the duty of civilization. You can not take the Philippines out of the operation of those forces which are binding all mankind into one vast and united intelligence. When Circumstance has raised our flag above them, we dare not turn these misguided children over to destruction by themselves or spoliation by others, and then make answer when the God of nations requires them at our hands, "Am I my brother's keeper?" ...

History establishes these propositions:

First: Every people who have become great, have become colonizers or administrators;

Second: Coincident with this colonization and administration, their material and political greatness develops;

Third: Their decline is coincident with the abandonment of the policy of possession and administration, or departure from the true principles thereof.

And as a corollary to these propositions is this self-evident and contemporaneous truth:

Every progressive nation of Europe to-day is seeking lands to colonize and governments to administer.

And can this common instinct of the most progressive peoples of the world—this common conclusion of the ablest statesmen of other nations—be baseless?

If the Opposition asks why this is the mission of the American people now more than heretofore, I answer that before any people assumes these great tasks it goes through a process of consolidation and unification, just as a man achieves maturity before he assumes the tasks of a man. ...

If France, Germany, Italy, Austria, would devote themselves to the world's great work of rescuing the wilderness, of planting civilization, of extending their institutions as England has done, as Germany is beginning to do, as the American Republic, under God, is going to lead the world in doing, the armaments of these European military powers would necessarily dissolve, because there would be no longer occasion for them; and because all their energies would be required in the nobler work to which they would thus set their hands.

To produce the same militarism in America that curses Europe, it would be necessary for Canada on the north to be an equal power with us, hostile with present rivalry and centuries of inherited hatred; and for Mexico to be the same thing on the south. And even then we should have only half the conditions that produce militarism in any European nation. Separate government in Cuba is the only proposed step that creates conditions of militarism in America. Militarism in extending American authority! No! No! The wider the dominion of the Stars and Stripes, the broader the reign of peace.

If we do our duty in the Philippines, it is admitted that we ought not to govern the Filipinos as fellow-citizens of the Republic. The Platform of the Opposition says that "to make the Filipinos citizens would endanger our civilization." To force upon Malays, who three hundred years ago were savages and who since that time have been schooled only in oppression, that form of self-government exercised by the citizens of the United States, would be to clothe an infant in the apparel of a giant and require of it a giant's strength and tasks. If we govern them, we must govern them with common sense. They must first be made familiar with the simplest principles of liberty—equal obedience to equal laws, impartial justice by unpurchasable courts, protection of property and of the right to labor—in short, with the *substance* of liberty which civilized government will establish among them.

The Filipinos must begin at the beginning and grow in the knowledge of free institutions, and, if possible, into the ultimate practice of free government by observing the operation of those institutions among them and by experiencing their benefits. They have experienced unjust, unequal and arbitrary taxation; this is the result of the institutions of tyranny. They must experience equal, just and scientific taxation; this is the result of free institutions. They have experienced arrest without cause, imprisonment without a hearing, and beheld justice bought and sold; these are the results of the institutions of tyranny. They must experience arrest only for cause publicly made known, conviction only after trial publicly conducted and justice impartial, unpurchasable and speedily administered; these are the results of free institutions.

They have experienced the violation of the home and robbery by public officers; these are the results of the institutions of tyranny. They must experience the sanctity of the fireside, the separation of Church and State, the punishment of soldier or public official practising outrage or extortion upon them; these are the results of free institutions. And these are the results which they will experience under the government of the American Republic....

We are a Nation. We can acquire territory. If we can acquire territory, we can govern it. If we can govern it, we can govern it as its situation may demand. If the Opposition says that power so broad is dangerous to the liberties of the American people, I answer that the American people's liberties can never be endangered at the hands of the American people; and, therefore, that their liberties can not be endangered by the exercise of this power, because this power is power exercised by the American people themselves.

"*Congress* shall have power to dispose of and make all needful rules and regulations respecting territory belonging to the United States," says the Constitution.

And what is Congress? The agent of the American people. The Constitution created Congress. But who created the Constitution? "We, the people," declares the Constitution itself.

The American people created the Constitution; it is their method. The American people established Congress; it is their instrument. The American people elect the members of Congress; they are the people's servants. Their laws are the people's laws. Their power is the people's power. And if you fear this power, you fear the people. If you want their power restricted, it is because you want the power of the people restricted; and a restriction of their power is a restriction of their liberty. So that the end of the logic of the Opposition is limitation upon the liberties of the American people, for fear that the liberties of the American people will suffer at the hands of the American people—which is absurd.

If the Opposition asserts that the powers which the Constitution gives to the legislative agents of the American people will not be exercised in righteousness, I answer that that can only be because the American people themselves are not righteous. It is the American people, through their agents, who exercise the power; and if those agents do not act as the people would have them, they will discharge those agents and annul their acts. The heart of the whole argument on the constitutional power of the government is faith in the wisdom and virtue of the people; and in that virtue and wisdom I believe, as every man must, who believes in a republic. In the end, the judgment of the masses is right. If this were not so, progress would be impossible, since only through the people is progress achieved.

Our Duty to the World*

BY WOODROW WILSON

1. *Address at New York, April 20, 1915*

Do you realize that, roughly speaking, we are the only great Nation at present disengaged? I am not speaking, of course, with disparagement of the greatness of those nations in Europe which are not parties to the present war, but I am thinking of their close neighborhood to it. I am thinking how their lives much more than ours touch the very heart and stuff of the business, whereas we have rolling between us and those bitter days across the water 3,000 miles of cool and silent ocean. Our atmosphere is not yet charged with those disturbing elements which must permeate every nation of Europe. Therefore, is it not likely that the nations of the world will some day turn to us for the cooler assessment of the elements engaged? I am not now thinking so preposterous a thought as that we should sit in judgment upon them—no nation is fit to sit in judgment upon any other nation—but that we shall some day have to assist in reconstructing the processes of peace. Our resources are untouched, we are more and more becoming by the force of circumstances the mediating Nation of the world in respect of its finance. We must make up our minds what are the best things to do and what are the best ways to do them. We must put our money, our energy, our enthusiasm, our sympathy into these things, and we must have our judgments prepared and our spirits chastened against the coming of that day.

So that I am not speaking in a selfish spirit when I say that our whole duty, for the present at any rate, is summed up in this motto, "America first." Let us think of America before we think of Europe, in order that America may be fit to be Europe's friend when the day of tested friendship comes. The test of friendship is not now sympathy with the one side or the other, but getting ready to help both sides when the struggle is over. The basis of neutrality,

---

* From *The Politics of Woodrow Wilson*, edited by August Heckscher. Harper & Brothers, 1956. Reprinted by permission. Woodrow Wilson (1856–1924) was the 28th President of the United States.

gentlemen, is not indifference; it is not self-interest. The basis of neutrality is sympathy for mankind. It is fairness, it is good will, at bottom. It is impartiality of spirit and of judgment.

We are the mediating Nation of the world. I do not mean that we undertake not to mind our own business and to mediate where other people are quarrelling. I mean the word in a broader sense. We are compounded of the nations of the world; we mediate their blood, we mediate their traditions, we mediate their sentiments, their tastes, their passions; we are ourselves compounded of those things. We are, therefore, able to understand all nations; we are able to understand them in the compound, not separately, as partisans, but unitedly as knowing and comprehending and embodying them all. It is in that sense that I mean that America is a mediating Nation. The opinion of America, the action of America, is ready to turn, and free to turn, in any direction. Did you ever reflect upon how almost every other nation has through long centuries been headed in one direction? That is not true of the United States. The United States has no racial momentum. It has no history back of it which makes it run all its energies and all its ambitions in one particular direction.

And America is particularly free in this, that she has no hampering ambitions as a world power. We do not want a foot of anybody's territory. If we have been obliged by circumstances, or have considered ourselves to be obliged by circumstances, in the past, to take territory which we otherwise would not have thought of taking, I believe I am right in saying that we have considered it our duty to administer that territory, not for ourselves but for the people living in it, and to put this burden upon our consciences—not to think that this thing is ours for our use, but to regard ourselves as trustees of the great business for those to whom it does really belong, trustees ready to hand it over to the cestui que trust at any time when the business seems to make that possible and feasible. That is what I mean by saying we have no hampering ambitions. We do not want anything that does not belong to us. Is not a nation in that position free to serve other nations, and is not a nation like that ready to form some part of the assessing opinion of the world?

My interest in the neutrality of the United States is not the petty desire to keep out of trouble. . . . I am interested in neutrality because there is something so much greater to do than fight; there is a distinction waiting for this Nation that no nation has ever yet got. That is the distinction of absolute self-control and self-mastery. Whom do you admire most among your friends? The irritable man? The man out of whom you can get a "rise" without trying? The man who will fight at the drop of the hat, whether he knows what the hat is dropped for or not? Don't you admire and don't you fear, if you have to contest with him, the self-mastered man who watches you with calm eye and comes in only when you have carried the thing so far that you must be disposed of? That is the man you respect. That is the man who, you know, has at bottom a much more fundamental and terrible courage than the irritable, fighting man. Now, I covet for America this splendid courage of reserve moral force. . . .

2. *Address at Mount Vernon, July 4, 1918*

Gentlemen of the Diplomatic Corps and My Fellow Citizens: I am happy to draw apart with you to this quiet place of old counsel in order to speak a little of the meaning of this day of our Nation's independence. The place seems very still and remote. It is as serene and untouched by the hurry of the world as it was in those great days long ago when General Washington was here and held leisurely conference with the men who were to be associated with him in the creation of a nation. From these gentle slopes they looked out upon the world and saw it whole, saw it with the light of the future upon it, saw it with modern eyes that turned away from a past which men of liberated spirits could no longer endure. It is for that reason that we cannot feel, even here, in the immediate presence of this sacred tomb, that this is a place of death. It was a place of achievement. A great promise that was meant for all mankind was here given plan and reality. The associations by which we are here surrounded are the inspiriting associations of that noble death which is only a glorious consummation. From this green hillside we also ought to be able to see with comprehending eyes the world that lies about us and should conceive anew the purposes that must set men free.

It is significant,—significant of their own character and purpose and of the influences they were setting afoot,—that Washington and his associates, like the barons at Runnymede, spoke and acted, not for a class, but for a people. It has been left for us to see to it that it shall be understood that they spoke and acted, not for a single people only, but for all mankind. They were thinking, not of themselves and of the material interests which centered in the little groups of landholders and merchants and men of affairs with whom they were accustomed to act, in Virginia and the colonies to the north and south of her, but of a people which wished to be done with classes and special interests and the authority of men whom they had not themselves chosen to rule over them. They entertained no private purpose, desired no peculiar privilege. They were consciously planning that men of every class should be free and America a place to which men out of every nation might resort who wished to share with them the rights and privileges of free men. And we take our cue from them,—do we not? We intend what they intended. We here in America believe our participation in this present war to be only the fruitage of what they planted. Our case differs from theirs only in this, that it is our inestimable privilege to concert with men out of every nation what shall make not only the liberties of America secure but the liberties of every other people as well. We are happy in the thought that we are permitted to do what they would have done had they been in our place. There must now be settled once for all what was settled for America in the great age upon whose inspiration we draw today. This is surely a fitting place from which calmly to look out upon our task, that we may fortify our spirits for its accomplishment. And this is the appropriate place from which to avow, alike to the friends who look on and to the friends with whom we have the happiness to be associated in action, the faith and purpose with which we act.

This, then, is our conception of the great struggle in which we are engaged. The plot is written plain upon every scene and every act of the supreme tragedy. On the one hand stand the peoples of the world,—not only the peoples actually engaged, but many others also who suffer under mastery but cannot act; peoples of many races and in every part of the world,—the people of stricken Russia still, among the rest, though they are for the moment unorganized and helpless. Opposed to them, masters of many armies, stand an isolated, friendless group of governments who speak no common purpose but only selfish ambitions of their own by which none can profit but themselves, and whose peoples are fuel in their hands; governments which fear their people and yet are for the time their sovereign lords, making every choice for them and disposing of their lives and fortunes as they will, as well as of the lives and fortunes of every people who fall under their power,—governments clothed with the strange trappings and the primitive authority of an age that is altogether alien and hostile to our own. The Past and the Present are in deadly grapple and the peoples of the world are being done to death between them.

There can be but one issue. The settlement must be final. There can be no compromise. No halfway decision would be tolerable. No halfway decision is conceivable. These are the ends for which the associated peoples of the world are fighting and which must be conceded them before there can be peace:

I. The destruction of every arbitrary power anywhere that can separately, secretly, and of its single choice disturb the peace of the world; or, if it cannot be presently destroyed, at the least its reduction to virtual impotence.

II. The settlement of every question, whether of territory, of sovereignty, of economic arrangement, or of political relationship, upon the basis of the free acceptance of that settlement by the people immediately concerned, and not upon the basis of the material interest or advantage of any other nation or people which may desire a different settlement for the sake of its own exterior influence or mastery.

III. The consent of all nations to be governed in their conduct towards each other by the same principles of honor and of respect for the common law of civilized society that govern the individual citizens of all modern states in their relations with one another; to the end that all promises and covenants may be sacredly observed, no private plots or conspiracies hatched, no selfish injuries wrought with impunity, and a mutual trust established upon the handsome foundation of a mutual respect for right.

IV. The establishment of an organization of peace which shall make it certain that the combined power of free nations will check every invasion of right and serve to make peace and justice the more secure by affording a definite tribunal of opinion to which all must submit and by which every international readjustment that cannot be amicably agreed upon by the peoples directly concerned shall be sanctioned.

These great objects can be put into a single sentence. What we seek is the reign of law, based upon the consent of the governed and sustained by the organized opinion of mankind.

These great ends cannot be achieved by debating and seeking to reconcile and accommodate what statesmen may wish, with their projects for balances of power and of national opportunity. They can be realized only by the determination of what the thinking peoples of the world desire, with their longing hope for justice and for social freedom and opportunity.

I can fancy that the air of this place carries the accents of such principles with a peculiar kindness. Here were started forces which the great nation against which they were primarily directed at first regarded as a revolt against its rightful authority but which it has long since seen to have been a step in the liberation of its own people as well as of the people of the United States; and I stand here now to speak,—speak proudly and with confident hope,—of the spread of this revolt, this liberation, to the great stage of the world itself! The blinded rulers of Prussia have roused forces they knew little of,—forces which, once roused, can never be crushed to earth again; for they have at their heart an inspiration and a purpose which are deathless and of the very stuff of triumph!

The American Dilemma: Democracy and Empire*

BY CARLOS P. ROMULO

[1955]

RECENTLY when the French delegation walked out of the General Assembly of the United Nations in protest against the latter's decision to debate the question of Algeria, some people blamed France for its precipitate action while others expressed the fear that the United Nations had lost prestige thereby. But the real loser was the United States of America. France accused the United States of being half-hearted in its support of the French position while the Asian-African states condemned it for siding with a colonial power.

This incident illustrates the nature of the American dilemma. It is a dilemma with a number of facets. The colonial powers of Europe can and often do appeal to the United States to support their colonial policies in the interest of a common cultural or even racial inheritance. At the same time, Americans realize that one of the most valuable elements of that inheritance is the historic American struggle for independence from colonial control. Therefore, they cannot remain indifferent to the struggle of the Asian and African peoples to throw off the colonial yoke.

Furthermore, the colonial powers often remind the United States that continued Western control of the rich or strategic colonial territories of Asia and Africa is essential to the defense of the free world against encroaching Communism. Yet again the American people know enough about the historical process and the mechanics of power in the modern world to realize that the defense of freedom requires not only the control of essential raw materials, military bases and armaments but, more important, the unwavering allegiance of men to the principles of freedom. In other words,

---

* From *Current History*, December 1955. Reprinted by permission. Carlos P. Romulo (b. 1899), Ambassador of the Philippines to the United States, has been President of the General Assembly of the United Nations, 1949–1950.

admitting that the danger of Communist encroachment in the colonial territories is one that must be met effectively, the American people know nevertheless that, in the long run, this danger cannot be neutralized by giving the Communists a virtual monopoly in supporting the struggle of the subject peoples for freedom and independence.

America is thus torn between its loyalty to the comity of Western civilization and its loyalty to the larger world community of free peoples of all races and creeds of which it is the recognized and honored leader. It is torn between the short-range consideration of maintaining its military power in relation to the Soviet Union and the long-range consideration of the need to establish its ultimate moral authority by waging a struggle against Communism for the minds of men.

This is the shape of the American dilemma: the dilemma of empire. The problem, in a nutshell, is whether American democracy can be reconciled with the new American "empire" of the Twentieth Century, and the first specific question to be asked is: Does American internationalism in the Twentieth Century have imperialistic connotations?

American imperialism was one of the last to enter the field and among the first to leave. There is no reason to suppose that it was more virtuous in its motives or more benign in its operations than the classical European imperialism that long preceded it. The American troops that followed Admiral Dewey to the Philippines were despatched by President McKinley under the most benevolent auspices and invariably with the avowed purpose of helping the Filipino people to defend their rights and safeguard their freedom. But when the Filipinos chose to resist the "civilizing mission" of the American soldiers, the latter thought themselves unjustly misunderstood and made up a famous song with a familiar refrain about "civilizing the Filipinos with the Krag."

Nevertheless, there is good reason to believe that the American overseas imperialism near the turn of the century was a somewhat enfeebled extension of the tremendous drive that in less than a century had resulted in the conquest of an entire continent. The original impulse that had pushed the American frontier to the very shores of the Pacific Ocean had become somewhat attenuated by the time it reached the sea, and there was a whole vast continent to engage the energies of a young and burgeoning nation. Moreover, most of the territories in Asia and the Far East had been pre-empted by the time America was ready to join the imperialist game; there were left only a few minor pieces of real estate, like Cuba in the Caribbean and the Philippines in the Far East, which the bankrupt Spanish Empire seemed ready to relinquish.

Thus it came about that the American conquest of the Philippines was carried out under a strange combination of crude self-interest and high-minded purpose. The well-known story about President McKinley going down on his knees in pious prayer to seek guidance from the Almighty on what ought to be done with the Philippines is symbolic of the mixed purposes of American imperialism. It may be that his understanding of the wishes of the Almighty did finally induce him to follow a policy of conquest and domination, though there is evidence to show that this policy happened also to have the enthusiastic support of the imperialist element who were inspired less by a concern for God's will than for national glory and the hope of material profit.

This is not to say that other imperialisms (the Spanish and the English, for example) did not try to disguise their crude motives of gain by pretending to equally high-minded purposes of converting the heathen or teaching them the ways of democracy. Thus, the sending of a boatload of American teachers to the Philippines in 1903 recalled the fact that nearly four hundred years before, the Spanish *conquistadores* in the Philippines were invariably accompanied by priests. It was necessary, in the view of both, to civilize the backward peoples even if they did not like it and showed their dislike by fighting the intruder.

Mixture of Motives

What distinguished American imperialism from all the rest was this: that being neither subtle in its motivations nor ruthless in its methods, it was eternally bothered with a conscience. Now, conscience becomes a luxury for any country that hopes to get the most out of an imperialist adventure. To succeed in this business, a nation must either be as unswervingly devoted to the single goal of material profit, resolutely ignoring all others, as the Dutch were in Indonesia, or as indefatigably pragmatic, refusing to be ruled by dogma, as the British were in India.

By contrast, the American regime in the Philippines was organized from a bewildering mixture of motives and conducted in varying moods and styles of government that included the high-mindedness of William Howard Taft, the tough-mindedness of Leonard Wood, and the tender-mindedness of Francis Burton Harrison. Underlying these varying motives and modes and styles was always to be felt the peculiar quality which characterized the American regime in the Philippines throughout: an abounding good nature and spirit of generosity triumphant in the end over every manifestation of selfishness or cupidity.

There is no doubt that at the turn of the century when America finally established control of the Philippines, there were Americans who had dreams of an empire extending over the two Americas and across the Pacific to the fabled East. Here was a young giant flexing his muscles and looking beyond the prairies and the oceans, and it is not difficult to understand that some Americans at the time would have entertained bright visions of national power and imperial glory.

But if there were arch-imperialists in America then, there were also dedicated anti-imperialists who from the beginning opposed the conquest of the Philippines as contrary to the American tradition and who, even when it had been decided that America would stay, were nevertheless determined that it would be for a temporary period, at the end of which the Filipinos would be granted self-government and independence.

Thus after the soldiers came the teachers, the doctors, the engineers, the administrators. They came in the earnest belief that the opening of schools, the building of hospitals, the construction of roads and bridges, and the establishment of a civil service were essential prerequisites to the enjoyment by the Filipinos of freedom and independence.

The history of Philippine-American relations, however, is not an idealized story without blemishes. Therefore it must be recorded that other Americans also came to the Philippines—businessmen who saw in the substantial human and material resources of the country an opportunity for gain through an expanding trade between an agricultural Philippines and industrial America. They saw that the Philippines produced some of the world's most important raw materials: copra, coconut oil, hemp, sugar, lumber, tobacco, minerals. They also saw that the Philippines was potentially one of the leading consumers of American manufactured products. What was more natural therefore than that a system of free trade should be established which would permit the unimpeded flow of goods between the two countries?

This was done by enactment of the United States Congress. The Filipinos objected to this as an artificial arrangement which would result in a hothouse economy for the Philippines, protected from the normal hazards of international trade by special free trade relations which were not likely to last forever. But the American view prevailed, and the Philippine economy was firmly riveted to the American economy for the next forty years.

Political and Economic Dichotomy

It is typical of the free-and-easy American rule in the Philippines that there was little or no attempt to harmonize and synchronize the economic policies and the political objectives. Almost from the very beginning it had been understood, first tacitly and later in a formal pledge by the United States Congress, that the Philippines would be given independence. All the political and social reforms were directed toward this end. Even if the American Government had not given a solemn promise to grant the independence of the Philippines (as it did in 1933), the inevitable net result of the improved health, education and livelihood of the Filipino people would have been the sharpening of their appetite for self-government and independence.

On the other hand, in the economic field, the policy of free trade had the effect of tying the Philippines closer to America. As a result of these conflicting tendencies, it became necessary, in 1935, at the beginning of the ten-year transitory period before independence, to devise a sliding system of quotas and tariffs which would gradually eliminate the preferential trade between the Philippines and the United States upon the advent of independence.

The war in the Pacific broke out in 1941, at the very middle of the ten-year period of transition. After liberation, on July 4, 1946, the United States, true to its pledged word, recognized the independence of the Philippines. Loyal to the principles of American democracy and grateful for the generosity of the American people, the Filipinos had seen their country ravaged by the invader. On the morrow after liberation, while the land lay prostrate from four years of war and occupation, its economy in total ruin, the Filipino people gathered enough strength to stand up proudly before the world and receive from America the priceless guerdon of independence.

Thus was enacted in the Philippines, at the end of the war, one of the most inspiring and heart-warming events in the unhappy history of colonialism.

Politically, the act was completed. However, the economic implications of independence needed to be worked out separately, for it was not an easy thing to untie the economic bonds between the two countries. Accordingly, a Trade Agreement was signed between the two countries designed to level off over a longer period the reduction of quotas and the increase of the tariff duties on Philippines goods entering the United States. This year the two governments agreed to amend this agreement by further prolonging the period of leveling off and removing certain features which the Filipinos have long considered as inequalities.

It is only fair to say that through all these transactions and arrangements, the point of view of the hard-headed American businessman who is quite unsentimental when it comes to questions of profit and loss, was always ably represented. Generous as the American attitude has always been in these matters, there has been no question of charity or a free handout. The Philippines was not asking for charity either.

It is pertinent to recall in this connection one of the less lovely aspects of the story of Philippine independence, namely, that the movement to achieve this independence from the United States had been greatly abetted by the powerful American dairy and sugar producers and investors who saw that the exclusion of duty-free Philippine coconut oil and sugar from the United States market would yield them many millions of dollars in profits.

This, in brief, is the American record in the Philippines. It is not, as I have indicated, a record without blemish. The Americans were no angels in the Philippines. In their relations with the

Filipinos they were not without motives of cupidity and pride. But judging the American record as a whole, it can be said beyond honest contradiction that the United States, by its conduct in the Philippines and by the sincerity and resolve with which it successfully brought off its great experiment there, rang the death-knell of colonialism in the Twentieth Century.

A New Empire?

This is the American record in a concrete instance of an adventure in imperialism that lasted nearly fifty years. We return to our original question: Does American internationalism in the Twentieth Century have imperialistic connotations?

One part of our answer would be: Yes, it unfortunately has such connotations, but they can hardly be helped. To be the richest and most powerful nation in the world is not necessarily to be universally admired, loved or trusted. America must pay a certain penalty for the power and influence it wields in the world, and the penalty is often paid in terms of the suspicion, fear and hate of others. This uncomfortable position is the result of certain factors beyond its control.

Ineluctable geography dictates the nature of the American policy to ensure its leadership of the free world. The Russian "empire" is a compact, virtually unbroken land mass that stretches from the Pacific Ocean westward to the Baltic Sea and the Stettin-Trieste line. When a monolithic political and economic ideology based on fraud and ruthless compulsion goes hand in hand with such a physical conformation, it becomes relatively easy to establish a unified system pursuing common ends in all vital fields of activity including foreign affairs. On the other hand, the free-world alliance led by the United States consists of countries widely scattered around the globe, with varying political and economic systems, and often pursuing conflicting objectives. With all its wealth and power, the United States cannot weld these countries into a semblance of unity by force or deceit but only by methods of free consent.

Thus, when Soviet Russia needs a military base in Bulgaria or desires certain economic changes in Poland, the monolithic principle of conformity within the "empire" operates to render quite unnecessary the type of difficult and often tortuous negotiations which the United States must undertake in order, for example, to establish a military base in the Philippines or Morocco, or to be able to give economic and financial assistance to France or Turkey under terms acceptable to these countries.

The connotations of imperialism in almost anything that America does to strengthen the sinews of the free world and to maintain its capacity for united action are thus virtually inevitable. Those connotations arise only because America is compelled by its own traditions and by the nature of the very principle of freedom which it seeks to defend, to use only the methods of consultation, discussion, and mutual consent. By contrast, Russian internationalism betrays no such connotations because it is in fact based on a system that is the most inherently imperialistic that the world has ever seen.

The second part of our answer to the question posed above is this: Yes, American internationalism in the Twentieth Century has imperialist connotations, but compared to the American record in the Philippines, in Cuba, in Puerto Rico, in Central America, they are greatly muted ones. Whatever their disappointments and disillusionments with American policy, the discontented Latin American, the suspicious Asian, and the disgruntled African know in their own hearts that they are infinitely better off with a sometimes slow, uncertain or mistaken America than they would be with the shining efficiency and certitude of Communist Russia. They know that America is a land with an unlimited promise of freedom for the human mind and the human spirit, and that its power will ever be used to strengthen the will to struggle for freedom. Following America we may stumble and hesitate and even occasionally lose our way, but lose sight of the final objective of human freedom, never. This is a nation with instincts far closer to the aspirations of our fallible and striving human nature, a nation whose heart is in the right place or, if it is not there at the moment, soon returns there.

By the nature of its birth and origin, by reason of the unique human conglomeration of which it is the splendid outcome, America has the opportunity to found and rule an empire greater than any the world has yet seen. But to become this, it must be an empire of the mind and heart.

Our Duty to Ourselves*

BY GEORGE F. KENNAN

[1951]

LET US TURN our thoughts to that something which we picture to ourselves by the phrase: "national interest." I say that "something" because this is one of those things that you know must exist—you can demonstrate it by the process of exclusion—but it is too vast, too rich in meaning, too many-sided, for any positive definition. And for that reason, I'm going to ask your indulgence if I try to make it clearer by talking—not about what it is—but about what it is *not*.

Concepts and ideas are sometimes like shy wild animals. You can never get near enough to touch them and make exact measurements of them, but you can round them up and gradually pen them in; you can mark out certain directions in which they are not permitted to move. In this way you can confine them from time to time and have some idea of what they are like.

And so, I would like to take this idea of national interest and try to box it in a little by telling you . . . this: The interest of the United States in international affairs is *not* a detached interest in our international environment *for its own sake,* independent of our own aspirations and problems here at home. It does *not* signify things we would like to see happen in the outside world primarily for the sake of the outside world, just because we feel altruistic or ashamed of our relative wealth or sorry for other people or are attached to some abstraction or other. It's something that begins at home. It is a function of our duty to ourselves in our domestic problems.

Why is this? It's because we do not live just for our relations with others—we do not live just in order to conduct foreign policy. It would be more correct to say that we conduct foreign policy in order to live. But not just to live "period"—not just to live in the

---

* Reprinted by special permission of the *Illinois Law Review* (Northwestern University School of Law), volume 45, number 6, 1951.

sense of waiting, individually, for death to come to us, but to live as a people, joined together in a social compact, for a purpose related primarily to ourselves and not to others.

It is not an expression of national selfishness to say that our first duty, as a nation, is to ourselves. It is an expression of self-respect. You all know how odious, in individual life, is the sloppy person who neglects his own affairs and his own appearance and runs around in a cloud of loose ends and unfulfilled obligations and at the same time busies himself with schemes for the improvement of others. The idealism may be commendable in principle, but we all find ourselves wishing that such people would hold in abeyance their desire to save humanity until they had gone home and cleaned themselves up and put their own affairs in order and ceased to be a problem to themselves and everybody around them. So it is with a nation. Its first obligation is to itself, to the cleanliness and orderliness and decency of its own national life. A nation which is meeting its problems, and meeting them honestly and creditably, is not apt to be a problem to its neighbors. And, strangely enough, having figured out what it wants to do about itself, it will find that it has suddenly and mysteriously acquired criteria, which it did not have before, for knowing what to do about its relations with others. But a nation which attempts to go in for world improvement as an end in itself is not going to be apt to do much good to anybody else and it is going to find itself in the realm of perpetual confusion in trying to make the day-by-day decisions which the conduct of foreign policy involves.

Whoever would determine national interest in the foreign field must begin, then, at home, and he must satisfy himself about the higher national purpose, of which foreign policy can only be a function. If he does that, he has something to go on. And it is for this reason that anyone who is pressed to define national interest in terms of our international relations has a right to say: "Tell me what we Americans are trying to be to ourselves, and I will tell you what we ought to try to be to other people."

Now I don't want to expatiate about this problem of the over-all national purpose. Who can say with any authority what it is? It is not what I think; it is not what you think individually; it is not even what the President thinks as an individual. It is the sum total of what all of us think—it is the consensus of our thought—and it finds expression in our behavior as citizens and in the behavior of the representatives we send to Washington.

But I think we might recall at this point certain things about our American system. This system was founded on the belief that the civil organization of society exists for the individual and not vice versa. Government exists, according to the Declaration of Independence, in order to secure to the individual citizen certain rights of life, liberty, and the pursuit of happiness which it is assumed he requires in order to pursue private purposes believed to be worth while. Government is regarded here as only the purveyor, and the guarantor, of these rights. It is not regarded as the channel through which the ultimate purposes are to be pursued. There is an assumption that the worth-while things of life will flow from the natural processes of private interest; it is the function of government to see to it that the flood gates are kept open and the channels clear through which the stream of private interest can flow and to do this by protecting the rights and the freedoms of the individual.

The Declaration of Independence says that man has the right to life. But it does not tell him how to live this life. The Declaration says he has a right to liberty—the right, that is, to be free. But there is no freedom in the abstract. There is only a freedom from something and to something. The Declaration does not specify what people shall be free to do, to what ends they should use their freedom. Finally, the Declaration says man has a right to the pursuit of happiness. But it does not attempt to define what constitutes happiness for the individual. Our system has been predicated thus far on the belief that the individual is capable of knowing these things for himself, and that if he is secured in his rights and guarded from the danger of infringement on the rights of others, he will know how to pursue these aims. In this way the purposes of society are expected to flow of themselves from the individual will and conscience and from the initiatives that the individual may take. And the duty of government, I reiterate, is primarily to stand guardian over the rights which are to make all this possible.

Now these philosophical concepts of Government are being sorely tested and buffeted by the realities of the time; and perhaps, since change is the immutable law of nature, they will have to undergo eventually some sort of adjustment to meet the requirements of the day—an adjustment whereby a higher value would be assigned to the collective activities of man, and a lower one to his purely private undertakings—an adjustment whereby the national purpose, as embodied and expressed in the work of government, would come to be more widely recognized as an independent and important theatre of man's strivings and achievements. Should such things come to pass, then I could imagine that our concept of what we are doing in international affairs might be somewhat altered.

But none of this has yet taken place—or very little. What we teach in our schools is still the rational-liberal political philosophy, not materially altered since the day of Jefferson: the philosophy of the watch-dog government, the guardian of private interest, operating by regulatory statute and held to this limited role by various checks and balances. It is not the philosophy of the purposeful government, operating by flexible administrative control, designed as the vehicle, rather than the mere guardian, of the vital processes of society.

I said these concepts of government were being tested and buffeted by the realities of the day. I want to emphasize this point. Let us never forget that our system of government is still an experiment. Lincoln described the civil war of his own day as a contest which was to determine whether any such system as our own—any system "conceived in liberty and dedicated to the proposition that all men are created equal"—could long survive. He did not regard that question as one which had been answered, even after 87 years of independence. I think he viewed the Civil War as something which would determine whether the system was to survive the crisis *of that particular time* or whether it was to break up then and there. I am sure he did not think of it as the last and final test. And I think he would agree, if he were alive now, that in this year 1951 the greatest of those remaining tests has not yet come, although it is not far off. He would still see it as Bryce foresaw it and described it fifty years ago—as something "looming on the horizon, and now no longer distant, a time of mist and shadows, wherein dangers may lie concealed whose form and magnitude America can scarcely yet conjecture." It is the test Tocqueville had in mind when, here in this country in the 1830's, he marvelled at the vitality of American institutions in the simple conditions of that day. And yet, he said: "One has to return into oneself, to struggle against the current, to perceive that such simple and logical institutions would not suit a great nation which needs a strong internal government and a fixed foreign policy."

On every side of us we see proof of this thesis that our American civilization is still something experimental, unfinished, not fully tested. We see it in our failure to bring our lives into balance with the natural resources of this continent; we see it in our failure, to date, to find a happier and more orderly answer to the problems of labor and wages and prices; we see it in the depressing and flimsy aspect of great portions of our sprawling big cities; we see it in the pathetic shallowness and passiveness of our recreational habits; we see it in our bewilderment as to how to handle the forces which modern technology has released among us—the telephone, the automobile, the television set, atomic energy—the forces which Henry Adams described in 1907 as ones which grasped man's wrists and "flung him about as though he had hold of a live wire or a run-away automobile." In all these things there stands written the reminder that our society has not yet passed its final test, and that we cannot yet claim for it, whatever may be our hopes or our faith, any final validity as the answer to the problems of political organization here or elsewhere on this planet.

That being the case, we must preserve a certain modesty about what we conceive to be our role on the stage of international affairs. We have no right to recommend our institutions to others or to expect others to understand entirely what it is we are doing here in this country. By the same token—not having yet finally demonstrated to ourselves the permanent validity of our own system—we have no right to be too emphatic or critical in our views about the validity of others. The whole nature of our national development here at home tells us not to become too ambitious in our ideas about intimacy with other peoples, but rather to lay upon ourselves the restraint of one who goes his own way out of his own conviction, realizing that he has not yet proven his reasons for doing so and that in the eyes of others he may appear a species of "queer duck," and asking only that others reserve judgment on what he is doing and leave him alone to do it, as long as he, in turn, minds his own business and does not step on their toes.

If you take this view, as I am free to say that I do, you conclude that the greatest thing America can do for this world is to make a success of what it is doing here on this continent and to bring itself to a point where its own internal life is one of harmony and stability and self-assurance. The fundamental interest of our Government in international affairs is, then, to assure that we should be permitted, as a people, to continue this Pilgrim's Progress toward a better America under the most favorable possible conditions, with a minimum of foreign interference, and also with a minimum of inconvenience or provocation to the interests of other nations. Concluding this, you do not hold that we have an obligation to assure the morality of domestic or international conduct elsewhere in the world, except where it affects our duty to ourselves. You do not hold that we have a responsibility to try to prevent economic hardship or to alter living standards elsewhere in the world, except, again, where the existence of poverty and misery might really interfere with what we are trying to do here. You do not regard this country as the policeman or the guardian of other people's liberties, except where the loss of those liberties would really threaten damage to our own. In particular, you distrust such generalizations as "world leadership"; and you shudder at the idea that this century should be considered, for the world at large, "the American century." It's a big world, bigger than we think. Our shoulders are broad, but they are not broad enough to bear all its miseries and its iniquities. Our own miseries and iniquities would be an ample load if we were to acknowledge them and shoulder them as we ought to.

Now I know what's in your minds at this point. You say: All right, we are supposed to go ahead with our domestic life here at home, and the rest will flow from that. That sounds fine in theory, but what about the Russians and world communism? Suppose they don't leave us any chance to do these things? Suppose their threatening behavior is already ripping the stuffings out of this domestic life of ours—twisting business out of shape, causing shortages, threatening financial stability, disrupting educational patterns? Are you telling us to ignore these things and to go our own way as though none of this existed? Is this isolationism?

The answer to these questions is "no." I recognize these things, and I think I know something about the bitterness and seriousness of the problem of Soviet Communism. I know that we are on the spot. Of course we have to react and defend our interests in such a time and such a situation. But what I am concerned with here is the spirit and concept within which we do all that, and what we expect to get out of it. What I am afraid of is that we may get unrealistic and exalted ideas about what it is we can expect to achieve. If our policies of resistance to Russia are successful, I think that the best we can hope is that there will be no major war and that within some years this will be a somewhat easier world in which to live—one in which we will not have to devote quite so much attention to defending ourselves from external dangers. I think all these things are relative. I do not think we can ever hope to have a situation in which such a country as Russia will not be a problem for us. That goes for other countries, too. I don't think we can ever expect to get this contrary and restless outside world out of our hair. My plea is that if we are going to resist threats to our security, we do just that: We take them as they come, and deal with them as they come. We don't kid ourselves into thinking that just because we coped successfully with one problem at one moment we are going to be immune from other ones rising in the next moment. We don't assume that the only world we can live in is a perfect one, and we recognize that no matter how impressive is the external threat, it is never impressive enough to absolve us from our duty to ourselves and to the improvement of our national life.

I would plead, then, for concepts of national interest more modest than those with which we are accustomed to flatter our sensibilities, but for a greater dignity and quietness and self-discipline in the implementation of those concepts. I would plead, particularly at this genuinely crucial moment in American history, for cool nerves and a clear eye, for the husbanding of our strength, and for an iron self-discipline in refusing to be provoked into using that strength where we cannot see some plausible and reasonably promising end to what we are beginning. I would plead for the restoration of a sense of comradeship and tolerance in our public life and public debates, and for a recognition of the fact that Americans may be wrong without being evil, and that those wrong ones may even conceivably be ourselves.

If we can achieve these things we need not be too exacting in our demands for a definition of national interest. We will then have done the best we can do to bring the world closer to that state of understanding based necessarily more on respect than on intimacy, but fortified by mutual restraint and moderation, and all the more durable and serviceable for its modesty of concept. Therein—not in the world of hatred or of intolerance or of vainglorious pretense—lies the true glory and the true interest of this nation.

X

STATESMANSHIP IN A FREE DEMOCRACY

Slavery and the Right to Self-Government*

BY STEPHEN A. DOUGLAS

[1858]

Ladies and Gentlemen:

It is now nearly four months since the canvass between Mr. Lincoln and myself commenced. On the sixteenth of June the Republican Convention assembled at Springfield and nominated Mr. Lincoln as their candidate for the United States Senate, and he, on that occasion, delivered a speech in which he laid down what he understood to be the Republican creed and the platform on which he proposed to stand during the contest.

The principal points in that speech of Mr. Lincoln's were: First, that this government could not endure permanently divided into free and slave states, as our fathers made it; that they must all become free or all become slave; all become one thing or all become the other, otherwise this Union could not continue to exist. I give you his opinions almost in the identical language he used. His second proposition was a crusade against the Supreme Court of the United States because of the Dred Scott decision** urging as an especial reason for his opposition to that decision that it deprived the Negroes of the rights and benefits of that clause in the Constitution of the United States which guarantees to the citizens of each state all the rights, privileges, and immunities of the citizens of the several states.

---

* From Douglas' Opening Speech in a Debate with Abraham Lincoln, Alton, Ill., October 15, 1858. Stephen A. Douglas (1813–1861), a leading statesman of the Democratic Party in the 1840's and 1850's, was United States Senator from Illinois from 1847 to his death. His main efforts were devoted to contrive a compromise between the Northern and the Southern States to preserve the Union.

** Editor's Note: In the case of *Dred Scott* v. *Sanford*, Chief Justice Taney's opinion ruled that the residence of a slave in a free state or territory did not make him a free man and a citizen; he upheld the right of property owners to move their property, including slaves, wherever they willed within the Union.

On the tenth of July I returned home and delivered a speech to the people of Chicago in which I announced it to be my purpose to appeal to the people of Illinois to sustain the course I had pursued in Congress. In that speech I joined issue with Mr. Lincoln on the points which he had presented. Thus there was an issue clear and distinct made up between us on these two propositions laid down in the speech of Mr. Lincoln at Springfield and controverted by me in my reply to him at Chicago.

On the next day, the eleventh of July, Mr. Lincoln replied to me at Chicago, explaining at some length, and reaffirming the positions which he had taken in his Springfield speech. In that Chicago speech he even went further than he had before and uttered sentiments in regard to the Negro being on an equality with the white man. He adopted in support of this position the argument which Lovejoy and Codding, and other abolition lecturers had made familiar in the northern and central portions of the state, to wit: that the Declaration of Independence having declared all men free and equal, by divine law, also that Negro equality was an inalienable right, of which they could not be deprived. He insisted, in that speech, that the Declaration of Independence included the Negro in the clause, asserting that all men were created equal, and went so far as to say that if one man was allowed to take the position that it did not include the Negro, others might take the position that it did not include other men. He said that all these distinctions between this man and that man, this race and the other race, must be discarded, and we must all stand by the Declaration of Independence, declaring that all men were created equal. . . .

. . . In my speeches I confined myself closely to these three positions which he had taken, controverting his proposition that this Union could not exist as our fathers made it, divided into free and slave states, controverting his proposition of a crusade against the Supreme Court because of the Dred Scott decision, and controverting his proposition that the Declaration of Independence included and meant the Negroes as well as the white men when it declared all men to be created equal. . . . I took up Mr. Lincoln's three propositions in my several speeches, analyzed them, and pointed out what I believed to be the radical errors contained in them. First, in regard to his doctrine that this government was in violation of the law of God, which says that a house divided against itself cannot stand, I repudiated it as a slander upon the immortal framers of our Constitution. I then said, I have often repeated, and now again assert, that in my opinion our government can endure forever, divided into free and slave states as our fathers made it—each state having the right to prohibit, abolish, or sustain slavery, just as it pleases. This government was made upon the great basis of the sovereignty of the states, the right of each state to regulate its own domestic institutions to suit itself, and that right was conferred with the understanding and expectation that, inasmuch as each locality had separate interests, each locality must have different and distinct local and domestic institutions, corresponding to its wants and interests. Our fathers knew when they made the government that the laws and institutions which were well adapted to the Green Mountains of Vermont were unsuited to the rice plantations of South Carolina. They knew then, as well as we know now, that the laws and institutions which would be well adapted to the beautiful prairies of Illinois would not be suited to the mining regions of California. They knew that in a republic as broad as this, having such a variety of soil, climate, and interest, there must necessarily be a corresponding variety of local laws—the policy and institutions of each state adapted to its condition and wants. For this reason this Union was established on the right of each state to do as it pleased on the question of slavery and every other question; and the various states were not allowed to complain of, much less interfere with, the policy of their neighbors. . . .

You see that if this abolition doctrine of Mr. Lincoln had prevailed when the government was made, it would have established slavery as a permanent institution, in all the states, whether they wanted it or not, and the question for us to determine in Illinois now as one of the free states is whether or not we are willing, having become the majority section, to enforce a doctrine on the minority which we would have resisted with our heart's blood had it been attempted on us when we were in a minority. How has the South lost her power as the majority section in this Union, and how have the free states gained it, except under the operation of that principle which declares the right of the people of each state and each territory to form and regulate their domestic institutions in their own way. It was under that principle that slavery was abolished in New Hampshire, Rhode Island, Connecticut, New York, New Jersey, and Pennsylvania; it was under that principle that one-half of the slaveholding states became free; it was under that principle that the number of free states increased until, from being one out of twelve states, we have grown to be the majority of states of the whole Union, with the power to control the House of Representatives and Senate, and the power, consequently, to elect a President by northern votes without the aid of a southern state. Having obtained this power under the operation of that great principle, are you now prepared to abandon the principle and declare that merely because we have the power you will wage a war against the southern states and their institutions until you force them to abolish slavery everywhere . . . ?

I answer specifically if you want a further answer and say that, while under the decision of the Supreme Court, as recorded in the opinion of Chief Justice Taney, slaves are property like all other property, and can be carried into any territory of the United States the same as any other description of property, yet when you get them there they are subject to the local law of the territory just like all other property. You will find in a recent speech delivered by that able and eloquent statesman, Hon. Jefferson Davis, at Bangor, Maine, that he took the same view of this subject that I did in my Freeport speech. He there said:

> If the inhabitants of any territory should refuse to enact such laws and police regulations as would give security to their property or to his, it would be rendered more or less valueless in proportion to the difficulties of holding it without such protection. In the case of property in the labor of man, or what is usually called slave property, the insecurity would be so great that the owner could not ordinarily retain it. Therefore, though the right would remain, the remedy being withheld, it would follow that the owner would be practically debarred, by the circumstances of the case, from taking slave property into a territory where the sense of the inhabitants was opposed to its introduction. So much for the oft-repeated fallacy of forcing slavery upon any community. . . .

The whole South are rallying to the support of the doctrine that, if the people of a territory want slavery, they have a right to have it, and, if they do not want it, that no power on earth can force it upon them. I hold that there is no principle on earth more sacred to all the friends of freedom than that which says that no institution, no law, no constitution, should be forced on an unwilling people contrary to their wishes; and I assert that the Kansas and Nebraska Bill contains that principle. . . . I will never violate or abandon that doctrine if I have to stand alone. I have resisted the blandishments and threats of power on the one side, and seduction on the other, and have stood immovably for that principle, fighting for it when assailed by northern mobs or threatened by southern hostility. I have defended it against the North and the South, and I will defend it against whoever assails it, and I will follow it wherever its logical conclusions lead me. I say to you that there is but one hope, one safety, for this country, and that is to stand immovably by that principle which declares the right of each state and each territory to decide these questions for themselves. This government was founded on that principle and must be administered in the same sense in which it was founded.

But the Abolition party really think that under the Declaration of Independence the Negro is equal to the white man and that Negro equality is an inalienable right conferred by the Almighty, and hence that all human laws in violation of it are null and void. With such men it is no use for me to argue. I hold that the signers of the Declaration of Independence had no reference to Negroes at all when they declared all men to be created equal. They did not mean Negro, nor the savage Indians, nor the Fiji Islanders, nor any other barbarous race. They were speaking of white men. They alluded to men of European birth and European descent—to white men and to none others—when they declared that doctrine. I hold that this government was established on the white basis. It was established by white men for the benefit of white men and their posterity forever and should be administered by white men and none others. But it does not follow, by any means, that merely because the Negro is not a citizen, and merely because he is not our equal, that, therefore, he should be a slave. On the contrary, it does follow that we ought to extend to the Negro race, and to all other dependent races all the rights, all the privileges, and all the immunities which they can exercise consistently with the safety of society. Humanity requires that we should give them all these privileges; Christianity commands that we should extend those privileges to them. The question then arises: What are those privileges and what is the nature and extent of them. My answer is that that is a question which each state must answer for itself. We in Illinois have decided it for ourselves. We tried slavery, kept it up for twelve years, and, finding that it was not profitable, we abolished it for that reason, and became a free state. We adopted in its stead the policy that a Negro in this state shall not be a slave and shall not be a citizen. We have a right to adopt that policy. For my part I think it is a wise and sound policy for us. . . . If the people of all the states will act on that great principle, and each state mind its own business, attend to its own affairs, take care of its own Negroes, and not meddle with its neighbors, then there will be peace between the North and the South, the East and the West, throughout the whole Union. Why can we not thus have peace? Why should we thus allow a sectional party to agitate this country, to array the North against the South, and convert us into enemies instead of friends, merely that a few ambitious men may ride into power on a sectional hobby? How long is it since these ambitious northern men wished for a sectional organization? Did any one of them dream of a sectional party as long as the North was the weaker section and the South the stronger? Then all were opposed to sectional parties; but the moment the North obtained the majority in the House and Senate by the admission of California, and could elect a President without the aid of southern votes, that moment ambitious northern men formed a scheme to excite the North against the South, and make the people be governed in their votes by geographical lines, thinking that the North, being the stronger section, would outvote the South, and consequently they, the leaders, would ride into office on a sectional hobby. I am told that my hour is out. It was very short.

Slavery and the Constitution*

BY ABRAHAM LINCOLN

[1858]

... IT IS NOT TRUE that our fathers, as Judge Douglas assumes, made this government part slave and part free. Understand the sense in which he puts it. He assumes that slavery is a rightful thing within itself—was introduced by the framers of the Constitution. The exact truth is that they found the institution existing among us, and they left it as they found it. But, in making the government, they left this institution with many clear marks of disapprobation upon it. They found slavery among them, and they left it among them because of the difficulty—the absolute impossibility—of its immediate removal. . . .

I confess, when I propose a certain measure of policy, it is not enough for me that I do not intend anything evil in the result, but it is incumbent on me to show that it has not a *tendency* to that result. I have met Judge Douglas in that point of view. I have not only made the declaration that I do not *mean* to produce a conflict between the states, but I have tried to show by fair reasoning, and I think I have shown to the minds of fair men, that I propose nothing but what has a most peaceful tendency. The quotation that I happened to make in that Springfield speech, that "a house divided against itself cannot stand," and which has proved so offensive to the Judge, was part and parcel of the same thing. He tries to show that variety in the domestic institutions of the different states is necessary and indispensable. I do not dispute it. I have no controversy with Judge Douglas about that. . . .

Now irrespective of the moral aspect of this question as to whether there is a right or wrong in enslaving a Negro, I am still in favor of our new territories being in such a condition that white men may find a home—may find some spot where they can better their condition—where they can settle upon new soil and better

---

* From Lincoln's Reply to Douglas in the Debate at Alton, Ill., October 15, 1858. Abraham Lincoln (1809–1865) was the 16th President of the United States.

their condition in life. I am in favor of this not merely (I must say it here as I have elsewhere) for our own people who are born amongst us, but as an outlet for *free white people everywhere,* the world over—in which Hans and Baptiste and Patrick, and all other men from all the world, may find new homes and better their conditions in life.

I have stated upon former occasions, and I may as well state again, what I understand to be the real issue in this controversy between Judge Douglas and myself. On the point of my wanting to make war between the free and the slave states, there has been no issue between us. So, too, when he assumes that I am in favor of introducing a perfect social and political equality between the white and black races. These are false issues, upon which Judge Douglas has tried to force the controversy. There is no foundation in truth for the charge that I maintain either of these propositions. The real issue in this controversy—the one pressing upon every mind—is the sentiment on the part of one class that looks upon the institution of slavery *as a wrong* and of another class that *does not* look upon it as a wrong. The sentiment that contemplates the institution of slavery in this country as a wrong is the sentiment of the Republican party. It is the sentiment around which all their actions—all their arguments circle—from which all their propositions radiate. They look upon it as being a moral, social, and political wrong; and, while they contemplate it as such, they nevertheless have due regard for its actual existence among us, and the difficulties of getting rid of it in any satisfactory way and to all the constitutional obligations thrown about it. Yet having a due regard for these, they desire a policy in regard to it that looks to its not creating any more danger. They insist that it should, as far as may be, *be treated* as a wrong, and one of the methods of treating it as a wrong is to *make provision that it shall grow no larger*. They also desire a policy that looks to a peaceful end of slavery at sometime as being wrong. These are the views they entertain in regard to it as I understand them; and all their sentiments—all their arguments and propositions—are brought within this range. I have said, and I repeat it here, that if there be a man amongst us who does not think that the institution of slavery is wrong in any one of the aspects of which I have spoken, he is misplaced and ought not to be with us. And if there be a man amongst us who is so impatient of it as a wrong as to disregard its actual presence among us and the difficulty of getting rid of it suddenly in a satisfactory way, and to disregard the constitutional obligations thrown about it, that man is misplaced if he is on our platform. We disclaim sympathy with him in practical action. He is not placed properly with us.

On this subject of treating it as a wrong, and limiting its spread, let me say a word. Has anything ever threatened the existence of this Union save and except this very institution of slavery? What is it that we hold most dear amongst us? Our own liberty and prosperity. What has ever threatened our liberty and prosperity save and except this institution of slavery? If this is true, how do you propose to improve the condition of things by enlarging slavery— by spreading it out and making it bigger? You may have a wen or cancer upon your person and not be able to cut it out lest you bleed to death; but surely it is no way to cure it, to engraft it and spread it over your whole body. That is no proper way of treating what you regard a wrong. You see this peaceful way of dealing with it as a wrong—restricting the spread of it, and not allowing it to go into new countries where it has not already existed. That is the peaceful way, the old-fashioned way, the way in which the fathers themselves set us the example.

On the other hand, I have said there is a sentiment which treats it as *not* being wrong. That is the Democratic sentiment of this day. I do not mean to say that every man who stands within that range positively asserts that it is right. That class will include all who positively assert that it is right, and all who like Judge Douglas treat it as indifferent and do not say it is either right or wrong. These two classes of men fall within the general class of those who do not look upon it as a wrong. . . .

The Democratic policy in regard to that institution will not tolerate the merest breath, the slightest hint, of the least degree of wrong about it. Try it by some of Judge Douglas' arguments. He says he "don't care whether it is voted up or voted down" in the territories. I do not care myself in dealing with that expression, whether it is intended to be expressive of his individual sentiments on the subject or only of the national policy he desires to have established. It is alike valuable for my purpose. Any man can say that who does not see anything wrong in slavery, but no man can logically say it who does see a wrong in it; because no man can logically say he does not care whether a wrong is voted up or voted down. He may say he does not care whether an indifferent thing is voted up or down, but he must logically have a choice between a right thing and a wrong thing. He contends that whatever community wants slaves has a right to have them. So they have if it is not a wrong. But if it is a wrong, he cannot say people have a right to do wrong. He says that, upon the score of equality, slaves should be allowed to go in a new territory, like other property. This is strictly logical if there is no difference between it and other property. If it and other property are equal, his argument is entirely logical. But if you insist that one is wrong and the other right, there is no use to institute a comparison between right and wrong. . . .

That is the real issue. That is the issue that will continue in this country when these poor tongues of Judge Douglas and myself shall be silent. It is the eternal struggle between these two principles—right and wrong—throughout the world. They are the two principles that have stood face to face from the beginning of time and will ever continue to struggle. The one is the common right of humanity and the other the divine right of kings. It is the same principle in whatever shape it develops itself. It is the same spirit that says, "You work and toil and earn bread, and I'll eat it." No matter in what shape it comes, whether from the mouth of a king who seeks to bestride the people of his own nation and live by the fruit of their labor, or from one race of men as an apology for enslaving another race, it is the same tyrannical principle. . . .

I suppose most of us (I know it of myself) believe that the people of the southern states are entitled to a congressional Fugitive Slave law—that is a right fixed in the Constitution. But it cannot be made available to them without congressional legislation. In the Judge's language, it is a "barren right" which needs legislation before it can become efficient and valuable to the persons to whom it is guaranteed. And as the right is constitutional I agree that the legislation shall be granted to it—and that not that we like the institution of slavery. We profess to have no taste for running and catching niggers—at least I profess no taste for that job at all. Why then do I yield support to a Fugitive Slave law? Because I do not understand that the Constitution, which guarantees that right, can be supported without it. And if I believed that the right to hold a slave in a territory was equally fixed in the Constitution with the right to reclaim fugitives, I should be bound to give it the legislation necessary to support it. I say that no man can deny his obligation to give the necessary legislation to support slavery in a territory who believes it is a constitutional right to have it there. No man can who does not give the abolitionists an argument to deny the obligation enjoined by the Constitution to enact a Fugitive Slave law. Try it now. It is the strongest abolition argument ever made. I say if that Dred Scott decision is correct, then the right to hold slaves in a territory is equally a constitutional right with the right of a slaveholder to have his runaway returned. No one can show the distinction between them. The one is express, so that we cannot deny it. The other is construed to be in the Constitution, so that he who believes the decision to be correct believes in the right. And the man who argues that by unfriendly legislation, in spite of that Constitutional right, slavery may be driven from the territories cannot avoid furnishing an argument by which abolitionists may deny the obligation to return fugitives and claim the power to pass laws unfriendly to the right of the slaveholder to reclaim his fugitive. I do not know how such an argument may strike a popular assembly like this, but I defy anybody to go before a body of men whose minds are educated to estimating evidence and reasoning and show that there is an iota of difference between the constitutional right to reclaim a fugitive and the constitutional right to hold a slave in a territory, provided this Dred Scott decision is correct. I defy any man to make an argument that will justify unfriendly legislation to deprive a slaveholder of his right to hold his slave in a territory that will not equally, in all its length, breadth, and thickness, furnish an argument for nullifying the Fugitive Slave law. Why, there is not such an abolitionist in the nation as Douglas, after all.

The Law of God and the Law of the Land*

BY HENRY DAVID THOREAU

[1854]

I WOULD remind my countrymen that they are to be men first, and Americans only at a late and convenient hour. No matter how valuable law may be to protect your property, even to keep soul and body together, if it do not keep you and humanity together.

I am sorry to say that I doubt if there is a judge in Massachusetts who is prepared to resign his office, and get his living innocently, whenever it is required of him to pass sentence under a law which is merely contrary to the law of God. I am compelled to see that they put themselves, or rather are by character, in this respect, exactly on a level with the marine who discharges his musket in any direction he is ordered to. They are just as much tools, and as little men. Certainly, they are not the more to be respected, because their master enslaves their understandings and consciences, instead of their bodies.

The judges and lawyers,—simply as such, I mean,—and all men of expediency, try this case by a very low and incompetent standard. They consider, not whether the Fugitive Slave Law is right, but whether it is what they call *constitutional.* Is virtue constitutional, or vice? Is equity constitutional, or iniquity? In important moral and vital questions, like this, it is just as impertinent to ask whether a law is constitutional or not, as to ask whether it is profitable or not. They persist in being the servants of the worst of men, and not the servants of humanity. The question is, not whether you or your grandfather, seventy years ago, did not enter into an agreement to serve the Devil, and that service is not accordingly now due; but whether you will not now, for once and at last, serve

---

* From an address "Slavery in Massachusetts" delivered on July 4, 1854, at Framingham, Mass. Henry David Thoreau (1817–1862), New England writer and philosopher, was a leading spokesman for the cause of Abolitionism. In this address, Thoreau was discussing the Fugitive Slave Law which was designed to enforce the return of fugitive slaves to their masters.

God,—inspite of your own past recreancy, or that of your ancestor, —by obeying that eternal and only just CONSTITUTION, which He, and not any Jefferson or Adams, has written in your being.

The amount of it is, if the majority vote the Devil to be God, the minority will live and behave accordingly,—and obey the successful candidate, trusting that, some time or other, by some Speaker's casting-vote, perhaps, they may reinstate God. This is the highest principle I can get out or invent for my neighbors. These men act as if they believed that they could safely slide down a hill a little way,—or a good way,—and would surely come to a place, by and by, where they could begin to slide up again. This is expediency, or choosing that course which offers the slightest obstacles to the feet, that is, a downhill one. But there is no such thing as accomplishing a righteous reform by the use of "expediency." There is no such thing as sliding up hill. In morals the only sliders are backsliders. . . .

Will mankind never learn that policy is not morality,—that it never secures any moral right, but considers merely what is expedient? chooses the available candidate,—who is invariably the Devil,—and what right have his constituents to be surprised, because the Devil does not behave like an angel of light? What is wanted is men, not of policy, but of probity,—who recognize a higher law than the Constitution, or the decision of the majority. The fate of the country does not depend on how you vote at the polls,—the worst man is as strong as the best at that game; it does not depend on what kind of paper you drop into the ballot-box once a year, but on what kind of man you drop from your chamber into the street every morning.

What should concern Massachusetts is not the Nebraska Bill, nor the Fugitive Slave Bill, but her own slaveholding and servility. Let the State dissolve her union with the slaveholder. She may wriggle and hesitate, and ask leave to read the Constitution once more; but she can find no respectable law or precedent which sanctions the continuance of such a union for an instant.

Let each inhabitant of the State dissolve his union with her, as long as she delays to do her duty.

The events of the past month teach me to distrust Fame. I see that she does not finely discriminate, but coarsely hurrahs. She considers not the simple heroism of an action, but only as it is connected with its apparent consequences. She praises till she is hoarse the easy exploit of the Boston tea party, but will be comparatively silent about the braver and more disinterestedly heroic attack on the Boston Court-House, simply because it was unsuccessful!

Covered with disgrace, the State has sat down coolly to try for their lives and liberties the men who attempted to do its duty for it. And this is called *justice!* They who have shown that they can behave particularly well may perchance be put under bonds for *their good behavior*. They whom truth requires at present to plead guilty are, of all the inhabitants of the State, preeminently innocent. While the Governor, and the Mayor, and countless officers of the Commonwealth are at large, the champions of liberty are imprisoned.

Only they are guiltless who commit the crime of contempt of such a court. It behooves every man to see that his influence is on the side of justice, and let the courts make their own characters. My sympathies in this case are wholly with the accused, and wholly against their accusers and judges. Justice is sweet and musical; but injustice is harsh and discordant. The judge still sits grinding at his organ, but it yields no music, and we hear only the sound of the handle. He believes that all the music resides in the handle, and the crowd toss him their coppers the same as before.

Do you suppose that that Massachusetts which is now doing these things,—which hesitates to crown these men, some of whose lawyers, and even judges, perchance, may be driven to take refuge in some poor quibble, that they may not wholly outrage their instinctive sense of justice,—do you suppose that she is anything but base and servile? that she is the champion of liberty?

Show me a free state, and a court truly of justice, and I will fight for them, if need be; but show me Massachusetts, and I refuse her my allegiance, and express contempt for her courts.

The effect of a good government is to make life more valuable,—of a bad one, to make it less valuable. We can afford that railroad and all merely material stock should lose some of its value, for that only compels us to live more simply and economically; but suppose that the value of life itself should be diminished! How can we make a less demand on man and nature, how live more economically in respect to virtue and all noble qualities, than we do? I have lived for the last month—and I think that every man in Massachusetts capable of the sentiment of patriotism must have had a similar experience—with the sense of having suffered a vast and indefinite loss. I did not know at first what ailed me. At last it occurred to me that what I had lost was a country. I had never respected the government near to which I lived, but I had foolishly thought that I might manage to live here, minding my private affairs, and forget it. For my part, my old and worthiest pursuits have lost I cannot say how much of their attraction, and I feel that my investment in life here is worth many per cent. less since Massachusetts last deliberately sent back an innocent man, Anthony Burns, to slavery. . . .

I feel that, to some extent, the State has fatally interfered with my lawful business. It has not only interrupted me in my passage through Court Street on errands of trade, but it has interrupted me and every man on his onward and upward path, on which he had trusted soon to leave Court Street far behind. What right had it to remind me of Court Street? I have found that hollow which even I had relied on for solid. . . .

I walk toward one of our ponds; but what signifies the beauty of nature when men are base? We walk to lakes to see our serenity reflected in them; when we are not serene, we go not to them. Who can be serene in a country where both the rulers and the ruled are without principle? The remembrance of my country spoils my walk. My thoughts are murder to the State, and involuntarily go plotting against her.

On Deceiving the Public for the Public Good*

BY LYMAN BRYSON

[1950]

THIS boldly stated question, "Should a leader deceive the public for the public good?" is not likely to get less important as time goes on, for leaders do not get less powerful or less ambitious and the means of deceiving the people increases. We have made political leaders more than ever dependent on the suffrage of their followers, but we have put mechanical messengers into their hands by which they can coerce public consent. The growth of an industrial culture has tended also to enlarge constantly the units of organization in which men work, and also the collectives in which they think and communicate with each other and act together. The great collectives take over more and more of our lives, in their economic and political aspects especially, and make it more and more difficult for us to see the relations between what we do and any general principles of truth or rightness. The leader is tempted by the tools of manipulation. Is it ever wise for him to use them?

We ought first to ask ourselves the more general question: What are leaders for? Why do we need leaders in a free country? I would answer that the leader's function is to help to determine, in any crisis, which of our possible selves will act. We are all multiple: our personalities are bundles of possible responses, each with the accent of our own peculiar self, but still all widely differing from each other. If we could always be counted on, each one of us, always to act in the same way, no matter how we are challenged, life and politics would be much simpler. It would also be dull and uninteresting: our unpredictability is part of our human charm. The leader is an embodied suggestion, and the combinations of causes and chances that determine the leader who will catch our attention

---

* From *Conflict of Loyalties*, edited by R. M. MacIver, Harper & Brothers, 1952. Reprinted by permission. Lyman Bryson (b. 1888), educator, is Professor at Teacher's College, Columbia University.

and our support at any time, are the casual chain of history.

We might suppose, for an example, that the people of a tragically unhappy country like Germany could have followed a more calm and righteous leader than Hitler, and that they had all reacted to the pressure of the situation in the same way. In that case there might never have been a world conflict. . . .

What did Hitler actually do? He called out the aggressive self that was latent in practically every normal German and made it the dominant active self. He made most of them as aggressive as they were individually capable of being. More than that, he called out and got into positions of power all the Germans who were even more than normally aggressive, and the nation was put in a generally aggressive and dangerous posture. Great crises make for great instability of selves, or of character, and Germany's crisis was catastrophic. The result was an enormous overdepelopment of a normal human trait, socially organized and expressed. This, in generally less damaging ways, sometimes to our great good, is the function of the leader. He cannot make us over; he can make us be our best or our worst within our range.

A free country, which is our ideal and our partial accomplishment, has the same need for leaders of the right kind as any other. But before we conclude this discussion we have to ask ourselves another question of general principle. What is the purpose of political life and active citizenship in a free country? I believe that the purpose of sharing in the political thought of my country, and of my own community as a part of it, is not ultimately the solution of political problems. Freedom for men to think and learn and act on their acquired wisdom is, I believe, more likely to get good solutions for political problems than any other system; our record would prove that. But even if there were a better way of getting the merely correct technical, or practical—shall we say, the material answers?—to a political problem, I would still hold fast to democracy and to decision by the people. Our ultimate judgment on a social or a political system should not be based, solely, on the criterion of practical success. It is based on the evidence of growth in the people. Sometimes free men make mistakes; they must be allowed that privilege. If they learn the lessons of politics and conduct from making mistakes, the principle of democracy is fulfilled. Whether or not there are other areas of living, outside the political, in which this principle cannot be followed, is another question. We shall stick to politics.

I can turn to Germany again for an example to make the point. Several years ago a small group of German women who were then holding political office at home were brought over here for some lessons in American ideas of democracy.... In the conversation we had, one of the most intelligent of the nine women struggled to understand the essential relation in a democracy between freedom and authority, political freedom and technical authority. She told of being in a city where the administration of municipal affairs was entrusted to a city manager. He was appointed because of his professional competence and experience. He was an excellent person and it was evident that he would know the right answers to most practical questions. And yet—this disturbed her deeply—the members of the city council, who were not elected for professional competence, had final authority to overrule his decisions. Even his right decisions! she said.

My answer was that the principle of democracy would always make it necessary for the directly elected representatives of the people to have the final word. Even when they were wrong? Even when, by any technical judgment, they were wrong. And I insisted that she would not understand democracy until she could see why that had to be so. I tried to explain to her that she was still thinking of political action as having no purpose other than to solve political problems, and that if this principle, which was ultimately not democratic but authoritarian, was followed in any country, the rule of the people would eventually be ended. The purpose of political action and the opportunity of free political life is for the people ultimately to determine their own destiny, and—after they have had the chance to learn—even to make their own mistakes. The great end they are serving is the development of their country through the development of themselves, not by authoritative interventions, no matter how competent or benevolent. It is not true that tyrants can never be benevolent. The trouble is that they go on being tyrants, and under tyranny individual men dry up for lack of spiritual exercise.

We too often forget that it means something real and important, and perhaps greatly daring, to say that the purpose of a democratic society is to make great persons, that the end is the development of the person by experience and not the technical answers to civic problems. In fact, political experience has been and still is, and probably will always be, one of the greatest of the educational factors in any person's life; it cannot count for much unless it includes free decisions and a chance to learn from the consequences. We have to decide for ourselves in the light of what we can learn of the facts, and then learn the great lessons from the results of our choice. Both collective organization and authority can interfere with this purpose and defeat us. Unless we learn from politics, then politics is not worth our time. Decisions should be made for the good of the state, but the ultimate value of all political decisions is the experience of bearing part of the responsibility of making them.

The leader, then, is a person who helps us to choose, and if he is a great leader he helps us also to learn. . . .

These are all preliminary considerations which must be taken into account, I think, before we can answer our question: which is, if you have forgotten, "Should a leader deceive the public for the public good?" But we still have to answer another prior question: "What kind of society do we want?" There seems to be good evidence in history and in our own times that only those societies that believe their own political slogans and in the myths of their national virtue can build successful empires. By this, I mean, of course, that nations which dominate others must first put themselves under spiritual domination.

This is not the same thing as to say that we should lose our democracy if we set out to run the world; I do not believe that the problem is best put in those terms. Certainly other nations in the remote past and in approximately modern times have built empires and still remained democracies. Britain is the best example; Britain built her empire and her own democracy through the same period. But it is still true that nations which set out more or less deliberately to dominate other nations must be ruled by men who pronounce some political slogans that are not to be questioned, and if the people have even a healthy skepticism toward the goals of empire or the myths of their country's unspecked virtue, the will to power is fatally weakened.

We need only look at the difference between the attitude of the Greeks, who brilliantly failed at empire, and the attitude of the

Romans who brilliantly succeeded. It is not now a question of their degrees of democracy at home, although we might have stopped to notice that Rome also built her empire while building her civil law and the civil rights of her people. The point is to mark the difference between a magnificently endowed but skeptical people like the Greeks, and a very different people, the Romans, who learned how to conquer. We can be quite sure, I think, that most of the Romans who brought their own version of the classical civilization to the Western world and gave the Mediterranean basin several centuries of magnificent peace, really believed that Roman ideas and Roman slogans were not only best but were not to be examined. The Greeks in their most insolent moments never seemed able quite to believe that their imperial depredations were for the good of subjected peoples. When the Romans began their career as a mean tribe with great ideas, in the little group of hills in the center of Italy, when they conquered and took more and more of their neighbors into the Roman state, they believed not only that it was good for the Romans but also good for the slaves and the conquered who might, of course, end up by being taken into citizenship but who were, in the meantime, to be made over in the Roman image. They believed their slogans and took over the Western world.

It helps to have the slogans and the political myths founded on facts, but the crucial point is that they are believed. Today we can study the British and find them much like the Romans in their early brutal innocence and faith, like them also perhaps in the late phase, their present refinement, their discovery that social welfare at home is more important than carrying the white man's burden, perhaps that the white man's burden can be best carried at home, their period of the Antonines. We can, if necessary, save them from the next stage of being overcome by barbarians. But when they were powerful, they were extraordinary for efficiency as rulers and for self-confidence. They believed their own stuff. The French, on the other hand, like the Greeks, have long since been skeptical and weak imperialists. The principle holds.

We do not know how much of our own destiny is ever in our own hands. This is a philosophic, not a practical, question we are dealing with and we can assume for ourselves the power of national choice. We do not need to destroy our own freedoms to save us from destroying the freedoms of others; that ironic crime has often been committed, to keep freedom at home and conquer abroad. Witness Britain and India. The cost of empire lies not in our political freedoms, but in the adjacent area of our freedom to know and to learn by free informed experience. To lead us into building an empire, provided it is possible to us in material terms, our leaders have got to convince us and keep us convinced that all slogans of imperialism are true, that we are destined because of our virtue as well as our strength to subdue the world for its own sake, to spread the unquestioned doctrines of our own political and economic commonplaces, to repeat again the sorry old pattern of dominion.

The purpose of empire building, however, is not the building of great citizens; it is the building of a great state. Great states are built by leaders who lead their people into sacrifice for dreams of power and glory, who lead them to believe that some kinds of doctrinally prescribed freedom may be good for them but that weaker peoples must be led. Great men and women are built by leaders who lead them into sacrifice for the right freely to seek the truth. . . .

. . . Great citizens can be built only when leaders will dare to let them use their own minds, when the state helps us to knowledge of alternative choices, conserving for us the essential democratic experience, which is to seek the truth by our own efforts, to know it as far as it can be known, to act on it freely, and to learn freely from free action. We can be strong and we can be free. If we want to dominate others and shove our doctrines down the throats of weaker peoples, then our leaders will have to deal often in lies in order to blind us to the true picture of our behavior.

This question, then, can be put in Kantian terms. We can follow Immanuel Kant's distinction: we can take men as means or as ends. It is as simple and as difficult as that. If men are to be used as means, if human beings are to be treated by their leaders as means to an end—and I am specifying nothing whatever as to the quality or character of that end, even the realization of national power, or the greatness of an institution, the realization of an ideal, or the sacrifice of the present to the future—if men are to be taken as means to an end, then it is inevitable that leaders will at certain times deceive them for the public good. But if we accept Kant's ethical principle and believe that men should never be used as means but always as ends in themselves, if their experience is the purpose of political life, then the leader, by interfering with the people's experience of free inquiry, no matter how bitter their discoveries may be, is defeating democracy.

The distinction is easy but to make it work is difficult, because a leader, even within my definition, is always also a teacher. Part of his job in every situation and on any scale of operation, social or political or intellectual, is to enlarge constantly the range of thought and the range of possible choices in the minds of his followers. A leader who believes in our kind of democracy, which we practice with considerable success, although we often get mixed up in trying to define it, will say, "My duty is also the duty of a teacher, to increase men's freedom by enlarging their knowledge." The biggest dimension of freedom is the dimension of knowing. As we have to repeat, over and over in these days of deliberate obfuscation and innocent confusions, the choice you never can make is the choice you never heard of. Ignorance is not only a chain on your mind, it is also a chain that binds your will.

Any kind of so-called freedom that protects men against making mistakes cannot be what we are talking about, because you cannot protect men from error except by protecting them against knowledge. Any government or institution or leader that sets out to protect men from making mistakes must act on the assumption that there is an ultimate truth that can be stated as final closed doctrine—and here I am still speaking of political and social truths because religious truth may present other problems—not only that there is an ultimate truth but that it can be exactly stated and that they can state it better than anyone else. It implies that this doctrine is of a kind that cannot maintain itself by its own character in the open market of ideas, and that it cannot do its work for men unless they are forcibly prevented from ever hearing anything else. . . .

. . . The most dangerous tyrant is the one who has succumbed to the ultimate corruption of power and believes in his own benevolence. He can believe that he is helping the people, that he is doing us good by keeping us from knowing anything but the official doctrine whatever it may be. He will in all honesty and zeal keep us from learning by our own choice.

You may be thinking that we do not learn much from experience and that making a mistake is not a step that leads always to wisdom. This is true enough; if we learned from our mistakes and never repeated them, the world would not be so full of defeated and frustrated people. But, however much we may fail to learn from experience, it is quite certain that there are many things we cannot learn from anything else. They cannot be handed down to us, nor handed out, and if there is any new truth in the world it will not be discovered by men who are protected from error.

So we come back again to the leader. His chief work is to help us make a choice. He can lead us to take great spiritual risks and be great citizens. Or, he can play safe, stick to the doctrine, and let us drift into the fallacies of power. In that case, we may possibly win an empire but we are almost certain to lose our own souls. I am making a clear contrast here, between greatness in nations and greatness in men, meaning power and domination on the one hand, and the expansion of the soul by the search for truth, on the other. These are different ideals, and it often happens in the practical affairs of men that nations attain to mixtures and combinations of the ideals that are held by various members of their national group. By my own preference among the definitions of freedom, I should have to say that I should prefer a nation in which there was a generous variety of different goals, on all levels of political opinion, except in those matters that would endanger freedom itself. And I do not believe that all the overt national actions of a national group will always serve the same ideal. What we are seeking is the answer to a question that proposes a choice between two kinds of effort and two kinds of demand to be made on our leaders, knowing that the results will be mixed. The true spirit of freedom and power are not antithetical, any more than righteousness and prosperity are antitheses. But the man who seeks prosperity above righteousness will probably lose both, and the nation that seeks power above truth will have the same double disaster.

The alternative that we are not quite willing to state in realistic terms can be put this way: Do we want our leaders to decide what shall be our national fate without letting us in on their secret, and then manipulate our ignorance to achieve what they think we ought to have, without our knowing what are the other choices? It might be said that a little license to a ruler is not a mandate for tyranny. The difficulty is that if we give willingly to a leader the right to deceive us, we give over to him also the right and the chance to decide when and whenever the deceit shall be practiced. It is certain that the fates of nations are mixed, evil with good, and we may seek virtue even while we have a concern for material power. But which value is to be sacrificed to the other? It is the habit of knowing what value is to be held to, even at the cost of the others, that gives us our national morality.

In the present crisis in American life, it is important to look closely at one assumption, an almost unconscious assumption that gets into much discussion of our possible future power and influence. It is taken for granted and it is false. The notion, which you will meet in all kinds of writing and talk and political polemics, is that we can justify our material power, even a career of material domination, by the cultural achievements that our power makes possible. I mean, of course, cultural achievements by ourselves, great art and thought and science and philosophy, the great expression of our own ideals. And the notion is false because, in spite of the commonplaces of the textbooks and careless historians, great cultural achievements have not been inevitably, or even generally, concurrent with great material power, and certainly are not the results of it. . . .

We are in a critical period and we have a choice to make among ideals. Whether or not we are in fact allowed to choose our course may be a debatable question, because we are deep in change and we can neither see clearly, nor control completely, the forces that we ride. We may want to be great in the material sense, dominant, imperialistic, oppressive, and also to do great things with our minds and spirits. But if we seek to gain the city, in the Biblical phrase, we may all too easily lose our own souls, without realizing that it is no better for a nation than it is for a man to want power at too great a cost, or to be unaware of what is paid for it.

We have said that this price is not our freedom at home. There is an evident paradox here. If we can have an empire without giving up our freedom as citizens of our home country, how are we paying for empire by giving up freedom of thought? The examples already given of Rome and Britain, where domestic citizenship got to be more equitable and free during much of the time when the empire was being conquered, ought to indicate an answer. And it seems to me certain that, in Britain in the nineteenth century, the citizen of Birmingham or Aberdeen, busy with his own political and business affairs, did not know what his government was doing in Burma or the Sudan. He believed in the imperial myth and was willing to support it, the myth that his government was bringing benevolent civilization to recalcitrant "niggers" the world over, and when his leaders lied to him as they consistently did, he believed them and was willing, when needed, to die for empire. . . .

We might have had, beginning with Theodore Roosevelt, the kind of leader who would manipulate the sources of public information and the sources of agencies of public opinion, as we saw done in Germany, and as is now being done in many other places. In our country the net of public communication is very wide and very difficult to manage from a central point, but it is also pervasive and influential. A persistent plan to deceive us might have succeeded. If that had been done, it is conceivable that our policies, especially in foreign relations, might have been less vacillating as well as less democratically criticized, and we might have been a mightier power. Having more honest leaders we have been less successfully aggressive than others, and against aggression we have fought back only at the last moment with great cost to ourselves. We have not been the big masterful nation that we sometimes think we ought to be. But if, as a nation, we had been more masterful, our people in this generation would be much less fully aware of what has been going on and would have been less challenged by a painfully real knowledge of the issues. In that case, we might have been a greater nation of lesser persons.

Our question then, "Should a leader deceive the public for the public good?" is not a simple question of good or evil. It is a real choice in political action. Shall we, on the one hand, follow the example of the nations that can set forth uncomplicated simple myths about themselves and their destiny in the world, and believe them, and go on to conquest? Shall we have our own version of the slogan, "Take up the white man's burden?" We, too, can have our imperial adventure, and there are leaders already in training to take us in hand with the right slogans and comforting reassuring myths about our destiny, if we are ready to respond.

But if we want above all things the wisdom and knowledge of free experience, the high privilege of searching for truth by our own powers, we had better give up campaigns against the liberties of others and also—this is harder to do—all campaigns to save other nations from their own errors. We can declare our own faiths and our own gospels, of course, but not with sanctions. Nations that encourage their citizens to be openminded, skeptical, questioning, free, are not good candidates for hegemony. But they have something else. They have a democracy of the spirit and the mind, and what they achieve may really be for the good of others, as well as for themselves.

The Predicament of Democracy*

BY ALEXIS DE TOCQUEVILLE

[1835]

IT IS INCONTESTABLE that, in times of danger, a free people display far more energy than any other. But I incline to believe that this is especially true of those free nations in which the aristocratic element preponderates. Democracy appears to me better adapted for the conduct of society in times of peace, or for a sudden effort of remarkable vigor, than for the prolonged endurance of the great storms that beset the political existence of nations. The reason is very evident; enthusiasm prompts men to expose themselves to dangers and privations; but without reflection they will not support them long. There is more calculation even in the impulses of bravery than is generally supposed; and although the first efforts are made by passion alone, perseverance is maintained only by a distinct view of what one is fighting for. A portion of what is dear to us is hazarded in order to save the remainder.

But it is this clear perception of the future, founded upon judgment and experience, that is frequently wanting in democracies. The people are more apt to feel than to reason; and if their present sufferings are great, it is to be feared that the still greater sufferings attendant upon defeat will be forgotten.

Another cause tends to render the efforts of a democratic government less persevering than those of an aristocracy. Not only are the lower less awake than the higher orders to the good or evil chances of the future, but they suffer more acutely from present privations. The noble exposes his life, indeed, but the chance of glory is equal to the chance of harm. If he sacrifices a large portion of his income to the state, he deprives himself for a time of some of the pleasures of affluence; but to the poor man death has no glory, and the imposts that are merely irksome to the rich often deprive him of the necessaries of life.

---

* Reprinted from *Democracy in America*, Volume I by permission of Alfred A. Knopf, Inc. Copyright 1945 by Alfred A. Knopf, Inc. Published by Vintage Books, Inc., 1955.

This relative weakness of democratic republics in critical times is perhaps the greatest obstacle to the foundation of such a republic in Europe. In order that one such state should exist in the European world, it would be necessary that similar institutions should be simultaneously introduced into all the other nations.

I am of opinion that a democratic government tends, in the long run, to increase the real strength of society; but it can never combine, upon a single point and at a given time, so much power as an aristocracy or an absolute monarchy. If a democratic country remained during a whole century subject to a republican government, it would probably at the end of that period be richer, more populous, and more prosperous than the neighboring despotic states. But during that century it would often have incurred the risk of being conquered by them.

The difficulty that a democracy finds in conquering the passions and subduing the desires of the moment with a view to the future is observable in the United States in the most trivial things. The people, surrounded by flatterers, find great difficulty in surmounting their inclinations; whenever they are required to undergo a privation or any inconvenience, even to attain an end sanctioned by their own rational conviction, they almost always refuse at first to comply. The deference of the Americans to the laws has been justly applauded; but it must be added that in America legislation is made by the people and for the people. Consequently, in the United States the law favors those classes that elsewhere are most interested in evading it. It may therefore be supposed that an offensive law of which the majority should not see the immediate utility would either not be enacted or not be obeyed.

In America there is no law against fraudulent bankruptcies, not because they are few, but because they are many. The dread of being prosecuted as a bankrupt is greater in the minds of the majority than the fear of being ruined by the bankruptcy of others; and a sort of guilty tolerance is extended by the public conscience to an offense which everyone condemns in his individual capacity. In the new states of the Southwest the citizens generally take justice into their own hands, and murders are of frequent occurrence. This arises from the rude manners and the ignorance of the inhabitants of those deserts, who do not perceive the utility of strengthening the law, and who prefer duels to prosecutions.

Someone observed to me one day in Philadelphia that almost all crimes in America are caused by the abuse of intoxicating liquors, which the lower classes can procure in great abundance because of their cheapness. "How comes it," said I, "that you do not put a duty upon brandy?" "Our legislators," rejoined my informant, "have frequently thought of this expedient; but the task is difficult: a revolt might be anticipated; and the members who should vote for such a law would be sure of losing their seats." "Whence I am to infer," replied I, "that drunkards are the majority in your country, and that temperance is unpopular."

When these things are pointed out to the American statesmen, they answer: "Leave it to time, and experience of the evil will teach the people their true interests." This is frequently true: though a democracy is more liable to error than a monarch or a body of nobles, the chances of its regaining the right path when once it has acknowledged its mistake are greater also; because it is rarely embarrassed by interests that conflict with those of the majority and resist the authority of reason. But a democracy can obtain truth only as the result of experience; and many nations may perish while they are awaiting the consequences of their errors. The great privilege of the Americans does not consist in being more enlightened than other nations, but in being able to repair the faults they may commit.

The Union is free from all pre-existing obligations; it can profit by the experience of the old nations of Europe, without being obliged, as they are, to make the best of the past and to adapt it to their present circumstances. It is not, like them, compelled to accept an immense inheritance bequeathed by their forefathers, *an inheritance of glory mingled with calamities,* and of *alliances conflicting with national antipathies.* The foreign policy of the United States is eminently expectant; it consists more in abstaining than in acting.

It is therefore very difficult to ascertain, at present, what degree of sagacity the American democracy will display in the conduct of the foreign policy of the country; upon this point its adversaries as well as its friends must suspend their judgment. As for myself,

I do not hesitate to say that it is especially in the conduct of their foreign relations that democracies appear to me decidedly inferior to other governments. Experience, instruction, and habit almost always succeed in creating in a democracy a homely species of practical wisdom and that science of the petty occurrences of life which is called good sense. Good sense may suffice to direct the ordinary course of society; and among a people whose education is completed, the advantages of democratic liberty in the internal affairs of the country may more than compensate for the evils inherent in a democratic government. But it is not always so in the relations with foreign nations.

Foreign politics demand scarcely any of those qualities which are peculiar to a democracy; they require, on the contrary, the perfect use of almost all those in which it is deficient. Democracy is favorable to the increase of the internal resources of a state; it diffuses wealth and comfort, promotes public spirit, and fortifies the respect for law in all classes of society: all these are advantages which have only an indirect influence over the relations which one people bears to another. But a democracy can only with great difficulty regulate the details of an important undertaking, persevere in a fixed design, and work out its execution in spite of serious obstacles. It cannot combine its measures with secrecy or await their consequences with patience. These are qualities which more especially belong to an individual or an aristocracy; and they are precisely the qualities by which a nation, like an individual, attains a dominant position.

If, on the contrary, we observe the natural defects of aristocracy, we shall find that, comparatively speaking, they do not injure the direction of the external affairs of the state. The capital fault of which aristocracies may be accused is that they work for themselves and not for the people. In foreign politics it is rare for the interest of the aristocracy to be distinct from that of the people.

The propensity that induces democracies to obey impulse rather than prudence, and to abandon a mature design for the gratification of a momentary passion, was clearly seen in America on the breaking out of the French Revolution. It was then as evident to the simplest capacity as it is at the present time that the interest of the Americans forbade them to take any part in the contest which was about to deluge Europe with blood, but which could not injure their own country. But the sympathies of the people declared themselves with so much violence in favor of France that nothing but the inflexible character of Washington and the immense popularity which he enjoyed could have prevented the Americans from declaring war against England. And even then the exertions which the austere reason of that great man made to repress the generous but imprudent passions of his fellow citizens nearly deprived him of the sole recompense which he ever claimed, that of his country's love. The majority reprobated his policy, but it was afterwards approved by the whole nation.

If the Constitution and the favor of the public had not entrusted the direction of the foreign affairs of the country to Washington, it is certain that the American nation would at that time have adopted the very measures which it now condemns.

Almost all the nations that have exercised a powerful influence upon the destinies of the world, by conceiving, following out, and executing vast designs, from the Romans to the English, have been governed by aristocratic institutions. Nor will this be a subject of wonder when we recollect that nothing in the world is so conservative in its views as an aristocracy. The mass of the people may be led astray by ignorance or passion; the mind of a king may be biased and made to vacillate in his designs, and, besides, a king is not immortal. But an aristocratic body is too numerous to be led astray by intrigue, and yet not numerous enough to yield readily to the intoxication of unreflecting passion. An aristocracy is a firm and enlightened body that never dies.

This ceaseless agitation which democratic government has introduced into the political world influences all social intercourse. I am not sure that, on the whole, this is not the greatest advantage of democracy; and I am less inclined to applaud it for what it does than for what it causes to be done.

It is incontestable that the people frequently conduct public business very badly; but it is impossible that the lower orders should take a part in public business without extending the circle of their ideas and quitting the ordinary routine of their thoughts. The humblest individual who cooperates in the government of society acquires a certain degree of self-respect; and as he possesses authority, he can command the services of minds more enlightened than his own. He is canvassed by a multitude of applicants, and in seeking to deceive him in a thousand ways, they really enlighten him. He takes a part in political undertakings which he did not originate, but which give him a taste for undertakings of the kind. New improvements are daily pointed out to him in the common property, and this gives him the desire of improving that property which is his own. He is perhaps neither happier nor better than those who came before him, but he is better informed and more active. I have no doubt that the democratic institutions of the United States, joined to the physical constitustitution of the country, are the cause (not the direct, as is so often asserted, but the indirect cause) of the prodigious commercial activity of the inhabitants. It is not created by the laws, but the people learn how to promote it by the experience derived from legislation.

When the opponents of democracy assert that a single man performs what he undertakes better than the government of all, it appears to me that they are right. The government of an individual, supposing an equality of knowledge on either side, is more consistent, more persevering, more uniform, and more accurate in details than that of a multitude, and it selects with more discrimination the men whom it employs. If any deny this, they have never seen a democratic government, or have judged upon partial evidence. It is true that, even when local circumstances and the dispositions of the people allow democratic institutions to exist, they do not display a regular and methodical system of government. Democratic liberty is far from accomplishing all its projects with the skill of an adroit despotism. It frequently abandons them before they have borne their fruits, or risks them when the consequences may be dangerous; but in the end it produces more than any absolute government; if it does fewer things well, it does a greater number of things. Under its sway the grandeur is not in what the public administration does, but in what is done without it or outside of it. Democracy does not give the people the most skillful government, but it produces what the ablest governments are frequently unable to create: namely, an all-pervading and restless activity, a superabundant force, and an energy which is inseparable from it and which may, however unfavorable circumstances may be, produce wonders. These are the true advantages of democracy.

In the present age, when the destinies of Christendom seem to be in suspense, some hasten to assail democracy as a hostile power while it is yet growing; and others already adore this new deity which is springing forth from chaos. But both parties are imperfectly acquainted with the object of their hatred or their worship; they strike in the dark and distribute their blows at random.

We must first understand what is wanted of society and its government. Do you wish to give a certain elevation to the human mind and teach it to regard the things of this world with generous feelings, to inspire men with a scorn of mere temporal advantages, to form and nourish strong convictions and keep alive the spirit of honorable devotedness? Is it your object to refine the habits, embellish the manners, and cultivate the arts, to promote the love of poetry, beauty, and glory? Would you constitute a people fitted to act powerfully upon all other nations, and prepared for those high enterprises which, whatever be their results, will leave a name forever famous in history? If you believe such to be the principal object of society, avoid the government of the democracy, for it would not lead you with certainty to the goal.

But if you hold it expedient to divert the moral and intellectual activity of man to the production of comfort and the promotion of general well-being; if a clear understanding be more profitable to man than genius; if your object is not to stimulate the virtues of heroism, but the habits of peace; if you had rather witness vices than crimes, and are content to meet with fewer noble deeds, provided offenses be diminished in the same proportion; if, instead of living in the midst of a brilliant society, you are contented to have prosperity around you; if, in short, you are of the opinion that the principal object of a government is not to confer the greatest possible power and glory upon the body of the nation, but to ensure the greatest enjoyment and to avoid the most misery to each of the individuals who compose it—if such be your desire, then equalize the conditions of men and establish democratic institutions.

But if the time is past at which such a choice was possible, and if some power superior to that of man already hurries us, without consulting our wishes, towards one or the other of these two governments, let us endeavor to make the best of that which is allotted to us and, by finding out both its good and its evil tendencies, be able to foster the former and repress the latter to the utmost.

Democratic Representation in an Age of Crisis*

BY DEAN ACHESON

[1956]

AMONG the manifold uncertainties of this life, I venture one prediction. If, at the present time, the limitation imposed by democratic political practice makes it difficult to conduct our foreign affairs in the national interest, this difficulty will increase, and not decrease, with the years. The problem is not merely the one which de Tocqueville pointed out, though that persists. "As for myself," he wrote, "I do not hesitate to say that it is especially in the conduct of their foreign relations that democracies appear to me decidedly inferior to other governments.... Foreign politics demand scarcely any of those qualities which are peculiar to a democracy; they require, on the contrary, the perfect use of almost all those in which it is deficient.... [A] democracy can only with great difficulty regulate the details of an important undertaking, persevere in a fixed design, and work out its execution in spite of serious obstacle. It cannot combine its measures with secrecy or await their consequences with patience. These are qualities which more especially belong to an individual or an aristocracy."[1]

To this something else is added. In the first place, the questions to be understood, and then solved, have vastly increased in complexity and dimension. They have taken on technical aspects requiring knowledge rare even among the intellectually elite. They involve the collection and analysis of information so current that it is changing in the very process of collection. How different was

---

* From *A Citizen Looks at Congress*. Copyright 1956, 1957 by Dean Acheson. Published by Harper & Brothers. Reprinted by permission. Dean Acheson (b. 1893) was Secretary of State from 1949 to 1952 and now practices law in Washington, D. C.

[1] De Tocqueville, *Democracy in America*, Vol. I, pp. 234–235.

the leisurely pace when Secretary of State Thomas Jefferson wrote to William Carmichael, the American Chargé d'Affaires at Madrid, in March, 1791: "Your letter of May 6. 1789. is still the last we have received, & that is now near two years old. . . . A full explanation of the causes of this suspension of all information from you, is expected in answer to my letter of Aug. 6 [1790]. It will be waited for yet a reasonable time, & in the mean while a final opinion suspended."

In the second place, the task of understanding and solving these questions requires the coordinated work of many minds with diverse training and competence, so that innumerable facts and considerations may be held in solution until the moment for the catalyst of decision. This sort of attack the Congress by its organization and history is incapable of making. . . .

The problems of today and tomorrow are different in their very nature from those of the heyday of democratic theory, the eighteenth century, just as the content of man's knowledge is different. It is, perhaps, not too great an illusion to believe that even at the end of the eighteenth century a man of capacity and zest could keep pretty well abreast of what was being done in the arts and sciences. . . .

A Franklin or a Jefferson was not a stranger in any field of learning. A gifted amateur could still hold his own with the professionals. And the legislative assembly was, above anything else, an assembly of amateurs. Politics was not yet a profession. If politicians had any profession it was likely to be law; and lawyers were the greatest jacks of all trades, the greatest amateurs, of them all. . . .

Nowhere is the age of the amateur more clearly reflected than in the military establishment where the main reliance of the nation's defense was placed on the militia. The standing army was so small that at the turn of the twentieth century annual maneuvers had to be cancelled since there weren't enough troops to maneuver. Moreover, it was not until the first years of the twentieth century that a General Staff was established.

But today how small is the comfortable and familiar area of our intellectual globe, how vast the unknown! As the amateur, whether in or out of Congress, begins to think of the nature and extent of dangers to our national life and ways to meet them, he runs into difficulty at the very start. The crudest danger, the easiest to understand as an idea, is that from foreign force. But is it easy to understand as a fact? If we cannot fully understand, at least we can dimly grasp the immensity of the destruction wrought by thermonuclear bombs. But the facts regarding capability of delivering these bombs on target and defense against this—both the active defense of intercepting and destroying the attacker and the passive defense of minimizing the effect of the blow—are more complicated and more problematical. Here enter such technical matters as early warning and interceptor systems, guided missiles launched from the ground and from the air and from surface craft and submarines, and, in the background, still to be developed, intercontinental ballistic missiles. By and large, the amateur, skeptical of extravagant claims, can gather that the development of the defense is well behind that of the offense, and not gaining appreciably; that whether the objective of the attack is the enemy's air force or his centers of population and industry, the latter are likely to suffer heavily and the former is not apt to be rendered incapable of retaliating strongly enough to cause destruction about equal to that of the attack, barring accidents. But this is small comfort to the Congressman whose basic role is control of the purse strings. How much money shall he grant to whom and for what? How much for offense? How much for defense? And who are the proper guardians of offense, the Air Force alone, or the Air Force and the Navy together? . . .

To the amateur the problems of nuclear war seem much more complicated and difficult than those of competitive coexistence. He is probably wrong about this, not because the complexities of the former are less than he thinks, but because the complexities of the latter are far greater and our knowledge of them less.

For here we leave whatever exactitude may reside in the physical sciences and launch into the imponderables of man's behavior. Suppose, for instance—as seems to be true—that the productive power of the Soviet Union is growing far faster than the more mature economic systems of the West, and perhaps faster than any society of which we have records, what do we conclude from this? We know many of the reasons—the jump from a peasant society into the latest industrial techniques developed by others, the forced austerity in consumption to permit production of capital goods, and so on. But will the rate of development, or anything like it, continue? When will the U.S.S.R.'s production exceed that of the U.S.A.? What will be the curve of China's industrialization? What will be the effect of such a shift in the ratio of productive power on military capabilities? On capability for economic and political maneuvering? On the attraction of the Communist bloc for the uncommitted peoples? Is it desirable to attempt to accelerate our own growth in productive power? How? For what immediate purposes? These questions, I submit, are quite as important and quite as difficult as those in the realm of nuclear physics. They are equally out of the field of competence of the amateur.

But to him they do not seem to be. Although the furthest marches of this territory may be trackless wilderness to him, it begins pretty close to the back pasture of the home farm. For twenty years as a trustee of a university I have noticed that the governing body, for the most part businessmen, lawyers, and clergymen—amateurs in education—wisely and passively accept the recommendations of the university administration, on faculty appointments and so on, in the scientific departments. But one can sense the restless consciousness of competence to criticize when attention turns to the departments of economics, history, law or religion. We may not know even the criteria for determining who is sound and who is flighty in biochemistry, but we have our own ideas on deficit financing—or think we have until the government shows signs of stopping it. Here we, and the legislator, whose position—frustrations and all—is very like that of a member of a board of trustees, incur grave risks. Our ignorance is far greater than our consciousness of it. . . .

How—to be concrete—can the non-expert member of Congress go to work on problems of military policy? First, by putting the tools of analysis to work so that a point for discussion may stand out. . . .

No American purpose, it could be pointed out, depends upon our using force against anyone. But we must be prepared to deter or meet the use of or the threat of force against our interests. When we speak of deterring the use of force against us, what do we mean? A deterrent is a threat under certain circumstances to do harm to another, which the other believes we will do and does not want to provoke. A threat is not believed, and therefore cannot deter, unless there is general conviction that the threatener has both the capacity and the intention to carry out the threat.

The deterrent to the use or threat of nuclear bombs today is the belief by the initiator that he would receive retribution about as devastating as the attack. This deterrent will be effective to protect us only as long as it is believed elsewhere that we are maintaining the capacity and the will to launch a crushing nuclear reply if our vital interests are attacked.

But nuclear capacity will not deter all use of force against a nation possessing it, because it is not credible that many occasions are serious enough to lead that nation to use atomic weapons, and thereby incur the risk of having them used against it.

Atomic capacity did not save Dienbienphu or prevent Mr. Nasser from seizing the Suez Canal. It was not credible that it would be used.

Therefore, to deter or meet force used or threatened on a local basis, capacity in what are called conventional forces is required —that is, forces which can conduct limited warfare and keep it limited. Even these will not act as a deterrent or moderating factor unless others believe that they will be used. When the British Labour party and the United States Government pronounced against the use of force to counter the forceful seizure of the Suez Canal, Mr. Nasser could obviously no longer regard its use as a practical possibility.

This analysis—and similar analysis of political or economic policy in the foreign field—is susceptible of rational discussion by people who are not experts. . . .

Can the Congress discipline itself to do its own job and not try to do the executive's job? Those who know it best will have their doubts. The virus of busyness has bitten deep. Impulses born of parochial interests at stake, of personal ambition, of partisan maneuver, all make for interference with administration and the attempt to control it. They are hard to resist in favor of more plain and unspectacular courses. But there have always been men in the Congress who have made this harder choice. . . .

The task of concentration upon what members of a popular assembly can do amid the complexities of the twentieth century, and of doing that well, is a hard one. But I venture to say that upon the measure of its achievement will depend the future of representative government, the power of control exercised by a responsible legislature, in the United States.

The Constitution of the United States

The Federal Constitution

AS AGREED UPON BY THE CONVENTION
SEPTEMBER 17, 1787

WE, THE PEOPLE OF THE UNITED STATES, in order to form a more perfect union, establish justice, insure domestic tranquillity, provide for the common defence, promote the general welfare, and secure the blessings of liberty to ourselves and our posterity, do ordain and establish this Constitution for the United States of America.

ARTICLE I. SECTION 1. All legislative powers herein granted shall be vested in a Congress of the United States which shall consist of a Senate and House of Representatives.

SECTION 2. The House of Representatives shall be composed of members chosen every second year by the people of the several States, and the electors in each State shall have the qualifications requisite for electors of the most numerous branch of the State legislature.

No person shall be a representative who shall not have attained to the age of twenty-five years, and been seven years a citizen of the United States, and who shall not, when elected, be an inhabitant of that State in which he shall be chosen.

Representatives and direct taxes shall be apportioned among the several States which may be included within this Union, according to their respective numbers, which shall be determined by adding to the whole number of free persons, including those bound to service for a term of years, and excluding Indians not taxed, three fifths of all other persons. The actual enumeration shall be made within three years after the first meeting of the Congress of the United States, and within every subsequent term of ten years, in such manner as they shall by law direct. The number of representatives shall not exceed one for every thirty thousand but each State shall have at least one representative; and until such enumeration shall be made, the State of New Hampshire shall be entitled to choose three; Massachusetts, eight; Rhode Island and Providence Plantations, one; Connecticut, five; New York, six; New Jersey, four; Pennsylvania, eight; Delaware, one; Maryland, six; Virginia, ten; North Carolina, five; South Carolina, five; and Georgia, three.

When vacancies happen in the representation from any State, the executive authority thereof shall issue writs of election to fill such vacancies.

The House of Representatives shall choose their Speaker and other officers, and shall have the sole power of impeachment.

SECTION 3. The Senate of the United States shall be composed of two Senators from each State, chosen by the legislature thereof, for six years; and each Senator shall have one vote.

Immediately after they shall be assembled in consequence of the first election, they shall be divided as equally as may be into three classes. The seats of the Senators of the first class shall be vacated at the expiration of the second year, of the second class at the expiration of the fourth year, and of the third class at the expiration of the sixth year, so that one-third may be chosen every second year; and if vacancies happen by resignation or otherwise during the recess of the legislature of any State, the executive thereof may make temporary appointments until the next meeting of the legislature, which shall then fill such vacancies.

No person shall be a Senator who shall not have attained to the age of thirty years, and been nine years a citizen of the United States, and who shall not, when elected, be an inhabitant of that State for which he shall be chosen.

The Vice-President of the United States shall be President of the Senate, but shall have no vote, unless they be equally divided.

The Senate shall choose their other officers and also a President *pro tempore* in the absence of the Vice-President, or when he shall exercise the office of President of the United States.

The Senate shall have the sole power to try all impeachments. When sitting for that purpose, they shall be on oath or affirmation. When the President of the United States is tried, the Chief Justice shall preside; and no person shall be convicted without the concurrence of two-thirds of the members present.

Judgment in cases of impeachment shall not extend further than to removal from office, and disqualification to hold and enjoy any office of honor, trust, or profit under the United States; but the party convicted shall, nevertheless, be liable and subject to indictment, trial, judgment, and punishment, according to law.

SECTION 4. The times, places, and manner of holding elections for Senators and Representatives shall be prescribed in each State by the legislature thereof; but the Congress may at any time by law make or alter such regulations, except as to the places of choosing Senators.

The Congress shall assemble at least once in every year, and such meeting shall be on the first Monday in December, unless they shall by law appoint a different day.

SECTION 5. Each House shall be the judge of the elections, returns, and qualifications of its own members, and a majority of each shall constitute a quorum to do business; but a smaller number may adjourn from day to day, and may be authorized to compel the attendance of absent members, in such manner, and under such penalties, as each House may provide. Each House may determine the rules of its proceedings, punish its members for disorderly behavior, and with the concurrence of two thirds, expel a member.

Each House shall keep a journal of its proceedings, and from time to time publish the same excepting such parts as may in their judgment require secrecy, and the yeas and nays of the members of either House on any question shall, at the desire of one-fifth of those present, be entered on the journal.

Neither House, during the session of Congress, shall, without the consent of the other, adjourn for more than three days, nor to any other place than that in which the two Houses shall be sitting.

SECTION 6. The Senators and Representatives shall receive a compensation for their services, to be ascertained by law and paid out of the Treasury of the United States. They shall, in all cases except treason, felony, and breach of the peace, be privileged from arrest during their attendance at the session of their respective Houses, and in going to and returning from the same; and for any speech or debate in either House, they shall not be questioned in any other place.

No Senator or Representative shall, during the time for which he was elected, be appointed to any civil office under the authority of the United States, which shall have been created, or the emoluments whereof shall have been increased during such time; and no person holding any office under the United States shall be a member of either House during his continuance in office.

SECTION 7. All bills for raising revenue shall originate in the House of Representatives; but the Senate may propose or concur with amendments as on other bills.

Every bill which shall have passed the House of Representatives and the Senate shall, before it become a law, be presented to the President of the United States; if he approve he shall sign it, but if not he shall return it, with his objections, to that House in which it shall have originated, who shall enter the objections at large on their journal and proceed to reconsider it. If after such reconsideration two-thirds of that House shall agree to pass the bill, it shall be sent, together with the objections, to the other House, by which it shall likewise be reconsidered, and if approved by two-thirds of that House it shall become a law. But in all such cases the vote of both Houses shall be determined by yeas and nays, and the names of the persons voting for and against the bill shall be entered on the journal of each House respectively. If any bill shall not be returned by the President within ten days (Sundays excepted) after it shall have been presented to him, the same shall be a law, in like manner as if he had signed it, unless the Congress by their adjournment prevent its return, in which case it shall not be a law.

Every order, resolution or vote to which the concurrence of the Senate and the House of Representatives may be necessary (except on a question of adjournment) shall be presented to the President of the United States; and before the same shall take effect, shall be approved by him, or being disapproved by him, shall be repassed by two-thirds of the Senate and House of Representatives, according to the rules and limitations prescribed in the case of a bill.

SECTION 8. The Congress shall have power

To lay and collect taxes, duties, imposts and excises; to pay the debts and provide for the common defence and general welfare of the United States; but all duties, imposts, and excises, shall be uniform throughout the United States;

To borrow money on the credit of the United States;

To regulate commerce with foreign nations, and among the several States, and with the Indian tribes;

To establish an uniform rule of naturalization, and uniform laws on the subject of bankruptcies throughout the United States;

To coin money, regulate the value thereof, and of foreign coin, and fix the standard of weights and measures;

To provide for the punishment of counterfeiting the securities and current coin of the United States;

To establish post offices and post roads;

To promote the progress of science and useful arts by securing for limited times to authors and inventors the exclusive right to their respective writings and discoveries;

To constitute tribunals inferior to the Supreme Court;

To define and punish piracies and felonies committed on the high seas and offences against the law of nations;

To declare war, grant letters of marque and reprisal, and make rules concerning captures on land and water;

To raise and support armies, but no appropriation of money to that use shall be for a longer term than two years;

To provide and maintain a navy;

To make rules for the government and regulation of the land and naval forces;

To provide for calling forth the militia to execute the laws of the Union, suppress insurrections, and repel invasions;

To provide for organizing, arming, and disciplining the militia, and for governing such parts of them as may be employed in the service of the United States, reserving to the States respectively the appointment of the officers, and the authority of training the militia according to the discipline prescribed by Congress;

To exercise exclusive legislation in all cases whatsoever over such district (not exceeding ten miles square) as may, by cession of particular States and the acceptance of Congress, become the seat of the government of the United States, and to exercise like authority over all places purchased by the consent of the legislature of the State in which the same shall be, for the erection of forts, magazines, arsenals, dock yards, and other needful buildings;

To make all laws which shall be necessary and proper for carrying into execution the foregoing powers, and all other powers vested by this Constitution in the government of the United States, or in any department or officer thereof.

SECTION 9. The migration or importation of such persons as any of the States now existing shall think proper to admit shall not be prohibited by the Congress prior to the year one thousand eight hundred and eight, but a tax or duty may be imposed on such importation, not exceeding ten dollars for each person.

The privilege of the writ of *habeas corpus* shall not be suspended, unless when in cases of rebellion or invasion the public safety may require it.

No bill of attainder or *ex post facto* law shall be passed.

No capitation or other direct, tax shall be laid, unless in proportion to the *census* or enumeration herein before directed to be taken.

No tax or duty shall be laid on articles exported from any State. No preference shall be given by any regulation of commerce or revenue to the ports of one State over those of another; nor shall vessels bound to or from one State be obliged to enter, clear or pay duties in another.

No money shall be drawn from the Treasury, but in consequence of appropriations made by law; and a regular statement and account of the receipts and expenditures of all public money shall be published from time to time.

No title of nobility shall be granted by the United States; and no person holding any office of profit or trust under them, shall, without the consent of the Congress accept of any present, emolument, office, or title of any kind whatever from any king, prince, or foreign state.

SECTION 10. No State shall enter into any treaty, alliance, or confederation; grant letters of marque and reprisal; coin money; emit bills of credit; make any thing but gold and silver coin a tender in payment of debts; pass any bill of attainder, *ex post facto* law, or law impairing the obligation of contracts, or grant any title of nobility.

No state shall, without the consent of the Congress, lay any imposts or duties on imports or exports, except what may be absolutely necessary for executing its inspection laws; and the net produce of all duties and imposts, laid by any State on imports or exports, shall be for the use of the Treasury of the United States; and all such laws shall be subject to the revision and control of the Congress. No State shall, without the consent of Congress, lay any duties of tonnage, keep troops and ships of war in time of peace, enter into any agreement or compact with another State, or with a foreign power, or engage in war, unless actually invaded, or in such imminent danger as will not admit of delay.

ARTICLE II. SECTION 1. The executive power shall be vested in a President of the United States of America. He shall hold his office during the term of four years, and together with the Vice-President, chosen for the same term, be elected as follows:

Each State shall appoint, in such manner as the legislature thereof may direct, a number of Electors, equal to the whole number of Senators and Representatives to which the State may be entitled in the Congress, but no Senator or Representative, or person holding an office of trust or profit under the United States, shall be appointed an Elector.

The Electors shall meet in their respective States and vote by ballot for two persons, of whom one at least shall not be an inhabitant of the same State with themselves. And they shall make a list of all the persons voted for, and of the number of votes for each; which list they shall sign and certify, and transmit sealed to the seat of government of the United States, directed to the President of the Senate. The President of the Senate shall, in the presence of the Senate and House of Representatives, open all the certificates, and the votes shall then be counted. The person having the greatest number of votes shall be the President, if such number be a majority of the whole number of electors appointed; and if there be more than one who have such majority, and have an equal number of votes, then the House of Representatives shall immediately choose by ballot one of them for President; and if no person have a majority, then from the five highest on the list the said House shall in like manner choose the President. But in choosing the President, the votes shall be taken by States, the representation from each State having one vote; a quorum for this purpose shall consist of a member or members from two-thirds of the States, and a majority of all the States shall be necessary to a choice. In every case, after the choice of the President, the person having the greatest numb·r of votes of the Electors shall be the

Vice-President. But if there should remain two or more who have equal votes, the Senate shall choose from them by ballot the Vice-President.

The Congress may determine the time of choosing the Electors, and the day on which they shall give their votes; which day shall be the same throughout the United States.

No person except a natural-born citizen, or citizen of the United States at the time of the adoption of this Constitution, shall be eligible to the office of President; neither shall any person be eligible to that office who shall not have attained to the age of thirty-five years, and been fourteen years a resident within the United States.

In case of the removal of the President from office, or of his death, resignation, or inability to discharge the powers and duties of the said office, the same shall devolve on the Vice-President, and the Congress may by law provide for the case of removal, death, resignation, or inability, both of the President and Vice-President, declaring what officer shall then act as President, and such officer shall act accordingly until the disability be removed or a President shall be elected.

The President shall, at stated times, receive for his services a compensation, which shall neither be increased nor diminished during the period for which he shall have been elected, and he shall not receive within that period any other emolument from the United States or any of them.

Before he enter on the execution of his office he shall take the following oath or affirmation:

"I do solemnly swear (or affirm) that I will faithfully execute the office of President of the United States, and will to the best of my ability, preserve, protect, and defend the Constitution of the United States."

SECTION 2. The President shall be Commander-in-Chief of the army and navy of the United States, and of the militia of the several States when called into the actual service of the United States; he may require the opinion, in writing, of the principal officer in each of the executive departments, upon any subject relating to the duties of their respective offices, and he shall have power to grant reprieves and pardons for offences against the United States, except in cases of impeachment.

He shall have power, by and with the advice and consent of the

Senate, to make treaties, provided two thirds of the senators present concur; and he shall nominate, and, by and with the advice and consent of the Senate, shall appoint ambassadors, other public ministers and consuls, judges of the Supreme Court, and all other officers of the United States whose appointments are not herein otherwise provided for, and which shall be established by law; but the Congress may by law vest the appointment of such inferior officers, as they think proper, in the President alone, in the courts of law, or in the heads of departments.

The President shall have power to fill up all vacancies that may happen during the recess of the Senate, by granting commissions which shall expire at the end of their next session.

SECTION 3. He shall from time to time give to the Congress information of the state of the Union, and recommend to their consideration such measures as he shall judge necessary and expedient; he may, on extraordinary occasions, convene both Houses, or either of them, and in case of disagreement between them, with respect to the time of adjournment, he may adjourn them to such time as he shall think proper; he shall receive ambassadors and other public ministers; he shall take care that the laws be faithfully executed, and shall commission all the officers of the United States.

SECTION 4. The President, Vice-President and all civil officers of the United States shall be removed from office on impeachment for and conviction of treason, bribery, or other high crimes and misdemeanors.

ARTICLE III. SECTION 1. The judicial power of the United States shall be vested in one Supreme Court, and in such inferior courts as the Congress may from time to time ordain and establish. The judges, both of the supreme and inferior courts, shall hold their offices during good behavior, and shall, at stated times, receive for their services a compensation which shall not be diminished during their continuance in office.

SECTION 2. The judicial power shall extend to all cases in law and equity, arising under this Constitution, the laws of the United States, and treaties made, or which shall be made, under their authority; to all cases affecting ambassadors, other public ministers, and consuls; to all cases of admiralty and maritime jurisdiction; to controversies to which the United States shall be a party; to controversies between two or more States; between a State and citizen of another State; between citizens of different States; between citizens of the same State claiming lands under grants of different States, and between a State, or the citizens thereof, and foreign states, citizens, or subjects.

In all cases affecting ambassadors, other public ministers and consuls, and those in which a State shall be party, the Supreme Court shall have original jurisdiction. In all the other cases before mentioned the Supreme Court shall have appellate jurisdiction, both as to law and fact, with such exceptions and under such regulations as the Congress shall make.

The trial of all crimes, except in cases of impeachment, shall be by jury; and such trial shall be held in the State where the said crimes shall have been committed; but when not committed within any State, the trial shall be at such place or places as the Congress may by law have directed.

SECTION 3. Treason against the United States shall consist only in levying war against them, or in adhering to their enemies, giving them aid and comfort. No person shall be convicted of treason unless on the testimony of two witnesses to the same overt act, or on confession in open court.

The Congress shall have power to declare the punishment of treason, but no attainder of treason shall work corruption of blood or forfeiture except during the life of the person attainted.

ARTICLE IV. SECTION 1. Full faith and credit shall be given in each State to the public acts, records, and judicial proceedings of every other State. And the Congress may by general laws prescribe the manner in which such acts, records, and proceedings shall be proved, and the effect thereof.

SECTION 2. The citizens of each State shall be entitled to all privileges and immunities of citizens in the several States.

A person charged in any State with treason, felony, or other crime, who shall flee from justice, and be found in another State, shall, on demand of the executive authority of the State from which he fled, be delivered up, to be removed to the State having jurisdiction of the crime.

No person held to service or labor in one State, under the laws thereof, escaping into another, shall, in consequence of any law or regulation therein, be discharged from such service or labor, but shall be delivered up on claim to the party to whom such service or labor may be due.

Section 3. New States may be admitted by the Congress into this Union; but no new State shall be formed or erected within the jurisdiction of any other State; nor any State be formed by the junction of two or more States, or parts of States, without the consent of the legislatures of the States concerned as well as of the Congress.

The Congress shall have power to dispose of and make all needful rules and regulations respecting the territory or other property belonging to the United States; and nothing in this Constitution shall be so construed as to prejudice any claims of the United States or of any particular State.

Section 4. The United States shall guarantee to every State in this Union a republican form of government, and shall protect each of them against invasion, and on application of the legislature, or of the Executive (when the legislature cannot be convened), against domestic violence.

Article V. The Congress, whenever two thirds of both houses shall deem it necessary, shall propose amendments to this Constitution, or, on the application of the legislatures of two thirds of the several States, shall call a convention for proposing amendments, which in either case shall be valid to all intents and purposes as part of this Constitution, when ratified by the legislatures of three-fourths of the several States, or by conventions in three-fourths thereof, as the one or the other mode of ratification may be proposed by the Congress; provided, that no amendment which may be made prior to the year one thousand eight hundred and eight shall in any manner affect the first and fourth clauses in the Ninth Section of the First Article; and that no State, without its consent shall be deprived of its equal suffrage in the Senate.

Article VI. All debts contracted and engagements entered into, before the adoption of this Constitution, shall be as valid against the United States under this Constitution as under the Confederation.

This Constitution, and the laws of the United States which shall be made in pursuance thereof, and all treaties made, or which shall be made, under the authority of the United States, shall be the supreme law of the land; and the judges in every State shall be bound thereby, any thing in the Constitution or laws of any State to the contrary notwithstanding.

The Senators and Representatives before mentioned, and the

members of the several State legislatures, and all executive and judicial officers both of the United States and of the several States, shall be bound by oath or affirmation to support this Constitution; but no religious test shall ever be required as a qualification to any office or public trust under the United States.

ARTICLE VII. The ratification of the conventions of nine States shall be sufficient for the establishment of this Constitution between the States so ratifying the same.

DONE in convention, by the unanimous consent of the States present, the seventeenth day of September, in the year of our Lord one thousand seven hundred and eighty-seven, and of the independence of the United States of America the twelfth. In witness whereof, we have hereunto subscribed our names.

GEORGE WASHINGTON, *President, Deputy from Virginia.*

New-Hampshire:
John Langdon
Nicholas Gilman

Massachusetts:
Nathaniel Gorham
Rufus King

Connecticut:
William Samuel Johnson
Roger Sherman

New York:
Alexander Hamilton

New Jersey:
William Livingston
David Brearley
William Paterson
Jonathan Dayton

Pennsylvania:
Benjamin Franklin
Thomas Mifflin
Robert Morris
George Clymer
Thomas Fitzsimons
Jared Ingersoll
James Wilson
Gouverneur Morris

Delaware:
George Read
Gunning Bedford, *Junior*
John Dickinson
Richard Bassett
Jacob Broom

Maryland:
James M'Henry
Daniel Jenifer, *of St. Thomas*
Daniel Carroll

Virginia:
John Blair
James Madison, *Junior*

North Carolina:
William Blount
Richard Dobbs Spaight
Hugh Williamson

South Carolina:
John Rutledge
Charles Cotesworth Pinckney
Charles Pinckney
Pierce Butler

Georgia:
William Few
Abraham Baldwin

Attest. WILLIAM JACKSON, *Secretary*

AMENDMENTS TO THE CONSTITUTION*

ARTICLE THE FIRST. Congress shall make no law respecting an establishment of religion, or prohibiting the free exercise thereof; or abridging the freedom of speech or of the press; or the right of the people peaceably to assemble, and to petition the government for a redress of grievances.

ARTICLE THE SECOND. A well-regulated militia being necessary to the security of a free State, the right of the people to keep and bear arms shall not be infringed.

ARTICLE THE THIRD. No soldier shall, in time of peace, be quartered in any house without the consent of the owner, nor in time of war, but in the manner prescribed by law.

ARTICLE THE FOURTH. The right of the people to be secure in their persons, houses, papers, and effects, against unreasonable searches and seizures, shall not be violated, and no warrants shall issue but upon probable cause, supported by oath or affirmation, and particularly describing the place to be searched, and the persons or things to be seized.

ARTICLE THE FIFTH. No person shall be held to answer for a capital, or otherwise infamous crime, unless on a presentment or indictment of a grand jury, except in cases arising in the land or naval forces, or in the militia when in actual service in time of war or public danger; nor shall any person be subject for the same offence to be twice put in jeopardy of life or limb; nor shall be compelled in any criminal case to be a witness against himself, nor be deprived of life, liberty or property, without due process of law; nor shall private property be taken for public use without just compensation.

ARTICLE THE SIXTH. In all criminal prosecutions the accused shall enjoy the right to a speedy and public trial, by an impartial jury of the State and district wherein the crime shall have been committed, which district shall have been previously ascertained

---

* The first *ten* amendments were proposed in Congress during its *first* session, and on the 15th of December, 1791, were ratified. The *eleventh* amendment was proposed during the *first* session of the *third* Congress, and was announced by the President of the United States in a message to it, of date January 8th, 1798, as having been ratified. The *twelfth* amendment originated with Hamilton, and was proposed during the *first* session of the *eighth* Congress, and was adopted in 1804.

by law, and to be informed of the nature and cause of the accusation; to be confronted with the witnesses against him; to have compulsory process for obtaining witnesses in his favor, and to have the assistance of counsel for his defence.

ARTICLE THE SEVENTH. In suits at common law, where the value in controversy shall exceed twenty dollars, the right of trial by jury shall be preserved, and no fact tried by a jury shall be otherwise re-examined in any court of the United States than according to the rules of the common law.

ARTICLE THE EIGHTH. Excessive bail shall not be required, nor excessive fines imposed, nor cruel and unusual punishments inflicted.

ARTICLE THE NINTH. The enumeration in the Constitution of certain rights shall not be construed to deny or disparage others retained by the people.

ARTICLE THE TENTH. The powers not delegated to the United States by the Constitution, nor prohibited by it to the States, are reserved to the States respectively, or to the people.

ARTICLE THE ELEVENTH. The judicial power of the United States shall not be construed to extend to any suit in law or equity, commenced or prosecuted against one of the United States by citizens of another State, or by citizens or subjects of any foreign State.

ARTICLE THE TWELFTH. The electors shall meet in their respective States and vote by ballot for President and Vice-President, one of whom, at least, shall not be an inhabitant of the same State with themselves; they shall name in their ballots the person voted for as President, and in distinct ballots the person voted for as Vice-President, and they shall make distinct lists of all persons voted for as President and of all persons voted for as Vice-President, and of the number of votes for each; which lists they shall sign and certify, and transmit sealed to the seat of the government of the United States, directed to the President of the Senate. The President of the Senate shall, in the presence of the Senate and House of Representatives, open all the certificates and the votes shall then be counted. The person having the greatest number of votes for President shall be the President, if such number be a majority of the whole number of Electors appointed; and if no person have such majority, then from the persons having the highest numbers not exceeding three on the list of those voted for as President, the

House of Representatives shall choose immediately, by ballot, the President. But in choosing the President the votes shall be taken by States, the representation from each State having one vote; a quorum for this purpose shall consist of a member or members from two-thirds of the States, and a majority of all the States shall be necessary to a choice. And if the House of Representatives shall not choose a President whenever the right of choice shall devolve upon them, before the fourth day of March next following, then the Vice-President shall act as President as in the case of the death or other constitutional disability of the President.

The person having the greatest number of votes as Vice-President shall be the Vice-President, if such number be a majority of the whole number of Electors appointed; and if no person have a majority, then from the two highest numbers on the list the Senate shall choose the Vice-President; a quorum for the purpose shall consist of two-thirds of the whole number of Senators, and a majority of the whole number shall be necessary to a choice.

But no person constitutionally ineligible to the office of President shall be eligible to that of Vice-President of the United States.

The following amendment was ratified by Alabama, December 2, 1865, which filled the requisite complement of ratifying States, and was certified by the Secretary of State to have become valid as a part of the Constitution of the United States, December 18, 1865.

ARTICLE THE THIRTEENTH. SECTION 1. Neither slavery nor involuntary servitude, except as a punishment for crime whereof the party shall have been duly convicted, shall exist within the United States, or any place subject to their jurisdiction.

SECTION 2. Congress shall have power to enforce this article by appropriate legislation.

The following amendment was certified by the Secretary of State to have become valid as a part of the Constitution of the United States, July 28, 1868.

ARTICLE THE FOURTEENTH. SECTION 1. All persons born or naturalized in the United States, and subject to the jurisdiction thereof, are citizens of the United States and of the State wherein they reside. No State shall make or enforce any law which shall abridge the privileges or immunities of citizens of the United States; nor shall any State deprive any person of life, liberty or property without due process of law; nor deny to any person within its jurisdiction the equal protection of the laws.

SECTION 2. Representatives shall be apportioned among the several States according to their respective numbers, counting the whole number of persons in each State, excluding Indians not taxed. But when the right to vote at any election for the choice of Electors for President and Vice-President of the United States, representatives in Congress, the executive and judicial officers of a State, or the members of the legislature thereof, is denied to any of the male inhabitants of such State, being twenty-one years of age, and citizens of the United States, or in any way abridged except for participation in rebellion or other crime, the basis of representation therein shall be reduced in the proportion which the number of such male citizens shall bear to the whole number of male citizens twenty-one years of age in such State.

SECTION 3. No person shall be a senator or representative in Congress, or elector of President and Vice-President, or hold any office, civil or military, under the United States or under any State, who, having previously taken an oath as a member of Congress, or as an officer of the United States, or as a member of any State legislature, or as an executive or judicial officer of any State, to support the Constitution of the United States, shall have engaged in insurrection or rebellion against the same, or given aid or comfort to the enemies thereof. But Congress may, by a vote of two-thirds of each House, remove such disability.

SECTION 4. The validity of the public debt of the United States, authorized by law, including debts incurred for payment of pensions and bounties for services in suppressing insurrection or rebellion, shall not be questioned. But neither the United States nor any State shall assume or pay any debt or obligation incurred in aid of insurrection or rebellion against the United States, or any claim for the loss or emancipation of any slave; but all such debts, obligations, and claims shall be held illegal and void.

SECTION 5. The Congress shall have power to enforce, by appropriate legislation, the provisions of this article.

The following amendment was proposed to the legislatures of

the several States by the fortieth Congress, on the 27th of February, 1869, and was declared, in a proclamation of the Secretary of State, dated March 30, 1870, to have been ratified by the legislatures of twenty-nine of the thirty-seven States.

ARTICLE THE FIFTEENTH. SECTION 1. The right of citizens of the United States to vote shall not be denied or abridged by the United States, or by any State, on account of race, color, or previous condition of servitude.

SECTION 2. The Congress shall have power to enforce this article by appropriate legislation.

ARTICLE THE SIXTEENTH. The Congress shall have the power to lay and collect taxes on incomes, from whatever source derived, without apportionment among the several States, and without regard to any census or enumeration.

ARTICLE THE SEVENTEENTH. SECTION 1. The Senate of the United States shall be composed of two Senators from each State, elected by the people thereof, for six years; and each Senator shall have one vote. The electors in each State shall have the qualifications requisite for electors of the most numerous branch of the State legislatures.

SECTION 2. When vacancies happen in the representation of any State in the Senate, the executive authority of such State shall issue writs of election to fill such vacancies: Provided, that the legislature of any State may empower the executive thereof to make temporary appointments until the people fill the vacancies by election as the legislature may direct.

SECTION 3. This amendment shall not be so construed as to affect the election or term of any Senator chosen before it becomes valid as part of the Constitution.

ARTICLE THE EIGHTEENTH. SECTION 1. After one year from the ratification* of this article the manufacture, sale or transportation of intoxicating liquors within, the importation thereof into, or the exportation thereof from the United States and all territory subject to the jurisdiction thereof, for beverage purposes, is hereby prohibited.

SECTION 2. The Congress and the several States shall have concurrent power to enforce this article by appropriate legislation.

---

* Jan. 16, 1919.

SECTION 3. This article shall be inoperative unless it shall have been ratified as an amendment to the Constitution by the legislatures of the several States, as provided in the Constitution, within seven years from the date of the submission hereof to the States by the Congress.

ARTICLE THE NINETEENTH. SECTION 1. The right of citizens of the United States to vote shall not be denied or abridged by the United States or by any State on account of sex.

SECTION 2. Congress shall have power to enforce this article by appropriate legislation.

ARTICLE THE TWENTIETH. SECTION 1. The terms of the President and Vice-President shall end at noon on the twentieth day of January, and the terms of Senators and Representatives at noon on the third day of January, of the years in which such terms would have ended if this article had not been ratified; and the terms of their successors shall then begin.

SECTION 2. The Congress shall assemble at least once in every year, and such meeting shall begin at noon on the third day of January, unless they shall by law appoint a different day.

SECTION 3. If, at the time fixed for the beginning of the term of the President, the President-elect shall have died, the Vice-President-elect shall become President. If a President shall not have been chosen before the time fixed for the beginning of his term or if the President-elect shall have failed to qualify, then the Vice-President-elect shall act as President until a President shall have qualified; and the Congress may by law provide for the case wherein neither a President-elect nor a Vice-President-elect shall have qualified, declaring who shall then act as President, or the manner in which one who is to act shall be selected, and such persons shall act accordingly until a President or Vice-President shall have qualified.

SECTION 4. The Congress may by law provide for the case of the death of any of the persons from whom the House of Representatives may choose a President whenever the right of choice shall have devolved upon them, and for the case of the death of any of the persons from whom the Senate may choose a Vice-President whenever the right of choice shall have devolved upon them.

SECTION 5. Sections 1 and 2 shall take effect on the fifteenth day of October following the ratification of this article.

SECTION 6. This article shall be inoperative unless it shall have been ratified as an amendment to the Constitution by the legislatures of three-fourths of the several States within seven years from the date of its submission.

ARTICLE THE TWENTY-FIRST. SECTION 1. The eighteenth article of amendment to the Constitution of the United States is hereby repealed.

SECTION 2. This article shall be inoperative unless it shall have been ratified as an amendment to the Constitution by the legislators of three-fourths of the several States within seven years from the date of its submission to the States by the Congress.

SECTION 3. This article shall be inoperative unless it shall have been ratified as an amendment to the Constitution by conventions in the several States, as provided in the Constitution, within seven years from the date of the submission hereof to the States by the Congress.

ARTICLE THE TWENTY-SECOND. SECTION 1. No person shall be elected to the office of the President more than twice. No person who has held the office of the President or acted as President for more than two years of a term to which some other person was elected President shall be elected to the office of the President more than once. But this article shall not apply to any person holding the office of President when this article was proposed by the Congress and shall not prevent any person who may be holding the office of President or acting as President during the term within which this article becomes operative from holding the office of President or acting as President during the remainder of such a term.

SECTION 2. This article shall be inoperative unless it shall have been ratified as an amendment to the Constitution by the legislators of three-fourths of the several States within seven years from the date of its submission to the States by the Congress.

John Lux